CAMBRIDGE MONOGRAPHS ON PHYSICS

GENERAL EDITORS

M. M. WOOLFSON, D.SC.
Professor of Theoretical Physics, University of York

J. M. ZIMAN, D.PHIL., F.R.S.
Professor of Theoretical Physics, University of Bristol

ELECTRON OPTICS

ELECTRON OPTICS

O. KLEMPERER

Formerly Assistant Professor and Reader
at Imperial College, University of London

THIRD EDITION

in collaboration with

M. E. BARNETT

Lecturer at Imperial College
University of London

CAMBRIDGE

AT THE UNIVERSITY PRESS

1971

Published by the Syndics of the Cambridge University Press
Bentley House, 200 Euston Road, London N.W.1
American Branch: 32 East 57th Street, New York, N.Y.10022

This edition © Cambridge University Press 1971

Library of Congress Catalogue Card Number: 74–118065

ISBN: 0 521 07928 4

First edition, 1939
Second edition, 1953
Third edition, 1971

Printed in Great Britain
at the University Printing House, Cambridge
(Brooke Crutchley, University Printer)

CONTENTS

CHAPTER 7

Electronic aberrations

CHAPTER 8

Electron optics and space-charge

CHAPTER 9

Emission systems and electron guns

CHAPTER 10

Deflecting fields

PREFACE TO THE THIRD EDITION

We have rewritten nearly half of the contents of the previous edition. However, we have no longer attempted to present an encyclopædic report, but a text suitable for students of the specialized subject. We have taken care not to omit any item of fundamental importance, and we have brought every section of the book up to date. Moreover, we have added some sections on new branches which have recently become important in electron optics: in particular we have described methods of trajectory tracing by digital computer and we have treated the optics of quadrupole lenses. To make room for the new additions, discussions of earlier investigations which seemed to us to be of minor importance have been eliminated. Chapters 10 and 12 of the second edition have been cancelled. Line focus lenses including quadrupoles have not been treated separately but their treatment has been added wherever corresponding properties (e.g. focal length, aberrations, etc.) were discussed for lenses of circular symmetry. Again, practical applications of electron optics have no longer been described in detail but have been joined to the appropriate sections, serving only as illustrations of the principles under discussion. In accordance with the experimental standpoint of the book we found it preferable to give the best available experimental data or the results of the most accurate computations, rather than deductions from inexact mathematical models. Furthermore, we have not included more advanced theoretical topics, such as the analytical computation of aberration coefficients. The bibliography at the end of the book, though it is very extensive, contains by no means a complete list of references; however, it is intended to guide the research worker to the important original literature and to provide the student with further reading of books and articles.

O. KLEMPERER

M. E. BARNETT

FROM THE PREFACE TO
THE SECOND EDITION (1953)

... The present book is intended to serve a twofold purpose: first, to introduce the student to a specialized subject, and secondly, to present the research worker or the designer of electron optical gear with all the basic principles and the most essential quantitative information on the subject and to be a guide to that literature where all available details can be found. The two purposes are complementary, since the specialized parts of the book which will be needed by the designer, serve as illustrations of the more general expositions which are addressed to the student. The expert will find many items which have been described in patent specifications only, and various other items which have not been published at all in the literature.

Selection, arrangement and presentation of the material have been made according to the personal experience of the author, who has worked for 11 years in charge of electron optical developments in industry and who has spent another 20 years at various universities engaged in teaching and research on electron physics. The book is concerned primarily with the experimental aspect of the subject. It should be intelligible both to the advanced student of general physics and to the research worker who has not specialized either in geometrical optics or in electron physics. ...

O. KLEMPERER

ACKNOWLEDGEMENTS

Thanks are due to the following for permission to reproduce figures from books and journals. Academic Press Inc. for figs. 8.11, 8.12, 9.21; *Adv. Electronics Electron Phys.*, 4.15; *Ann. Phys. Lpz.*, 9.17; *Arch. Elektrotech.*, 5.10, 6.13, 6.19; *Ark. Mat. Ark. Fys.* (Stockholm), 10.15; Butterworth & Co., 4.10, 10.1, 10.2, 10.16; *C.R. Acad. Sci., Paris*, 4.14; the Clarendon Press, 10.19; *Electl. Commun.*, 4.1; *Electron. Radio Engr.*, 9.22; Controller, H.M.S.O., 4.17; *J. appl. Phys.*, 3.4, 5.17, 6.2, 8.7, 9.20; *J. Br. Instn elect. Engrs*, 9.12, 9.18; *J. Electron. Control*, 8.6; *J. Instn elect. Engrs*, 9.15, 9.16; *J. scient. Instrum.*, 4.7, 6.12; *Jenaer Jb.*, 9.10; Macmillan & Co. Ltd., 10.8; *Naturwiss.*, 9.3; *Optik*, 6.27, 7.1, 7.14, 9.30; Pergamon Press Ltd., 7.7; *Philips tech. Rev.*, 9.6; *Proc. Instn elect. Engrs*, 2.6; *Revue Opt. théor. instrum.*, 5.7, 5.8, 5.9, 5.11, 5.12; Spon Ltd., 9.24; Springer Verlag, 4.8, 4.13, 6.16, 6.26; D. C. Swift & W. C. Nixon, 9.25; John Wiley & Sons Inc., 3.4, 6.13, 6.21; *Z. angew. Math. Phys.*, 4.2, 9.31; *Z. angew. Phys.*, 9.28; *Z. Phys.*, 6.7, 9.9, 10.18; *Z. tech. Phys.*, 10.20.

CHAPTER I

HISTORICAL INTRODUCTION.
THE FUNDAMENTAL PRINCIPLES OF
ELECTRON OPTICS

§1.1. Historical development

The birth of the subject of electron optics may be said to have occurred in 1926, when H. Busch showed that the action of a short axially symmetrical magnetic field on electron rays was similar to that of a glass lens on light rays. In 1931 and 1932, C. J. Davisson and C. J. Calbick, and independently E. Brüche and H. Johannson, recognized that axially symmetrical electrostatic fields could also be used as 'electron lenses'. The use of electron lenses for the projection of extended electron images was developed, first by M. Knoll and E. Ruska (1931), who worked mainly with magnetic lenses, then by E. Brüche and his collaborators (see Brüche and Scherzer, 1934), who worked mainly with electrostatic lenses and subsequently by many others. Up to 1939, electron optics experienced a rapid development, induced by a strong industrial need. Large companies which were especially interested in building up and producing a television technique of electronics employed considerable staffs of scientific workers who were engaged to an appreciable extent with fundamental research on electron optics. Thus the research laboratories of Electric and Musical Industries in England, of the RCA in USA, of Telefunken and AEG in Germany, and of Philips in Holland, have played a prominent part in the early development of electron optical science.

A detailed study of electron optical problems has also been taken up in connexion with electron microscopes of great magnification. The first working model of such an instrument was built up by Knoll and Ruska (1932). Then the technical development of the electron microscope was taken up by industrial firms; first in the field were Siemens and AEG in Germany, Radio Corporation and General Electric in USA and Metropolitan Vickers (now with GEC) in England.

During the 1939–45 war, electron optics again received strong stimulation through practical requirements. Thus, new types of cathode-ray tubes were developed for radar purposes and image-converter tubes were developed for infra-red vision. Most important, however, was the evolution of klystron and magnetron tubes for production of high-frequency electromagnetic waves. The need for improvement of these tubes stimulated the development of electron optics in the region of great space-charge densities.

The growth of electron optics has also been much stimulated by the demands of purely scientific research. The electron microscope has become indispensable to progress in many fields of study in biology and metallurgy. Investigation of energy losses of electron beams in atoms and in solids has required the construction of highly refined electron beam spectrometers capable of resolving energy levels within some millivolts. Investigation of nuclear energy levels led to the electron optical development of high energy particle spectrometers of great resolving and collecting power. Again the technique of particle acceleration by betatrons, synchrotrons, linear accelerators, etc., favoured the development of electron optics in the relativistic region of highest electron velocities.

While the problems of electron optics are very successfully attacked with methods which have been developed in geometrical light optics, the complementary method of tracing the paths of single electrons by studying the motion of point charges in the electromagnetic field is often more convenient.

§1.2. Variation principles and the refractive index of an electron*

The fundamental theory of electron optics is based on Hamilton's conceptions of the identity of the optical description of the path of a light ray through refractive media and the mechanical description of the motion of a mass point through a potential field. These general conceptions were presented as early as 1828–37; they originally started from a comparison of Fermat's principle of least

* A reader who is primarily interested in the practical aspect of electron optics may omit §§1.2–1.7 and continue his perusal from §2.1.

time (about 1650) as applied to the path of a light ray, with Maupertuis's principle of least action (1744) as applied to any mechanical movement.

For conservative forces, the action a is given by the line integral of the momentum p of the particle along its path s between two points P and P', namely, by

$$a = \int_P^{P'} p \, ds. \tag{1.1}$$

For an electron, the momentum is given by

$$p = mu_{el},$$

where u_{el} is the velocity of the electron and m its inertial mass which for small velocities approximates the rest mass m_0. According to Maupertuis's principle, the action integral must be a minimum, i.e. its variation must vanish. In this case the electron proceeds along a line of least action, and the number of joule seconds (compare with the everyday term 'number of man-hours') is a minimum.

On the other hand, the time of transit for a light ray travelling from a point P to a point P' can be expressed by

$$t = \int_P^{P'} (\bar{N}/c) \, ds, \tag{1.2}$$

where the refractive index, $\bar{N}$, is usually defined as

$$\bar{N} = c/u_{ph}, \tag{1.3}$$

i.e. as the ratio of the velocity *in vacuo*, c, and the phase velocity u_{ph}, of the light wave. According to Fermat's principle, expression (1.2) must be a minimum, i.e. its variation must vanish. In other words, the ray proceeds along a line of 'fastest arrival'.

Hamilton developed his theory from the close relationship between wave and particle that follows from the comparison of (1.1) and (1.2), but he did not, of course, know the concept of an electron. The subject of electron motion did not come into existence until 60 years later (J. J. Thomson, 1897), and yet another 30 years had to elapse before the development of the optics of electron rays was started.

A comparison of Fermat's and Maupertius's principles suggests that the path of the electron is identical with the path of a wave if $\overline{N} \propto p$, and it is convenient to define a dimensionless refractive index

$$\overline{N} = p/m_0 c. \tag{1.4}$$

If the electron is identified with a wave group, it can be shown that the relation between its phase velocity and particle velocity is

$$u_{ph} \times u_{el} = c^2. \tag{1.5}$$

Combining (1.3), (1.4) and (1.5) it is seen that the phase velocity of the electron wave is given by

$$u_{ph} = \frac{m_0 c^2}{p} = \frac{\text{rest energy}}{\text{momentum}}.$$

In practice it is useful to express the refractive index of an electrostatic field by the electron energy in electron volts. The potential V is connected with the momentum of the electron of rest mass m_0 and charge* $-e$ by the principle of conservation of energy, V being defined to be zero when $p = 0$. For $u_{el} \ll c$, this can be written

$$\frac{p^2}{2m_0} - eV = 0. \tag{1.6}$$

Hence, with (1.4) one obtains

$$\overline{N} = k_1 \sqrt{V}, \tag{1.7}$$

where $k_1 = (1/c)\sqrt{2e/m} = 1.978 \times 10^{-3}$ [s m^{-1}] is a constant.

The relativistic formula for the refractive index may be given here as

$$\overline{N} = k_1 \sqrt{V_r}, \tag{1.8}$$

where $\quad V_r = V\left(1 + \frac{eV}{2mc^2}\right) = V(1 + 0.987 \times 10^{-6} V) \tag{1.9}$

is known as the relativistically corrected accelerating voltage. For small voltages (1.8) and (1.7) are identical.

Since the presence of constants does not change the significance of the variation principles (1.1) and (1.2), the constant k_1 may be

---

* Note the negative sign: in the greater part of electron optical literature and throughout this book the symbol e is used for the absolute value of the electronic charge.

disregarded and a practical refractive index of the electron may be expressed directly in electron volts, namely by

$$N = \frac{\overline{N}}{k_1} = \sqrt{V_r}. \tag{1.10}$$

In the following we shall confine the discussion mainly to low electron energies, say up to 10 keV where the error $(\sqrt{V_r} - \sqrt{V})$ is about 0·5 per cent and can be disregarded for most practical purposes. Care has to be taken in the labelling of the zero voltage level. For problems connected with the cathode-ray tubes for instance, it is quite usual to assume the cathode potential as the zero level. If the cathode is actually connected to earth, the voltage V of any equipotential as measured against earth potential leads very approximately to the right refractive index $\sqrt{V}$. Greater accuracy, however, would be obtained if the cathode potential were taken not as zero but as $V_c = V_{em}$, i.e. equal to the emission energy of the electron which, for an oxide cathode, is of the order of 0·1 eV and for a tungsten cathode is about 0·25 eV. For electrons from a radioactive source, the emission energy V_{em} is considerable. If the source is at earth potential, the refractive index at any equipotential is given by

$$N = \sqrt{(V_r + V_{em})}, \tag{1.11}$$

where V_r is the voltage difference of the equipotential as measured against earth potential.*

Equation (1.10) may be illustrated by the elementary example shown in fig. 1.1. There, a beam travelling through a medium with a refractive index N passes into another medium with a refractive index N'. The angle of incidence α and the angle of refraction α' are connected by Snell's law:

$$N \sin \alpha = N' \sin \alpha'. \tag{1.12}$$

This optical law may be interpreted mechanically in the way which Newton used to explain the refraction of light at a glass surface.

---

* In electron diffraction theory, the refractive index of a crystal is usually written as
$$N = \sqrt{[E/(E - W_i)]},$$
where E is the kinetic energy of the electron in vacuo and W_i is the 'internal potential' of the crystal. In our notation, the potential outside the crystal with respect to the electron source is $V_o = E/e$, hence $W_i = e(V_o - V_i)$ and
$$N = N_o/N_i = \sqrt{(V_o/V_i)}.$$

Suppose an electron travelling with uniform speed u through a space of constant potential V passes a potential step into a space with another homogeneous potential V', so that the path of the electron suddenly changes its direction. Assuming, as in fig. 1.1,

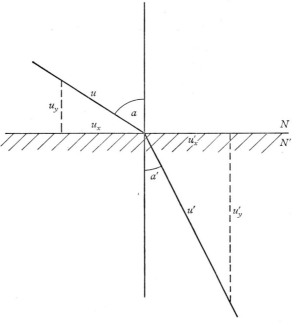

Fig. 1.1. Refraction of an electron beam.

that the potential V' is greater (positively) than V, the normal velocity component of the electron u_y is increased and the electron is accelerated. The tangential component u_x remains unchanged, so that $u_x = u'_x$. Now, at moderate energies, according to (1.6) the velocity of the electron is proportional to the square root of the potential. Moreover,

$$\sin \alpha = u_x/u,$$

$$\sin \alpha' = u'_x/u'.$$

Therefore
$$\frac{\sqrt{V}}{\sqrt{V'}} = \frac{\sin \alpha'}{\sin \alpha}, \qquad (1.13)$$

which is consistent with (1.10) and (1.12).

One principal difference between the great majority of light-optical and electron-optical refractions should be mentioned. Light rays are generally refracted by a finite number of refracting surfaces where the refractive index changes suddenly, whereas in electron-optical arrangements the refractive index changes continuously along the path of the electron. However, this difference is not an important obstacle in the application of practical light-optical methods to electron-optical problems. It is always possible to divide the range of continuously changing potential into a finite number of steps of an assumed constant potential corresponding to a finite number of sheets, each of constant refractive index. It can, for instance, be shown that the path of an electron may be traced through an electrostatic electron lens by a repeated application of Snell's law. In this case, a theoretical splitting up of the continuously varying field into a finite number of steps is sufficient to obtain results of high accuracy.

The optical description of the motion of an electron in a magnetic field is not as obvious as it is in an electrostatic field. In the presence of magnetic fields, Maupertuis's principle ceases to be applicable since the force on the electron is non-conservative, i.e. it cannot be expressed as the derivative of the potential energy of the electron in this field. The magnetic flux density B exerts a force at right angles to the electron velocity u and so has no effect upon the kinetic energy. Schwarzschild (1903) has formulated a variation principle according to which the expression

$$\int_{P}^{P'} (p - eA \cos \gamma)\, ds \qquad (1.14)$$

has to become a minimum. A is the magnetic vector potential defined* by the vector equation

$$B = \operatorname{curl} A, \qquad (1.15)$$

* The mathematical definition of the vector potential may be found in textbooks on vector theory. A physical picture for the distribution of the vector potential A in a magnetic field in which the flux density B is known (say, from search-coil measurements) can be gained from the fact that A is related to B in the same way as B is related to the generating current density i. Thus the B-lines may be represented by electric currents. The magnetic lines of flux density produced by these currents will show the direction and magnitude of A.

The determination of numerical values for A is complicated. In order to show how particular numerical values for A may be determined from a and γ is the angle between A and the path element ds. Comparing Schwarzschild's and Fermat's principles, it appears that the refractive index in the magnetic field is given by

$$\bar{N} = \frac{e}{m_0 c}\left[\sqrt{\left(\frac{2V}{e/m}\right)} - A\cos\gamma\right].$$ (1.18)

Again, since only relative values matter, one can take

$$N = \sqrt{V} - \sqrt{(e/2m)}\,A\cos\gamma$$

$$= \sqrt{V} - 2\cdot966\times 10^5\,A\cos\gamma,$$ (1.19)

where V is measured in volts and A in webers/metre. According to (1.19), N depends upon $A\cos\gamma$, i.e. the scalar product of the vectors A and ds. Thus N depends not only on the position, but also upon the direction of the motion of the electron. Under the influence of the magnetic field, the electron performs a spiralling motion in space so that the optical analogy appears to be rather complicated. (For a full discussion of the refractive index in the magnetic field, cf. Glaser (1933a, 1951), Opatowski (1944) and Ehrenberg and Siday (1949).)

In axially symmetrical magnetic fields the refraction of electron rays can be treated as a two-dimensional problem. The conservation of total angular momentum in the magnetic field can be expressed by the equation

$$r(mu_\psi - eA_\psi) = C,$$ (1.20)

where u_ψ is the tangential velocity component and r the distance of

simple flux-density distribution, (1.15) may be written in cylindrical co-ordinates (z, r, ψ) as follows:

$$\left.\begin{aligned}B_z &= \frac{1}{r}\left[\frac{\partial(rA_\psi)}{\partial r} - \frac{\partial A_r}{\partial \psi}\right],\\[4pt]B_r &= \frac{1}{r}\frac{\partial A_z}{\partial \psi} - \frac{\partial A_\psi}{\partial z},\\[4pt]B_\psi &= \frac{\partial A_r}{\partial z} - \frac{\partial A_z}{\partial r}.\end{aligned}\right\}$$ (1.16)

Thus for the homogeneous field $B_z = \text{const.}$, (1.16) can yield

$$A = A_\psi = \tfrac{1}{2}B_z r.$$ (1.17)

In this case, the numerical value of the vector potential equals half the electron-momentum value of an electron rotating about the z-axis. Further explanations will be given in chapter 5.

the electron from the axis. The constant C vanishes if the electron starts on the axis or crosses the axis where $r = 0$.

There,

$$mu_\psi = eA_\psi \tag{1.21}$$

or

$$\frac{mu_\psi^2}{2} = \frac{e^2}{m}\frac{A_\psi^2}{2}. \tag{1.22}$$

Now, the total velocity u is composed of the tangential velocity vector u_ψ and the velocity vector u_m in the rotating meridional plane, i.e.

$$u^2 = u_\psi^2 + u_m^2$$

or

$$\frac{mu^2}{2} = \frac{mu_\psi^2}{2} + \frac{mu_m^2}{2} = eV,$$

where V is the constant electron energy in electron volts. Substituting (1.22) gives

$$\frac{mu_m^2}{2} = e\left(V - \frac{e}{2m}A_\psi^2\right). \tag{1.23}$$

(1.23) is equivalent to an 'energy equation of the electron in the meridional plane'. The term in brackets is called the 'meridional potential' V_m. The integration constant in A is chosen to make $A_\psi = 0$ on the axis. Thus on the axis $V_m = V$, while with increasing distance from the axis V_m decreases. Analogously to (1.8) a 'meridional refractive index'

$$N_m = \sqrt{V_m} = \sqrt{\left(V - \frac{e}{2m}A_\psi^2\right)} \tag{1.24}$$

can be defined and electron rays can be traced through the meridional plane according to Snell's law.

§1.3. Image formation by collinear projection

At this stage it may be asked whether there is any advantage in discussing the action of potentials on electrons from the optical point of view, since at first sight, such a discussion might appear to be no more than a modern analogy for talking about an old problem. To take a simple example, there is certainly not much point in speaking of refractive indices instead of electric potentials, nor is there any advantage in calling a homogeneous electric or magnetic deflecting field an 'electric prism' or a 'magnetic prism'.

The advantages of the optical approach as compared with the ballistic or with the electrodynamic treatment of electron motion become evident when we deal with axially symmetric fields. Every electric or magnetic field of circular symmetry has the properties of a lens, i.e. it can project an optical image. This implies that we can obtain important parts of the electron path by considering two regions outside the particular refracting fields, the 'object space' and the 'image space'. A knowledge of only six points of the field, i.e. of the 'cardinal points' of the lens (which may be obtained, for example, by simple experimental procedure), is sufficient for predicting the paths of the electrons after their passage through even the most complicated electric and magnetic fields.

Object and image space are quite generally connected by a mathematical relationship of three linear equations (cf. Drude, 1925). This collinear relationship reduces to a relatively simple form in the case of circular symmetry. There, we need investigate the image formation only in a meridional section. Taking y and z as the radial and the axial coordinates respectively, we have the following relations between the object space (y, z) and the image space (y', z'):

$$\left. \begin{aligned} z' &= \frac{az+b}{cz+d}, \\[2mm] y' &= \frac{ey}{cz+d}. \end{aligned} \right\} \tag{1.25}$$

The five constants a, b, c, d, e determine the relative positions of the six cardinal points. The plane $cz+d = 0$ of the object space is, according to (1.25), conjugate to an infinitely distant plane in the image space. It is called the focal plane and cuts the axis in the focus F. Similarly, a focus F' is defined in the image space. By placing the origins of the z and z' coordinates at the foci F and F' respectively, new axial coordinates

$$x = z + \frac{d}{c},$$

$$x' = z' - \frac{a}{c}$$

are obtained. Moreover, if we introduce the abbreviations

$$f = e/c,$$
$$ff' = \frac{cb - ad}{c^2},$$
(1.26)

we obtain
$$xx' = ff',$$
(1.27)

and
$$\frac{y'}{y} = \frac{f}{x} = \frac{x'}{f'}.$$
(1.28)

f and f' are the focal lengths in the object and in the image space respectively, (1.27) is Newton's lens equation and the ratio y'/y is the magnification. For all points of the plane $x = f$ of the object space to which the plane $x' = f'$ in the image space is conjugate, this ratio is unity according to (1.28). These planes are the principal planes. Every ray passes these two planes at equal radial distances from the axis. Consequently if a ray intersects the axis at x and again at x' making the angles Θ and Θ' with it in object and image space respectively,

$$\frac{\tan \Theta'}{\tan \Theta} = \frac{x+f}{x'+f'},$$

or from (1.27),
$$= \frac{f+x}{f'+ff'/x} = \frac{x(f+x)}{f'(x+f)} = \frac{x}{f'},$$
(1.29)

and from (1.28)
$$fy \tan \Theta = f'y' \tan \Theta'.$$
(1.30)

The conjugate points on the axis for which $\Theta = \Theta'$ are called the nodal points; they are the third pair of cardinal points in addition to the foci and to the principal points. The necessity for considering the general type of object–image relation is given by the fact that all electron lenses are fundamentally 'thick lenses'. A full discussion of the use of the cardinal points in the construction of optical paths and of their determination in practice will be found in chapter 2.

§1.4. Line focus lenses

Of much importance in electron optics are line focus lenses, the focal length of which have different values in two mutually perpendicular planes. The simplest types of such lenses are obtained by two-dimensional arrangements of electrodes or magnets. For instance, a simple two-dimensional electrostatic lens is produced by a long slot in a diaphragm which separates two spaces of different electrostatic field strength (cf. §4.9). A simple magnetic line focus lens is, for instance, produced between the poles of two coaxial magnetized iron cylinders, the opposing ends of which are of equal polarity. The electron beam which is focused, passes between these ends at right angles to the cylinder axis.

Of great practical interest, however, are the quadrupole electron lenses (§2.6, §4.9 and §5.10). A magnetic quadrupole consisting of four opposing magnetized cylinders had been used already in 1935 by Hehlgans for focusing the beam in a cathode-ray tube, but only after 1952 did quadrupoles find much application, when they began to be used for the focusing of rapid nuclear particles in high energy accelerators.

Line focus lenses have been thoroughly investigated in glass optics (see for instance M. von Rohr, 1920, pp. 182–96) and many of the results apply equally to electron optics. The glass optical lenses which correspond to the electron optical quadrupoles have surfaces with different curvatures in two mutually perpendicular cross-sections, namely the two principal sections which are called tangential and sagittal section respectively. The refracting surfaces of double curvature are for all rays (excepting skew rays) subject to the same formulae and laws which have been discussed here for an axially symmetrical surface, the single radius of curvature now being replaced by two radii of curvature in the two principal sections. For each object point, two conjugate, but independent image lines are projected, namely one in the tangential and the other in the sagittal plane. There is however one important difference between all glass-optical line focus lenses and many electron optical line focus systems. The glass-optical line focus lenses represent 'congradient' systems, in which the rays travel approximately in the direction of the gradient of the refractive index. On the other hand, we will show in §2.5 that quadrupole lenses are 'transgradient' with rays travelling at right angles to the field gradient.

Again, owing to the transverse nature of their active field these lenses have an advantage over electron lenses of circular symmetry where the central component of the field is longitudinal with respect to the beam axis. For the same energy of incident electrons, the convergence produced by quadrupole fields is much stronger; for this reason they are sometimes called 'strong-focusing lenses'.

Quadrupole electron lenses, unfortunately, have one inconvenient feature, they converge the rays in one principal section only, in the other principal section they always diverge the rays. One can, however, combine two or more quadrupoles, each turned about the axis by $\pi/2$, and with potentials and geometric dimensions chosen in such a way that the same converging effect results in the two principal sections (cf. §2.5). Such a system then works like a circular symmetrical lens.

§1.5. The physical problem of image formation

It is a very important fact that (1.30), which is derived from purely mathematical principles, can be applied in physics to paraxial rays only, i.e. only to rays which are close to the axis and make with it a very small angle Θ. For wide beams, the image formation for an element of area is no longer collinear. A correct point-by-point imaging of objects of any size by widely diverging beams is physically impossible. This statement applies to electron optics in the same sense as it applies to light optics since, as we have already shown, electron paths obey the same basic laws as light rays provided the appropriate values of refractive index (cf. (1.10) and (1.19)) are assumed.

An important relationship which must hold if correct point-to-point imaging of finite objects is to occur is expressed by Abbe's sine law. In its light optical form this can be written

$$Ny \sin \Theta = N'y' \sin \Theta', \qquad (1.31)$$

where y and y' are the distances from the axis of an object point and of an image point respectively. Θ and Θ' are the semi-aperture angles of the bundles of rays at object and image and N, N' are the refractive indices of object space and image space. If the object space is a region of constant potential V and the image space is a region of potential V', then the electron optical form of the sine law is given from (1.7) and (1.31) by

$$\sqrt{V}y \sin \Theta = \sqrt{V'}y' \sin \Theta'. \qquad (1.32)$$

The validity of the sine law for electromagnetic rays can be demonstrated in various ways. A physical interpretation of the theorem for electron rays can be obtained by noting that $\sqrt{V} \sin \Theta$ is proportional to the maximum transverse velocity u_r of electrons leaving the object, while $\sqrt{V'} \sin \Theta'$ is proportional to the maximum transverse velocity of electrons reaching the image. This implies the invariance of the product $u_r y$ for conjugate planes. This is consistent with Liouville's theorem of statistical mechanics, which states that if the motion of a bunch of particles is represented by the motion of their corresponding 'phase points' in 'phase space', then the density of phase points in the bunch remains invariant through the motion. If the electron-optical channel is non-absorbing, it follows that the volume of phase space occupied by the bunch does not change, since the number of particles remains constant. The product of $u_r y$ of the maximum transverse velocity and the maximum transverse displacement in a bunch of particles at the object plane is related to the phase volume occupied by the bunch, since the phase-space coordinates for the transverse motion are the transverse momentum and the transverse displacement. Thus (1.32) expresses the equality of phase-space densities at object and image.

In electron optics, just as in glass optics, Abbe's sine law implies that the image formation for an area element can be collinear for narrow bundles of rays with small aperture Θ only. Now, since

$$\sin \Theta = \Theta - \frac{\Theta^3}{3!} + \frac{\Theta^5}{5!} - ..., \qquad (1.33)$$

(1.32) may be reduced by neglecting higher order terms to

$$Ny\Theta = N'y'\Theta'. \qquad (1.34)$$

This is called Lagrange's law. A comparison of (1.34) with (1.30) leads to the conclusion that

$$\frac{f}{f'} = \frac{N}{N'}. \qquad (1.35)$$

Moreover, according to this comparison of (1.34) with (1.30) the collinear projection provides a satisfactory approximation to the physical image formation if Θ is small so that

$$\sin \Theta \approx \tan \Theta \approx \Theta,$$

i.e. if sin Θ is approximated accurately enough by the first term of (1.33). The treatment based on this approximation is called 'Gaussian optics' after its originator. Alternatively, it is called 'first-order theory' as distinguished from a more rigorous 'third-order theory' which attempts to include the second term of the sine series (1.33).

§1.6. Electron optical brightness and the directional intensity of a beam

The angular distribution of emission of electrons at the surface of the cathode has been shown by Langmuir (1937b) to follow Lambert's law which can be written

$$i_\Theta = \frac{i_{em}}{\pi} \cos \Theta, \qquad (1.36)$$

where i_Θ is the emission density in any direction making an angle Θ with the normal and i_{em} is the total emission per unit area into the hemisphere. Hence, referring to fig. 1.2, the emission from the

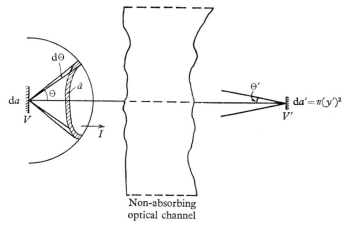

Fig. 1.2. Derivation of brightness law.

area da into the angular interval between Θ and $\Theta + d\Theta$, i.e. towards a ring area $\bar{a} = 2\pi \sin \Theta \, d\Theta$ on the unit sphere is given by

$$dI(\Theta) = i_\Theta \, da \, \bar{a} = \frac{i_{em}}{\pi} \cos \Theta \, da \, 2\pi \sin \Theta \, d\Theta. \qquad (1.37)$$

The amount of radiation emitted from a cathode between the angles Θ_c and $(\Theta_c + d\Theta_c)$ with the normal to the surface of area πy_c^2 can be written

$$dI(\Theta_c) = 2\pi\beta\,\pi y_c^2 \sin\Theta_c \cos\Theta_c\, d\Theta_c, \tag{1.38}$$

where the quantity $\beta = i_{em}/\pi$ (1.39)

is the electron optical brightness of emission which is analogous to the luminance of a light optical source.

At the electron optical image—and analogous considerations apply to the cross-over in an electron gun (cf. §9.4 and §9.9)—the radiation $dI(\Theta')$ will intersect the axis at angles Θ' which are connected to the corresponding quantities at the cathode by Abbe's sine law (1.32), namely,

$$\sin\Theta' = \frac{Ny_c}{N'y'} \sin\Theta_c. \tag{1.40}$$

On differentiation Abbe's law gives

$$\cos\Theta'\, d\Theta' = \frac{Ny_c}{N'y'} \cos\Theta_c\, d\Theta_c. \tag{1.41}$$

Substituting the last two equations into (1.38) yields

$$dI = 2\pi\beta\left(\frac{N'}{N}\right)^2 \pi(y')^2 \sin\Theta' \cos\Theta'\, d\Theta',$$

or taking $\beta' = \beta(N'/N)^2,$ (1.42)

we obtain $dI = 2\beta'(\pi y')^2 \sin\Theta' \cos\Theta'\, d\Theta'.$ (1.43)

The angular distribution of the radiation is thus the same in (1.38) and in (1.43). That is, if Lambert's law is obeyed by the cathode it will be also obeyed by the cross-over and by the image, provided the emission system is free from aberrations.

There are, however, the two significant differences between the cathode and the other cross-sections under consideration. First, at the cathode, electrons are emitted in all directions, i.e. $\Theta_c \leqq \frac{1}{2}\pi$. Hence, from Abbe's law, follows

$$\sin\Theta' \leqq Ny_c/N'y', \tag{1.44}$$

i.e. the rays pass through the cross-over or through the image at a limited angle of divergence. Secondly, the brightness of the cross-over or of the image is different from that of the emitting surface and, due to (1.42) with $N \sim \sqrt{V}$, the ratio of brightness

$$\frac{\beta'}{\beta} = \frac{V + V_{em}}{V_{em}} \qquad (1.45)$$

is given by the ratio of electron energies at the cathode (V_{em}) and at the cross-over or image respectively $(V_{em} + V)$. If $N = N'$, (1.42) becomes the familar law for ordinary optical instruments that the apparent brightness of a source of light cannot be changed by any focusing process.

The brightness law (1.45) together with Abbe's sine law (1.31) is sufficient to solve many problems in the design of electron beams (cf. Helmer, 1966) in particular, as a consequence of these two laws, there follows the existence of an upper limit for the current density obtainable in the focused spot of an electron gun; this will be shown in §7.1.

Considering the reduced angle Θ' of divergence of the beam indicated by (1.40), an expression for the current density per unit solid angle, namely (dropping the prime from Θ')

$$b = i/\Omega = i/2\pi(1 - \cos\Theta) \; [\text{A m}^{-2}\,\text{sterad}^{-1}] \qquad (1.46)$$

is found to be of much practical use as a measure of the concentration of current density i in a beam. b is the 'directional intensity' or 'brightness of the beam'.* For small Θ,

$$b = i/\pi\Theta^2. \qquad (1.47)$$

If the semi-aperture of the beam at a cross-over or image is Θ, then from (1.43) the current in the beam I is given by

$$I = 2\beta'(\pi y')^2 \int_0^\Theta \sin\Theta' \cos\Theta' \; d\Theta' = \beta'(\pi y')^2 \sin^2\Theta.$$

Thus from (1.42) the current density at the cross-over is given by

$$i = \frac{I}{\pi y'^2} = \pi\beta_c \left(\frac{N'}{N}\right)^2 \sin^2\Theta. \qquad (1.48)$$

* i/Ω was introduced in the electron optical literature by Dosse (1940) as the 'Richtstrahlwert', a term which also has been used sometimes in the English literature.

So that for small Θ, the directional intensity

$$b = \frac{i}{\pi\Theta^2} = \beta_c \left(\frac{N'}{N}\right)^2 = \beta_c \left(\frac{V+V_{em}}{V_{em}}\right).$$

Expressing the mean energy of electron emission (eV_{em}) in terms of the cathode temperature T and the Boltzmann constant k, we have

$$eV_{em} = kT. \tag{1.49}$$

Hence, the directional intensity is given by

$$b = \frac{i_{em}}{\pi} \left(1 + \frac{eV}{kT}\right). \tag{1.50}$$

In this form, the brightness law (1.50) shows that for a given emission density i_{em} from a cathode at the temperature T, the directional intensity b of the focused beam appears to depend only on its potential V relative to the cathode.

We should, however, mention that some reduction of b below its optimal value given by (1.50) will be caused by electron optical errors in the focusing system (cf. chapter 6) or by interactions between the electrons in beams of high space charge density (cf. Spanner, 1967). As a matter of practical interest, we compare here in table 1.1 some photometric and electron optical quantities.

TABLE 1.1. *Photometry and Electron Optics*

Light optics	Electron optics	Symbol
Luminous flux [lumen]	Electron current	I [A]
Emittance ⎫ ⎡lumen m^{-2}⎤ Luminous ⎬ ⎢ = lux ⎥ flux density ⎭ ⎣ ⎦ (illumination)	Emission density ⎫ Current density ⎬	i [A m^{-2}]
Luminous intensity [lumen sterad^{-1}]	Angular current density	I/Ω [A sterad^{-1}]
Luminance Brightness of surface, [lumen m^{-2} sterad^{-1}]	Electron optical ⎫ brightness ⎬	$\beta = i/\pi$ $\left[\dfrac{\text{A m}^{-2}}{\text{sterad}^{-1}}\right]$
Beam intensity ⎫ Brightness of beam ⎭ [lumen m^{-2} sterad^{-1}]	Directional beam intensity	$b = i/\Omega$ [A m^{-2} sterad^{-1}]

§1.7. Conclusion

The explanations given in this chapter illustrate the essential character of the optical method for the study of electron trajectories. It is seen that the optical method consists in the investigation of systems of rays, i.e. a manifold or collective of paths, while the individual paths of the electrons are derived from the properties of the electromagnetic field. Optical properties apply to this collective, e.g. foci, principal planes, etc., while mechanical properties apply to individual paths.

It is, however, immaterial how the optical properties of a system are derived, whether with the help of optical, or ballistic, or electrodynamic, or thermodynamic methods, and in the following chapters, we shall make use of all available methods in a complementary manner and according to best convenience.

CHAPTER 2

THE CARDINAL POINTS OF
AN ELECTRON LENS

§2.1. Location of cardinal points by ray tracing with a sliding target

A true optical image is formed if every point of the image space corresponds to a conjugate point of the object space. Straight lines, connecting two points, are in the ideal case reproduced as straight lines, and planes are reproduced as planes. According to §1.3, such image formation can be effected near the axis of a refracting system of circular symmetry. There, the relationship between object and image space can be defined completely by the location of six points. These are the two focal points, the two principal points and the two nodal points. We shall start with an explanation of these points in the case of the simplest electron lens. We shall also discuss how these points can be located experimentally and how the knowledge of their position can be used for practical purposes.

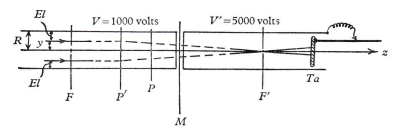

Fig. 2.1. Investigation of an electron lens by means of a sliding target.

The simplest arrangement acting as an electron lens consists of two co-axial metal tubes which are kept at different electrostatic potentials. In fig. 2.1 are shown two tubes of equal diameter, the left at the lower electric potential V, the right at a higher potential V'. In order to specify a concrete example, we shall assume that

$V = 1000$ volts and $V' = 5000$ volts. Both tubes lie symmetrically about an ideal plane, the so-called midplane of the system,
cutting the axis of circular symmetry (the z-axis) at the point M.
By a special arrangement, which will not be described at the moment,
electron beams El, travelling exactly parallel to the z-axis, are
transmitted through the tubes. These electrons are gradually
accelerated from the lower potential V to the higher potential V',
and in passing through the electric field they are also gradually
deflected. The path of the beams can be investigated by means of a
small target Ta, which is covered with a fluorescent substance. This
target is kept at the potential V' of the second tube and can be
moved along the axis of the tube. All points where the electron
beams hit the target are visible by fluorescence, so that the distances
of the beam from the z-axis can be everywhere determined. To
avoid disturbances of the field of the lens, the target should be
kept outside the lens, i.e. in the field-free space where the electrons
travel along perfectly straight lines. The path of the beams inside
the lens is actually curved. The straight, broken lines drawn in
fig. 2.1 are obtained by extrapolation of the measured parts of the
path.

It will be noticed that the electron beams intersect at one point,
the focus F'. Extrapolating the straight part of a traced electron
beam backwards until it cuts the prolongation of the original
parallel beam one finds the principal plane, which intersects the
axis in the principal point P'. Assuming parallel beams starting
from the high-voltage side of the lens (i.e. coming from the right
side of fig. 2.1), the electrons would be decelerated and focused at
a point F. The intersection of the extrapolated original parallel
beam and the extrapolated final direction would show the other
principal point P. Thus, by this simple experiment, there can be
found four cardinal points, P, P' and F, F', of the electron lens.
The two remaining points are the nodal points K and K'. They
lie in the same direction as the principal points with reference to
the respective focal points, but are shifted by the difference between
the numerical values of the two focal lengths, thus

$$KF = PF + |P'F' - PF| = P'F',$$
$$K'F' = PF.$$

$$(2.1)$$

Therefore if the focal points and the principal points are obtained experimentally, the nodal points can be located by an easy procedure. For a general estimate of the position of the cardinal points it may be helpful to notice that for all two-electrode lenses, both principal points lie on the same side of the midplane, namely, at the low-voltage side where the refractive index is small. The principal points as well as the nodal points are crossed over with respect to their appropriate focal points.

§2.2. Geometrical relations between object and image space

The location of the cardinal points for any axially symmetrical electric field is of the greatest practical importance. Their position controls the size and position of the electron image of any source of electron rays. This may be explained by means of fig. 2.2. There,

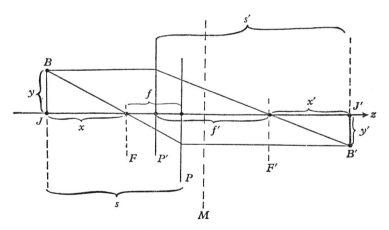

Fig. 2.2. Location of an optical image. Lateral magnification.

F and F' are the two focal points, P and P' the two principal points and JB the electron source. At J' there can be detected (for instance, by means of the sliding fluorescent target) a sharp electron image. In the example of fig. 2.2 this image $J'B'$ is found to be slightly reduced in comparison with the object JB. By the following simple construction this image can be found graphically. A beam, parallel to the z-axis drawn through a point B of the object, is refracted at the principal plane through P' so as to pass through the focus F'. Another beam emerging from B and travelling through the focus F is refracted at the principal plane through P and leaves the lens in a direction parallel to the axis. The two rays intersect at B', where the image is formed.

The general lens formulae which were derived in §1.3 from the collinear relationship can be obtained here at once geometrically from fig. 2.2. Taking $FJ = x$ and $J'F' = x'$ as object and image distances measured from the foci, and taking $FP = f$ and $F'P' = f'$ as the focal lengths of the system and $JB = y$ and $J'B' = y'$ as object and image size, it can be seen from fig. 2.2 that the following relation holds for the lateral magnification y'/y:

$$\frac{y'}{y} = \frac{f}{x} = \frac{x'}{f'}, \tag{2.2}$$

from which Newton's formula

$$xx' = ff' \tag{2.3}$$

is derived. Moreover, taking $s = f + x$ and $s' = f' + x'$ as the object and image distances from the appropriate principal points, there follows from (2.3)

$$(s-f)(s'-f') = ff'$$

or

$$\frac{f}{s} + \frac{f'}{s'} = 1. \tag{2.4}$$

The angular magnification or ratio of conjugate divergences is defined as the ratio of the two angles which two conjugate beams make with the axis. The angles may be so small that their tangents are substantially equal to their radian measure. In fig. 2.3 object and image points are represented by J and J', and the two conjugated divergences are Θ and Θ'. From fig. 2.3 it follows immediately (cf. (1.29))

$$\frac{\Theta'}{\Theta} = \frac{JP}{J'P'} = \frac{s}{s'} = \frac{f+x}{f'+x'}, \tag{2.5}$$

and using (2.3)

$$\frac{\Theta'}{\Theta} = \frac{x}{f'}. \tag{2.6}$$

The angular magnification Θ'/Θ and the lateral magnification y'/y are connected by Lagrange's law which, according to (1.34) and (1.10) can be written

$$\Theta y \sqrt{V} = \Theta' y' \sqrt{V'}. \tag{2.7}$$

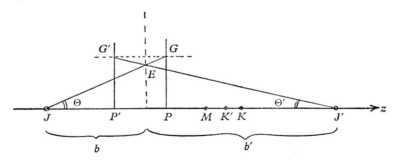

Fig. 2.3. Divergence of a beam. Angular magnification.
Thin equivalent lens.

Combining (2.7) with (2.6) and replacing there x by $(y/y')f$
from (2.2) leads to the relation (cf. also (1.35) and (1.7))

$$\frac{f}{f'} = \sqrt{\frac{V}{V'}}. \tag{2.8}$$

The ratio of the two focal lengths of the electron lens is equal to
the square root of the ratio of the voltages at the electrodes.

The principal planes are defined as planes of unit lateral magnification, i.e. a ray arriving at one of the principal planes at a given
distance from the axis will leave the other principal plane at the
same axial distance. Nodal planes, that is, planes perpendicular to
the axis through the nodal points, are correspondingly defined as
planes of unit angular magnification, i.e. a ray passing one of these
planes at an angle Θ to the axis, passes the other nodal plane with
the same inclination $\Theta' = \Theta$. Thus from Lagrange's law, (2.7), it
follows that the lateral magnification in the nodal planes is given
by $y'/y = \sqrt{V}/\sqrt{V'}$, or combining this result with (2.2) and (2.8)
the expression (2.1) for locating the nodal points is proved. The
location of the nodal points is of practical interest for the determination of the lateral magnification. Since the angular magnification
in the nodal points is unity, $\tan \Theta' = \tan \Theta$ or

$$\frac{y'}{K'J'} = \frac{y}{KJ}. \tag{2.9}$$

Hence the lateral magnification is equal to the ratio of the nodal
distances for any conjugate object and image.

Electron lenses are generally 'thick' lenses, since the field produced by the one electrode penetrates appreciably into the other electrode, so that the region of varying refractive index extends over a distance which is considerable in comparison with the focal length of the lens. However, for many practical considerations it is a useful fact that there can always be found a 'thin equivalent lens', i.e. an infinitely thin lens that will produce an image of the same size and in the same place as the image produced by the thick lens. The distances b and b' of the object and image from this thin equivalent lens depend on the magnification and on the refractive indices (Hess, 1934); they are given by:

$$\frac{y'}{y} = \frac{b'}{b} \sqrt{\frac{V}{V'}}. \qquad (2.10)$$

The position of the thin equivalent lens E is indicated in fig. 2.3. It may be noticed, however, that this position is not fixed, but changes with the position of the object and image. The thin equivalent lens always lies between the two principal planes P and P', it moves gradually from P' to P as the object J moves gradually from an infinite distance to the focus F (corresponding to a movement of the image J' from the focus F' to infinity). If the object J and the image J' are connected by straight lines with any point E of the thin equivalent lens (see fig. 2.3), the ratio of the object and image distances gives the correct angular magnification

$$\frac{\Theta}{\Theta'} = \frac{b'}{b}. \qquad (2.11)$$

This can be seen by introducing Lagrange's equation (2.7) in (2.10). Combining (2.5) with (2.11) one obtains an expression for locating the thin equivalent lens

$$\frac{b}{b'} = \frac{s}{s'}. \qquad (2.12)$$

The position of the thin equivalent lens divides the object-image distance JJ' in the ratio $JP/J'P'$, which is given by the distance of object and image from their appropriate principal planes.

§2.3. Combination of lenses

Of some interest are the effects of a combination of lenses. For instance, it is well known in glass optics that a sequence of two thin lenses of focal lengths f_1 and f_2 respectively will have the combined focal length given by

$$\frac{1}{f} = \frac{1}{f_1} + \frac{1}{f_2} - \frac{D}{f_1 f_2}. \qquad (2.13)$$

In the special case when $f_2 = -f_1$, it follows from (2.13) that $f > 0$. Thus the combination of a converging lens with an equally strong diverging lens will always result in a system which converges the electron beam. This result is of much importance for the so-called 'strong focusing lenses' which will be discussed in §4.9.

Another practically important problem concerning the combination of lenses occurs in the design of a chain of lenses which is able to channel a long beam, such that the transverse dimensions of it will always remain smaller than the aperture stops of these lenses. This problem has been discussed for instance for the special case of a chain of thick symmetrical lenses by Grivet (1965, p. 252). All these lenses are supposed to be equal and spaced at the same mutual distances. It is shown by Grivet (1952) that the outermost trajectory of the beam will be stable as long as

$$D < 2f, \qquad (2.14)$$

where f are the focal lengths for all lenses and D are the distances measured from the second focus of one lens to the first focus of the following lens.

§2.4. Practical methods for locating cardinal points

The theory of the location of cardinal points presented so far applies in the same way to electron lenses and to glass lenses. The technique of measuring the positions of these points, however, is substantially different in the two cases. In electron optics special experimental methods are required, since the measurements have to be taken in evacuated tubes and special precautions are necessary to avoid the occurrence of spurious electric or magnetic fields which easily would modify the path of charged particles.

The accurate, but rather elaborate, method by Klemperer and Wright (1939) is a direct application of the electron pencil tracing already mentioned at the beginning of this chapter. A series of initially parallel, very narrow pencils (diameter $\approx$ o·1 mm) emerges through the fine holes of a 'pepperpot-diaphragm', and then passes through the electron lens under investigation. The pencils are detected with a sliding fluorescent target. This target may be made from glass covered with Willemite (1 mg cm^{-2}) shielded by a fine wire gauze, to prevent distortion of the electron trajectories, owing to the accumulation of charges on the screen. The fluorescent spots are observed from the back of the target, and their mutual distances are measured by a scale in the eyepiece of a calibrated microscope of large object-to-objective distance.

A more rapid but less accurate method for the location of the cardinal points has been described by Epstein (1936). Lateral magnifications have to be measured at two different object and image distances. A wire mesh 'illuminated' with electrons may be used as a suitable object; suppose the mesh is arranged at a distance p from the midplane. The image can be found by means of the sliding fluorescent target; let it be distant q from the midplane. Using again the same notation as above, we obtain

$$
\left.
\begin{aligned}
p &= JM = x + MF = \frac{y}{y'} f + MF, \\[2mm]
q &= J'M = x' + MF' = \frac{y'}{y} f' + MF'.
\end{aligned}
\right\} \tag{2.15}
$$

If p_1, q_1 and y_1, y_1' correspond to one position of the object, p_2, q_2 and y_2, y_2' to another position of the object, it follows that

$$
f = \frac{p_1 - p_2}{\dfrac{y_1}{y_1'} - \dfrac{y_2}{y_2'}}, \tag{2.16}
$$

$$
MF = \left(p_1 \frac{y_1'}{y_1} - p_2 \frac{y_2'}{y_2} \right) \Big/ \left(\frac{y_1'}{y_1} - \frac{y_2'}{y_2} \right). \tag{2.17}
$$

The conjugate values of f' and MF' are obtained if in (2.16) and (2.17) the experimental values of p, y and y' are interchanged with those of q, y' and y. Since, however, the image has to be measured with paraxial rays only, the depth of focus is very large, and therefore the accuracy of locating the image and of measuring the distance q is low.

Klemperer and Wright (1939) pointed out that Epstein's method may be easily modified by replacing lateral by angular magnifications using Lagrange's law (2.7). In (2.16) and (2.17) therefore, every ratio of y/y' has to be replaced by $\dfrac{\Theta'}{\Theta}\sqrt{\dfrac{V'}{V}}$. The required pairs of values of p, q, Θ and Θ' are found experimentally by tracing divergent pencils emitted from a point-source object (see chapter 9) and focused to a point image.

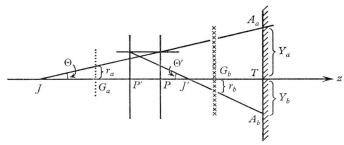

Fig. 2.4(a). Shadow method for location of cardinal points.

This line of attack has been followed up by Spangenberg and Field (1942a, b) and by other workers. They measured the angular magnification by their ingenious 'shadow method', which has the great advantage of avoiding the need of parts which have to be moved relative to one another in the vacuum. The method is illustrated in fig. 2.4(a). A cone of rays of fixed semi-vertical angle Θ is emitted from a point source J and detected by a fluorescent target T, both at a fixed position. At the object and at the image side respectively there are wire grids G_a and G_b, each casting a shadow on the target. Both these measuring grids consist of parallel equidistant

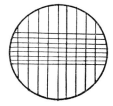

Fig. 2.4(b). Shadows of measuring grids G_a and G_b on the target.

wires at right angles to the axis. The wires of G_a are arranged in a perpendicular direction to those of G_b, so that their shadows on the target can be distinguished from one another as shown in fig. 2.4(b).

The radius of the illuminated area on the target is Y_a if no voltage difference is applied to the lens electrodes; it changes into Y_b if a voltage difference is applied. The radii of the beam at G_a and G_b are r_a and r_b respectively; they can be determined by counting the number of wire shadows in the image at the target. Now if the positions of the points G_a, G_b and T at which the two grids and the target respectively intercept the axis (z) are known, the angles of divergence of the beam at the object J and at the image J' are given by

$$\left. \begin{array}{l} \Theta = \dfrac{r_a}{JG_a} = \dfrac{Y_a}{JT}, \\[2ex] \Theta' = \dfrac{Y_b - r_b}{G_b T} = \dfrac{Y_b}{J'T}. \end{array} \right\} \qquad (2.18)$$

These equations are sufficient for the determination (i) of the angular magnification Θ'/Θ and (ii) of object and image positions J and J' respectively, i.e. of p and q. Two measurements have to be taken at two different positions of J giving p_1, p_2 and q_1, q_2 also:

$$\frac{y_1'}{y_1} = \frac{\Theta_1}{\Theta_1'} \sqrt{\frac{V_1}{V_1'}},$$

and

$$\frac{y_2'}{y_2} = \frac{\Theta_2}{\Theta_2'} \sqrt{\frac{V_2}{V_2'}}.$$

These six values for p_1, p_2, y_1, y_2, y_1' and y_2' can be substituted into (2.16) and (2.17) leading to the determination of the cardinal points.

A special case of practical importance is met in electron microscopy where lenses are operated under extreme conditions of high magnification, say $y'/y = 10^2$ to 10^3. There the object distance is very nearly equal to the focal length and the image distance s' is relatively so large that, with sufficient accuracy, it equals the mid-image distance q or the distance x' of the image from the focus, namely

$$s' \simeq q \simeq x'.$$

Moreover, the refractive index is generally the same in the object space and in the image space. Hence, using (2.8) and (2.15) one obtains

$$f = f' \simeq \frac{q}{y'/y} \simeq \frac{q}{(y'/y)+1}. \qquad (2.19)$$

In this way, the focal lengths of microscope lenses are obtained in practice (Dosse, 1941; Liebmann, 1952 a) by measuring, with an object of known size y, the image size y' and the image distance q from the centre of the lens.

§2.5. Cardinal points of line focus lenses*

We have discussed so far the cardinal points of lenses with circular symmetry. In electron lenses with translational symmetry, i.e. in 'two dimensional' fields we have corresponding cardinal lines.

The simplest type of such fields can be imagined to be produced by replacing the two tubes, shown in cross-section in fig. 2.1, by two pairs of plates, the axis of the tubes being replaced by a plane of symmetry in the middle between and parallel to the four plates.

The focal and principal lines of these plate lenses can be traced with the sliding fluorescent target in a similar way as explained in §2.1, however, the pepperpot-diaphragm which for the tubes contains co-axial circles of holes must now have the holes arranged in the form of squares, the sides of which are parallel or perpendicular respectively and symmetrically arranged with respect to the plates.

The focal lengths of four-plate lenses must be expected to be much shorter (by a factor ~ 2) than the focal lengths of two-tube lenses of the same cross-section. This is easily concluded by observing the decay of potential from the midplane M (cf. fig. 2.1) along the z-axis. The results of measurements of this potential decay for tubes, plates and a box of rectangular cross-section are shown in fig. 2.5. The lens electrodes are drawn in cross-section at the bottom of the figure; for a circular structure, A and B together represent a tube of radius R with axis z. For a two-dimensional structure, however, each, A and B, represent conducting plates extending to infinity in the x-direction, i.e. at right angles to the plane of the drawing. The mutual distance of the plates is $2R$, where R may be called the 'semi-aperture' of the 'plate lens'. The broken curve in fig. 2.5 corresponds to the case of a line focus lens in which the plates do not extend to infinity but are chopped off at distances $\pm x = \pm 2R$ from the xy-plane of the drawing and connected by another plate parallel to the yz-plane so that a box-

---

* Cf. Bernard (1954), Bullock (1955) and Septier (1961).

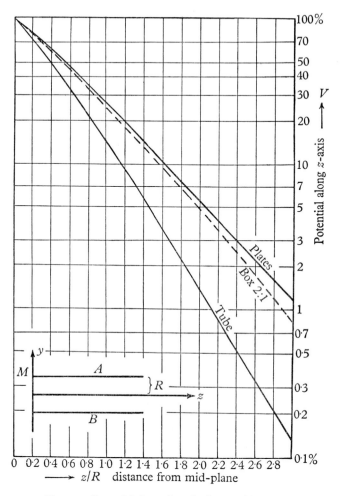

Fig. 2.5. Potential decay in tube lens and in
corresponding line focus lenses.

like electrode structure is obtained, the cross-section being a
rectangle of side ratio 1:2 in the *xy*-plane. It can be seen in
fig. 2.5 that the potential decay is less rapid in the more open
planar structures than in the circular structure. The conclusion of
a correspondingly greater focusing power of the planar structure is
illustrated by some experimental curves of Lubszynski *et al.* (1969)

shown in fig. 2.6. There mid-image distances q are plotted against mid-object distances p for a fixed voltage ratio $V_2/V_1 = 6\cdot0$ applied to both lenses. The two diameters of the tubes for the two-tube lens and the two-plate separations for the four-plate lens

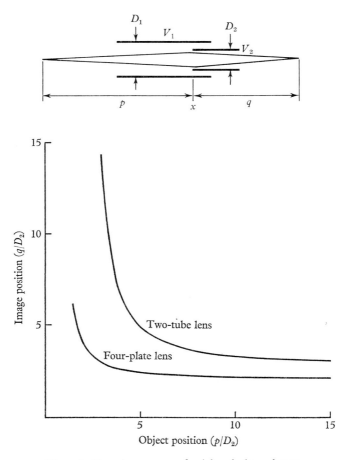

Fig. 2.6. Focusing powers of axial and planar lenses.
$V_1/V_2 = 6\cdot0, D_1/D_2 = 1\cdot6$.

are in each case D_1 and D_2. p and q are plotted in terms D_2, where $D_2 < D_1$.

As another example for the greater focusing power of a line focus lens compared with the corresponding lens of circular symmetry, we mention here that the focal length of a slit-aperture is just half as long as the focal length of a circular aperture; this will be derived in §4.4.

Of greatest interest are the quadrupole lenses which we have introduced already in §1.4. We will show now that the potential distribution in many of these lenses and also the position of their cardinal lines can be described by simple formulae.

The ideal electrostatic quadrupole field is formed by four electrodes, the cross-section of their surfaces are hyperbolae as shown in fig. 2.7. The axis (z) of the lens is at right angles to the

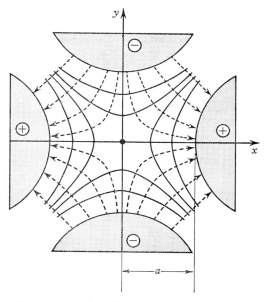

Fig. 2.7. Cross-section through an electrostatic quadrupole lens.

drawing and the electrodes are long (in the z-direction) in comparison to their distances $\pm x = a$ and $\pm y = a$ from the z-axis. One pair of opposite electrodes (that on the x-axis of the drawing) is positively charged, the other pair (on the y-axis) is negative. Electrons travelling in the z-direction are attracted by the positive electrodes, i.e. they are diverged in the x-direction away from the z-axis but they are repelled by the negative electrodes, i.e. they are converged in the y-direction towards the z-axis.

The electrostatic lines of force (shown in the figure as broken lines) form 4 systems of hyperbolae, the x- and y-axis being their asymptotes. The equipotentials which are drawn as solid lines, are again hyperbolae, they are orthogonal to the lines of force.

The potential at any point can be written

$$V = \tfrac{1}{2}k(x^2 - y^2) \tag{2.20}$$

at the electrode surfaces, the voltage will be

$$V_a = \pm \tfrac{1}{2}ka^2. \tag{2.21}$$

The field E has the components

$$\left.\begin{aligned} E_x &= -\frac{\partial V}{\partial x} = -kx, \\ E_y &= -\frac{\partial V}{\partial y} = +ky, \end{aligned}\right\} \tag{2.22}$$

and the equations of motion for the electron are

$$\left.\begin{aligned} \frac{-eE_x}{m} &= \ddot{x} = \omega^2 x, \\ \ddot{y} &= -\omega^2 y, \end{aligned}\right| \tag{2.23}$$

where $\omega^2 = -k\dfrac{e}{m}$, and substituting for k from (2.21) we obtain

$$\omega^2 = \pm\frac{e}{m}\frac{2V_a}{a^2}, \tag{2.24}$$

where the $+$ and $-$ signs refer to the x- and y-electrodes respectively. The solutions to the differential equations (2.23) are

$$x = x_0 \cosh(\omega t), \tag{2.25}$$

$$y = y_0 \cos(\omega t). \tag{2.26}$$

These equations refer to initial conditions of an electron entering the field with an initial velocity u in the z-direction and at an initial distance (x_0, y_0) from the z-axis.

Using (2.24) for ω^2 and taking the energy equation $u^2 = 2V_0(e/m)$ for the initial velocity of the electrons, we can now introduce the 'excitation constant'

$$\beta = \frac{\omega}{u} = -\frac{1}{a}\left(\frac{V_a}{V_0}\right)^{\frac{1}{2}}. \tag{2.27}$$

Moreover, if L is the length of the electrodes in the z-direction, the electron will spend the time

$$t_1 = L/u \qquad (2.28)$$

in the lens. The fringing fields are neglected here in the so-called 'rectangular model'.

Now for an electron with initial velocity in the z-direction only and with the initial distance y_0 from the z-axis, the displacement y_1 upon leaving the field is given by (2.26), hence the y-component of the velocity at this point is

$$\dot{y}_1 = -\omega y_0 \sin(\omega t). \qquad (2.29)$$

The trajectory will intersect the z-axis at a distance z_f from the lens, while the emergent angle Θ is given by

$$\tan \Theta = \frac{y_1}{z_f} = \frac{\dot{y}_1}{u}. \qquad (2.30)$$

Substituting in (2.30) from (2.26) and (2.29) for y_1 and $\dot{y}_1$ respectively, and also from (2.27) and (2.28) for $\omega = L/t_1$, we obtain

$$z_f = \frac{1}{\beta} \cot(\beta L). \qquad (2.31)$$

z_f is independent of y_0 which means that all trajectories initially parallel to the z-axis will intersect at the same point. This is just the property of the focus of a lens.

The focal length of this thick lens is obtained by producing the emergent ray back, to intersect the initial ray, as it was parallel to z, hence

$$\tan \Theta = y_0/f_y = \dot{y}_1/u \qquad (2.32)$$

and finally, from (2.32) using (2.26) for y_1 with (2.27), (2.28), (2.30) and (2.31), we obtain

$$f_y = 1/\beta \sin(\beta L), \qquad (2.33)$$

or for $\beta L \ll 1$ (i.e. $\sin \beta L \simeq \beta L$),

$$f_y = 1/(L\beta^2). \qquad (2.34)$$

As the potential of the electrons is the same in front $(+z)$ and behind $(-z)$ the quadrupole, and since its geometry is perfectly symmetrical about the xy-plane in the middle of the lens, the two focal lengths in front and behind ($f_x = f'_x$) must be equal and the two principal planes must coincide in the middle of the lens, i.e.

$$z(P_y) = \tfrac{1}{2}L. \tag{2.35}$$

With the specification of the last two equations we can apply to the quadrupole all the thick lens formulae given in §2.2.

For an electron path at an initial distance x_0 from the z-axis, the displacement x_1 upon leaving the field is given by (2.25), and the x-component of the velocity becomes

$$\dot{x}_1 = \omega x_0 \sinh (\omega t). \tag{2.36}$$

Analogously to the calculation for the y-components we obtain here that the two focal lengths owing to the E_x-field are (for $\beta L \ll 1$),

$$f_x = f'_x = -1/(L\beta^2) \tag{2.37}$$

and that the two principal planes again coincide with the centre of the lens, namely

$$z(P_x) = \tfrac{1}{2}L. \tag{2.38}$$

The minus sign in (2.37) indicates a diverging lens.

Thus we have a converging lens in the yz-cross-section and a diverging lens in the xz-section. The absolute value of the four focal lengths being the same. The line foci of f_y are parallel to the x-axis while the virtual line foci of f_x are parallel to the y-axis.

The results so far derived apply for $\beta \ll 1/L$, i.e. for long quadrupoles or weak excitation. The ratio f_x/f_y exceeds unity and the principal planes move away from the centre, unless

$$|\beta| = \frac{1}{a}\sqrt{\frac{V_a}{V_0}} \lesssim 0.5/L. \tag{2.39}$$

This has been shown in detail by Septier (1961, p. 107) and we shall specify his results in §4.9 for various quadrupoles.

It now remains to consider a doublet. As a simple example we take two identical quadrupoles arranged coaxially, separated by a small distance D and rotated by 90° with respect to each other. By substitution of (2.33) and (2.37) into (2.13) we obtain for the focal length of the doublet

$$f_{Dx} = f_{Dy} = \frac{1}{L^2\beta^4 D_1}, \tag{2.40}$$

where $D_1 = (D+L)$. We have treated here the two quadrupoles as two thin lenses. These lenses are located at the principal planes of the quadrupoles, i.e. at a distance $L/2$ from the actual end of the electrodes. According to Septier (1961, p. 116) this approximation is valid as long as $\beta L \lesssim 0.7$. As the focal lengths in the xz-plane and in the yz-plane respectively are equal and both positive, the doublet acts like an electron lens of circular symmetry.

CHAPTER 3

FIELD PLOTTING AND RAY TRACING: ANALOGUE AND COMPUTATIONAL METHODS

§3.1. Potential distribution and Laplace's equation

The equipotentials are the refracting surfaces of electrostatic electron optics. They correspond to the material refracting surfaces of glass lenses in ordinary optics. The electric field E is given in terms of the potential V by the general relation

$$E = -\operatorname{grad} V,$$

and the direction of the field is always perpendicular to the equipotentials.

A rough sketch of the field plot can be obtained in the following way: first, some equipotentials in the neighbourhood of the electrodes can be drawn parallel to the electrode surfaces. Then an equipotential between two electrodes can be drawn bisecting everywhere the angle between the conducting surfaces. Moreover, after drawing some field lines at right angles through the above equipotentials, the rest of the equipotential lines may be filled in duly regarding the conservation principle of flux, so that

$$\Phi = \epsilon \frac{\Delta V}{\Delta s} \Delta a = \text{const.}, \tag{3.1}$$

where Φ is the flux through a tube bounded by two field lines, ϵ is the dielectric constant, ΔV is the potential increment between two successive equipotentials of distance Δs at the cross-sectional area Δa of the tube. In two-dimensional fields, for instance, the area Δa is proportional to the distance Δb of the field lines bounding the tube. In fields of circular symmetry, however, $a \propto r \Delta b$, where r is the mean distance of the elementary volume $\Delta a \Delta s$ from the axis of symmetry.

More exact methods for the determination of potential distributions follow from the theory of the electrostatic field (see, for instance, Weber, 1950; Moon and Spencer, 1961). The potential distribution in space is entirely defined by the geometry and the potentials of the electrodes. Neglecting space charges, the potential distribution can be completely calculated from Laplace's differential equation

$$\nabla^2 V = 0. \tag{3.2}$$

But, although the theoretical possibility of obtaining analytical solutions for the field is very interesting in principle, an integration of (3.2) under the proper conditions as given by geometry and potentials of the electrodes is generally very complicated. In the majority of practically important cases, analogue or numerical methods have to be used.

There are a few practical cases, however, where analytical solutions for the field are of interest. We mention here:

(i) The potential distribution along the axis of the symmetrical two-tube lens with a small gap (fig. 3.3). This is, according to Gray (1939) given by

$$V(z) = \frac{V_1 + V_2}{2}\left[1 + \frac{1 - V_1/V_2}{1 + V_1/V_2}\tanh 1.32z\right], \tag{3.3}$$

the unit of length being taken as the radius of the tubes.

(ii) The potential along the axis of a circular aperture separating two regions of uniform field (cf. fig. 4.4) is given by Fry (1932a) namely:

$$V(z) = V_c + \tfrac{1}{2}(E + E')\, z - 1/\pi(E - E')\, |z|\left(\tan^{-1}\frac{z}{R} + \frac{R}{z}\right), \tag{3.4}$$

E and E' being the fields at some distance from the aperture and V_c being the potential of the diaphragm. The fields may be considered accurately uniform except within the region

$$-3/2R < z < 3/2R.$$

(iii) The potential distribution of an einzel lens (cf. fig. 4.5) consisting of three diaphragms with circular apertures. This is conveniently presented by a frequently used approximate formula of Glaser and Schiske (1954) namely:

$$V(z) = V_o + V_i\bigg/\left[1 + \frac{z^2}{w^2}\right], \tag{3.5}$$

where V_o is the potential of the two external diaphragms, V_i is the negative potential of the inner diaphragm and $2w$ is the half width of the 'bell-shaped' distribution curve.

An important conclusion can be drawn from the validity of Laplace's equation throughout the lens field, namely, that in an axially symmetrical system the potential on the axis uniquely determines the potential distribution outside it. It can be shown (see appendix A) that the solution of (3.2) can be expressed by the power series:

$$V(r, z) = V_0(z) - \frac{r^2}{2^2} \frac{d^2V_0(z)}{dz^2} + \frac{r^4}{2^2 4^2} \frac{d^4V_0(z)}{dz^4} \dots \qquad (3.6)$$

Hence, for the determination of a lens field, a measurement of the field $V_0(z)$ along the axis only should be sufficient, the rest being obtainable by calculation.

Taking the first two terms of (3.6), it can be seen that

$$d^2V_0(z)/dz^2$$

determines the radial field strength in the region close to the axis, since

$$E_r = -\frac{\partial V(r, z)}{\partial r} \approx +\frac{r}{z} \frac{d^2V_0(z)}{dz^2}. \qquad (3.7)$$

(3.7) demonstrates that an electron experiences a force towards or away from the axis according as whether d^2V_0/dz^2 is positive or negative.

The fact that the potential distribution in space is completely determined by the axial potential distribution brings out a fundamental difference between ordinary optics and electron optics which complicates the process of computing electron lenses. In light optics, the curvatures and refractive indices of lenses can be selected independently and so, for instance, by skilful choice, lens errors can be compensated. On the other hand, in electron optics, the radii of the refracting surfaces seem to be already determined by the change of refractive index along the axis. Referring to the practical problem, however, it must be stressed that insignificant variations of the axial potential can yield decisive changes in the potential distribution outside it. Hence from the practical point of view of obtaining reliable results for ray tracing the axial potential distribution should be used for paraxial tracing only. For zonal and marginal ray tracing, a measurement of the whole potential field is always advisable.

§3.2. Relaxation methods

Though the analytical calculation of the potential distribution for a given electrode configuration from Laplace's equation is impracticable in many cases, complete numerical computation can always be performed by a relaxation method.

For the two-dimensional case, Laplace's equation (3.2) reduces to

$$\frac{\partial^2 V}{\partial y^2} + \frac{\partial^2 V}{\partial z^2} = 0. \tag{3.8}$$

For a numerical method of extrapolating the potentials off the electrodes, let the yz-plane of translational symmetry be covered by a net of equidistant lines parallel to the two axes, as shown in fig. 3.1. The potential V_0 at any lattice point should be completely

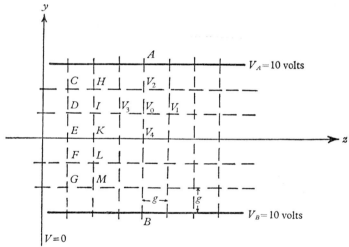

Fig. 3.1. Lattice for computational mapping.

determined by the potentials V_1, V_2, V_3, V_4 of its neighbouring four lattice points. Calling g the distance between the lattice lines, it is obvious that

$$\frac{V_2-V_0}{g} - \frac{V_0-V_4}{g} \approx \left(\frac{\partial V}{\partial y}\right)_{2,0} - \left(\frac{\partial V}{\partial y}\right)_{0,4} \approx g\left(\frac{\partial^2 V}{\partial y^2}\right)_0 \tag{3.9}$$

and

$$\frac{V_1-V_0}{g} - \frac{V_0-V_3}{g} \approx \left(\frac{\partial V}{\partial z}\right)_{1,0} - \left(\frac{\partial V}{\partial z}\right)_{0,3} \approx g\left(\frac{\partial^2 V}{\partial z^2}\right)_0. \tag{3.10}$$

According to Laplace's equation (3.8), the expressions (3.9) and (3.10) are equal, hence

$$V_0 \approx \tfrac{1}{4}(V_1 + V_2 + V_3 + V_4). \qquad (3.11)$$

With the help of (3.11) the potential distribution between electrodes of known potential can be extrapolated. This is done in practice by starting from an arbitrary distribution.

To give an example, let the two heavy lines A and B in fig. 3.1 represent two plate electrodes at a common potential $V = 10$ V., and let the y-axis be the section through an electrode at zero potential. The electrode arrangement corresponds practically to a four-plate lens, A and B being the cross-section of one pair of plates while the y-axis represents the section through the mid-plane.

If, for a start, all lattice points are assumed to be at $V = 10$, the potential distribution is not in disagreement with (3.11) excepting for those points which adjoin the y-axis. There, one has to begin with reducing the assigned V values. First, new values may be obtained for the points C and G, since these points are surrounded each by two neighbouring lattice points of the fixed potentials 0 and 10 respectively. From (3.11) follow the values

$$V_C = V_G \approx 30/4.$$

Hence we may try to assign the values 7 to these two points noting the 'residuals' ($+2$) for each. We then calculate with (3.11) new values for the neighbouring points D and F and we see that both will be changed to $V = 6$. The values of further neighbouring points can then be adjusted in the same way noting that V_E is doubly affected, i.e. both by the changes in D and in the symmetrical point F; this applies to all points on the z-axis. The procedure explained is further applied in turn to all other points, recording always the effect of a change at any point upon its immediate neighbours. Then the whole process is repeated until the residual for each potential is below a given value which may be ± 2 at a first trial. In this way an approximate potential distribution is obtained as shown in fig. 3.2 where the final residuals are given in brackets.

To obtain higher accuracy the potential at every point (and all the residuals) of the first solution (fig. 3.2) may be multiplied, for instance, by a factor 10 and the process of approximation continued, until the residuals are again reduced to (± 2). A comparison of such a refined numerical plot with the corresponding analytical solution (also compare the paraxial values of the rough plot of fig. 3.2 with the curve shown in fig. 2.5 of §2.5) indicates that an

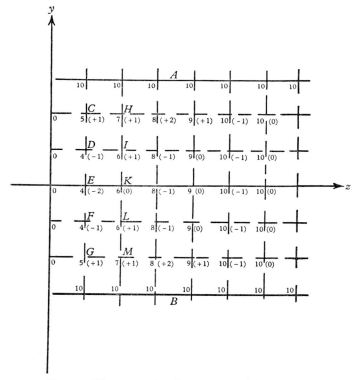

Fig. 3.2. First approximation to potential distribution.

error of a few per cent only can be reached by a few repetitions of the process. For greater accuracy, a lattice with smaller intervals may be chosen.

For the three-dimensional case of circular symmetry, the relaxation method is applied by using Laplace's equation in cylindrical coordinates (r, z) namely

$$\frac{\partial^2 V}{\partial r^2} + \frac{1}{r}\frac{\partial V}{\partial r} + \frac{\partial^2 V}{\partial z^2} = 0. \tag{3.12}$$

For the purpose of a numerical solution (3.12) may now be replaced by a system of difference equations. There, the values of V can be determined at the points of a square net of finite mesh length. For example, at any point P_0 of this net the potential V_0 can be expressed in terms of the potentials V_1, V_2, V_3 and V_4 of the four neighbouring mesh points P_1, P_2, P_3 and P_4. This follows from an expansion in Taylor series for the potentials of these points, namely,

$$\left.\begin{aligned}
V_1 &= V_0 + g\,\frac{\partial V}{\partial r} + \frac{g^2}{2}\frac{\partial^2 V}{\partial r^2} + \cdots, \\[1ex]
V_3 &= V_0 - g\,\frac{\partial V}{\partial r} + \frac{g^2}{2}\frac{\partial^2 V}{\partial r^2} - \cdots, \\[1ex]
V_2 &= V_0 + g\,\frac{\partial V}{\partial z} + \frac{g^2}{2}\frac{\partial^2 V}{\partial z^2} + \cdots, \\[1ex]
V_4 &= V_0 - g\,\frac{\partial V}{\partial z} + \cdots,
\end{aligned}\right\} \qquad (3.13)$$

where g is the length of the meshes. From (3.13) we obtain

$$V_1 + V_2 + V_3 + V_4 - 4V_0 \approx g^2\left(\frac{\partial^2 V}{\partial r^2} + \frac{\partial^2 V}{\partial z^2}\right). \qquad (3.14)$$

The right-hand side of this equation can, with the help of (3.12), be replaced by

$$-\frac{g^2}{r_0}\frac{\partial V}{\partial r}, \qquad (3.15)$$

where r_0 is the distance of point P_0 from the z-axis. Moreover, with the difference of the first two equations of (3.13) substituted into (3.15), (3.14) can be written

$$V_1\left(1 + \frac{g}{2r_0}\right) + V_3\left(1 + \frac{g}{2r_0}\right) + V_2 + V_4 - 4V_0 \approx 0. \qquad (3.16)$$

Hence the potential V_0 at any point P_0 of the net can be calculated if the potential of the surrounding four net points are known. The smaller the mesh size, the more exact does the relationship (3.16) become.

For a detailed account of the relaxation method reference may be made to the book by Southwell (1946) and to the paper by

Motz and Klanfer (1946) where the application of the method to the two-tube lens is described in detail.

Relaxation by digital electronic computer. The relaxation method as outlined above is not suitable for application when high accuracy is required, e.g. in the computation of the aberrations of electron lenses. Even when special procedures are adopted to speed the convergence of the method, the labour required to reach the desired accuracy in a 'paper-and-pencil' calculation is enormous. However, if a fast electronic computer is available, the relaxation technique can be satisfactorily applied to very accurate field determinations. Even so, in order to economize on computer time, the method needs to be systematized and steps taken to speed the convergence. Field determinations by computer have been discussed by Carré and Wreathall (1964), by Kulsrud (1967) and in more detail by Weber (1967).

If there are N points in the mesh, then for each point P_0 an equation of the form (3.16) can be written connecting the potential V_0 at the point with the potentials at surrounding points. The problem is then to solve by an iterative method the resulting set of N simultaneous equations for the potentials. The simplest systematic way of doing this is the procedure called 'simultaneous displacement'. This starts from a guessed set of potentials at the mesh points. For each point such as P_0, using (3.16), the potential V_0 is then recalculated from the assumed values at surrounding points, thus establishing an 'improved' set of potential values for the N points. In the next cycle of the iterative process, the procedure is repeated, the set of N 'improved' values being used to calculate a further set of potential values which will be a yet better approximation to the final solution. The process is repeated for as many cycles as are necessary to obtain a set of N potential values which satisfy the N finite difference equations.

The modified potentials obtained in each cycle are calculated from the potentials resulting from the previous cycle. Thus the memory of the computer must contain the potentials of the previous cycle and also store the modified values as they are calculated. By the end of a cycle, then, $2N$ potential values are stored. At the end of each cycle, the whole set of potentials of the previous cycle are displaced simultaneously by the potentials of the cycle just calculated.

It is only necessary to store N potential values if a routine of 'successive displacement' is used, i.e. if the previous potential at a point is displaced by the new potential as soon as it is calculated. This new potential is used in the difference equation of the neighbouring points that still have to be calculated, so that, in the 'successive displacement' method, the order in which the points are calculated is significant, and the points must be scanned in a prescribed order (Young, 1954).

The method of successive displacement converges more quickly than that of simultaneous displacement and is therefore to be preferred. The fastest convergence is obtained by using successive displacement in conjunction with the technique of 'over-relaxation'. The convergence is accelerated by using a modified version of the difference equation (3.16) into which a factor w, the 'acceleration factor', is introduced. The optimum value of this factor varies during the course of the iteration, and also depends on the boundary values. The speed of convergence is very critically dependent on the choice of w. The optimization of w is discussed by Carré (1961).

Weber (1967) has compared the simultaneous displacement method with the successive overrelaxation method for a simple electrode configuration, and shown that for a net of 5000 points, the latter method converges 100 times as fast as the former. Wreathall (1966) states that the difference equations for 1500 net points can be solved to an accuracy of 1 part in 10^5 in tens of seconds using a fast modern computer.

In choosing the number of such points, the required accuracy in various parts of the system must be taken into account. In regions in which the field is likely to vary quickly, a finer mesh may be needed. Furthermore, if the field values are needed for ray tracing, the field must be known more accurately where the electrons are travelling slowly, e.g. in the region of a cathode (cf. Francken and Dorrenstein, 1951, and also appendix D).

§3.3. Analogue methods of field plotting

If the highest accuracy is not required, then analogue simulation methods for field determination provide a useful alternative to digital computing. These methods had been used to investigate general electrostatic problems a long time before any electron optical research was started. The electrolytic trough method is based on the perfect physical analogy between the propagation of electric currents through electrolytes and the spreading of the lines of force through an electrostatic field. A good review on the electrolytic trough is given in the book by Karplus (1958).

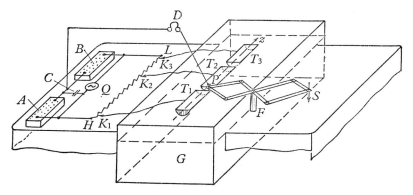

Fig. 3.3. Electrolytic field plotting trough.

The general features of the method are shown by fig. 3.3. G represents a large box or glass trough which is filled with a weakly conducting liquid. For instance, a solution of copper sulphate in distilled water (about 10 mg/l) or sometimes clear tap water is used as electrolyte. In the present example, T_1, T_2 and T_3 represent three electrodes of an electron lens; they are tubes, cut along their axes and fixed in such a way that the axis of the lens coincides with the surface of the liquid. An alternating-current source Q is connected to two large, variable precision resistances A and B, used as a potentiometer for the adjustment of a given potential at a probe P, the tip of which is just dipping into the surface of the liquid. A current-detecting device D will indicate that no current is taken by the probe as soon as its tip is in the particular equipotential line which corresponds to the adjustment of the ratio of the resistances A and B.

The source Q is also connected to the ends H and L of a potentio-meter supplying any desired voltage to the electrodes T_1, T_2 and T_3 through the sliding contacts K_1, K_2 and K_3. To set up any desired potential at an electrode T, this electrode may be touched with the probe of preadjusted potential, and the corresponding contact K may be adjusted to zero current in the detector D.

To avoid electrolysis, the source Q should supply alternating current. A valve oscillator of 400–1000 hertz is convenient. Head-phones are suitable for the detecting device D. In order to avoid disturbances due to a shift of the phase angle, it is often necessary to balance the capacity of the electrodes by a small variable condenser C which can be switched in parallel with either of the resistances A or B. The probe may be carried by a pantograph attached to the side of the trough and moving about the pivot F. This pantograph transmits the motion of the probe to a plotting pencil S which moves over a drawing board. The probe is fixed in an earthed metal barrel and it is connected by a screened flexible lead to the headphones.

The accuracy of electrolytic field plotting is limited mainly by electrolytic polarization and by the surface tension of the liquid producing a meniscus. According to Einstein (1951) the errors in the relative potential measurements can be reduced to 0·2 per cent. In a simple laboratory apparatus, in which no particular precautions are taken, the error in relative potential measurements could be as large as 5 per cent. At the other extreme Sander and Yates (1956) have described a special design of 'inverted trough' in which these problems have been minimized and in which potential distribution can be measured to an accuracy of 0·1 per cent.

The method of field plotting described can be applied to any electrode system possessing a plane of symmetry that can be placed in the surface of the liquid. In systems of circular symmetry, such a plane is given by any meridional plane, i.e. a plane which includes the axis. The reason for placing the liquid level in such a position is that in electron lenses all electric field vectors lie in meridional planes only. There are no field components tangential to a conductor. Analogously, in the electrolytic trough, there cannot exist any current vectors crossing meridional planes.

The wedge trough. A particular technical development of the method of electrolytic field plotting was initiated by the conclusion that an insulating plane may be placed in any meridional plane of the electrode arrangement. Manifold and Nicoll (1938) proposed the important design of the wedge trough, which implies an appreciable practical simplification. Such a trough is shown in fig. 3.4. There, a thin wedge of electrolyte is realized by a tilted trough containing electrolyte resting on a plane, insulating bottom.

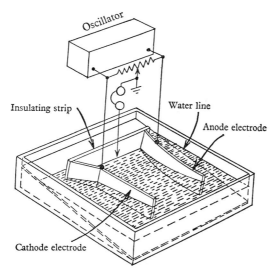

Fig. 3.4. Electrolytic wedge trough.

The thin edge of the wedge represents the axis of the electrode system. The bottom of the trough may be made from glass so that a coordinate system can be placed beneath it and seen from above. The tilt of the trough may be varied by foot screws. By their adjustment, the edge of the wedge can be moved into any desired position with respect to the electrodes. In order to obtain reliable plots, it is absolutely necessary for the liquid forming the wedge to have perfect contact with the glass bottom. The edge of the wedge must be really straight. To avoid disturbing effects of surface tension, the glass surface may be wiped perpendicularly to the edge with a fine abrasive (Klemperer, 1944).

If the angle of the wedge is small, the electrodes used need not be surfaces of revolution but can be replaced by strips of plane metal sheet, bent to the shape of a longitudinal section of the actual electrode. For instance, a two-tube lens is simply represented by two flat strips of metal placed in the liquid in a direction parallel to the thin edge (tube axis) of the wedge.

Electrolytic field plotting can also be applied to finding a potential distribution that is modified by the neighbourhood of an intense electron beam. According to Pierce (1940, 1949) and Samuel (1945), the edge of the beam may be represented by a strip of insulator in the electrolyte (see fig. 3.4). Since there can be no current flow normal to the insulator this ensures that the field at the edge of the beam normal to the beam surface will be zero. Field plotting in the presence of space charges will be fully discussed in §8.9.

Labelling of the equipotentials. In fig. 3.5 there is shown an actual potential plot of the symmetrical two-tube lens. The equipotentials

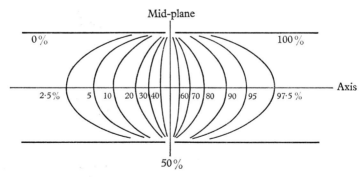

Fig. 3.5. Potential distribution in the symmetrical two-tube lens.

are marked as a percentage of the total potential difference between the two electrodes. For instance, if the bridge resistances are equal ($A = B$ in fig. 3.3) the probe is at the 50 per cent equipotential, or if $A/B = \frac{1}{9}$, the probe is at the 10 per cent equipotential, etc.

In this way one obtains a very general diagram, and the potential plot can easily be interpreted for all varieties of voltages or voltage ratios at the electrodes, since it is evident that the shape of the equipotentials does not change if the voltage at either of the two electrodes is altered. For instance, the two tubes may have the potentials $V = 1000$ V and $V' = 5000$ V. In this case the 10 per cent equipotential would be at

$$1000 + \frac{10}{100}(5000 - 1000) = 1400 \text{ V.}$$

In another example, the same two tubes may be at $V = 500$ V and $V' = 20000$ V; then the 10 per cent equipotential would be at 2450 V. The 1400 V equipotential of the first example would be in exactly the same place as the 2450 V equipotential of the second example.

The equipotentials in fig. 3.3 are given in steps of 10 per cent between 10 and 90 per cent, and to show the spreading of the field into the tubes, extra equipotentials are plotted at 2·5, 5, 95 and 97·5 per cent. One can see that the whole figure is perfectly symmetrical with respect to the midplane. Moreover, the potential gradient is greatest at the midplane, where the equipotentials are closest together, and it decays slowly towards the outer ends of the tubes, the 90 per cent equipotential having a distance of about one tube radius from the midplane and the 97·5 per cent equipotential being twice as far from that plane.

It is advisable to measure every geometrical distance in terms of the tube radius R as a unit, so that it is easy to transfer the results of a field plot to electron lenses of any size. In order to obtain good accuracy in plotting the equipotentials, the model of the electrodes should be made as large as the size of the trough allows. Other metal parts and insulators should, however, be sufficiently removed in order to avoid disturbances. It has been explained that insulators, such as the glass wall of the trough, act as mirrors towards the equipotentials. On the other hand, conductor surfaces are equipotentials. If, for example, in order to avoid external disturbances, the ends of the tubes in the electrolytic trough are closed by metal sheets, kept at the tube potential, these sheets must be sufficiently far away from the midplane in order to avoid any influence on the equipotentials to be investigated. According to practical experience, a plane sheet perpendicular to the axis and at a distance of three tube radii from the midplane would not modify markedly the shape or position of the 95 per cent equipotential.

When a part of the whole potential plot has to be investigated with increased accuracy, metal sheets could be shaped exactly like a natural equipotential measured in the electrolytic trough, and could be introduced in the trough. Two metal sheets replacing the equipotentials of n and m per cent may be connected to the ends of the resistances A and B. The right designation of any measured equipotential is then given by

$$n + \frac{(m-n)A}{100(A+B)}. \tag{3.17}$$

By this artifice of replacing two equipotentials by metal sheets, a relatively small fraction of the whole field could be investigated separately on a scale enlarged as much as the dimensions of the electrolytic trough would allow.

Resistor network. For the purpose of field plotting, the continuous distribution of the conducting medium of the electrolytic trough can be replaced by a network of discrete resistance units connecting a square array of studs. The electrodes of the electron-optical system are simulated by short-circuiting corresponding studs. The operation of the network can be seen from Kirchhoff's law

$$\Sigma I_n = 0.$$

The currents I_n, with $n = 1, 2, 3, 4$, towards any stud O are given by

$$I_n = \frac{V_n - V_0}{\Omega}, \tag{3.18}$$

where V_n are the potentials of the four studs adjoining the stud O of the potential V_0 and where Ω is the value of the resistance units. If all these units are equal, one obtains

$$\Sigma V_n - 4V_0 = 0. \tag{3.19}$$

This equation describes a two-dimensional field; it is identical with (3.11). Such a two-dimensional network analogue has been described first by Hogan (1943). On the other hand, Packh (1947) and Liebmann (1949b) have shown the application of the network method to three-dimensional potential distributions of circular symmetry. There, the resistances have to be graded logarithmically in the radial direction. Liebmann (1950 a) has shown experimentally that with a graded network a high degree of accuracy (10^{-3} to 10^{-4}) can be obtained, which compares favourably with that of the best wedge troughs. However, he used large arrays of precision resistances with 20 and 60 meshes in radial and in axial direction respectively; in addition, his board was surrounded by an 'infinity strip' network for moving the reflecting boundary farther away from the axis. In this version, the need for some 3 000 accurate resistors makes the expense prohibitive for the ordinary laboratory.

Preuss and Bas (1966) have discussed the relationship between the accuracy of the resistor values and the error in the measured potential. Statistical calculation shows that the mean error of the measured potentials is given by

$$\overline{\delta V} = \frac{1}{\sqrt{6}} \frac{\overline{\delta R}}{R} \Delta V_{max},$$

where $\overline{\delta R}$ is the mean error in resistance value and ΔV_{max} is the largest measured potential difference between any two mesh points. In practice, $\frac{1}{2}$ per cent tolerance in R can lead to an accuracy of better than 10^{-4} for V.

When simulating problems of rotational symmetry, the resistance values must decrease in the radial direction, so that the highest resistance values (typically 10000) are found at the axis. This highlights a fundamental difficulty in this application of the resistance network analogue. The lowest permissible values of resistance are set by the necessity to avoid heat dissipation. Thus, resistances near the axis will normally be far higher than this minimum value. However, the higher the resistance values, the higher is the required input impedance for the instrument which measures the voltage. The problem of constructing a network with lower resistances at the axis than normal so that the input impedance of the measuring instrument can be lower has been solved by Weber (1963). The same author (Weber, 1967) has developed a more accurate type of network analogue based on a nine-point difference equation, rather than the five-point equation of the type (3.19).

§3.4. Ray tracing by Snell's law

After having obtained an exact potential plot of the investigated electron lens with all the values of the equipotentials calculated in volts, the path of any electron could be traced in principle immediately from Snell's law. The spaces between the equipotentials V_1 and V_2, V_2 and V_3, V_3 and V_4, etc., are assumed to be media of constant refractive indices $N_1 = \sqrt{V_1}$, $N_2 = \sqrt{V_2}$, $N_3 = \sqrt{V_3}$, etc. The angle of incidence α_1 of the electron at the first equipotential V_1 may be measured with a protractor. The corresponding angle of refraction α_1' may be calculated with Snell's law (1.12) and plotted with the protractor at the other side of the equipotential. The path of the electron between the equipotentials V_1 and V_2 is drawn as a straight line. The same procedure is applied for the angles α_2 and α_2' at the equipotential V_2, and so on.

For practical purposes the tracing procedure by Snell's law is simplified if a field plot with geometrical progression of voltages of the equipotentials is available. There, the ratio of refractive indices is the same for each step, i.e.

$$\sqrt{\frac{V_1}{V_2}} = \sqrt{\frac{V_2}{V_3}} = \sqrt{\frac{V_3}{V_4}} = \dots = \sqrt{\frac{V_n}{V_{n+1}}}. \qquad (3.20)$$

This may be illustrated by a simple example in which—for the sake of argument—we want to plot in a two-tube lens, three equipotentials only as shown in fig. 3.6. Let the voltage ratio of this lens be $V'/V = 5$, the two tubes being at 1000 and at 5000 volts respectively. The three equipotentials divide the lens into $r = 4$ regions of constant voltages $V_0, V_1, \dots, V_r$. The voltages of the regions above the voltage V_0 of the incident electrons are calculated as

$$V_1 = V_0 \left(\frac{V_r}{V_0}\right)^{1/r} = 1495 \text{ volts,}$$

$$V_2 = V_0 \left(\frac{V_r}{V_0}\right)^{2/r} = 2236 \text{ volts,}$$

$$V_3 = V_0 \left(\frac{V_r}{V_0}\right)^{3/r} = 3343 \text{ volts,}$$

$$V_4 = V_0 \left(\frac{V_r}{V_0}\right)^{4/r} = 5000 \text{ volts.}$$

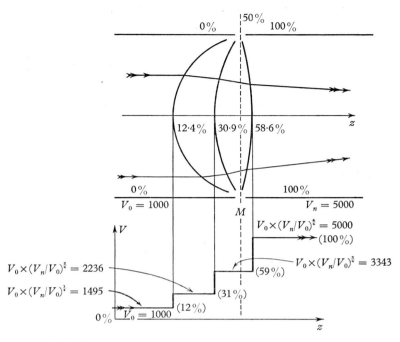

Fig. 3.6. Equipotentials on two-tube lens: voltage increase
in geometric progression.

The voltage ratio between consecutive regions is in every case

$$V_n/V_{n+1} \simeq 1 \cdot 5.$$

To find the refracted rays, it is convenient to make a stencil from
paper, or celluloid, etc., as shown by $CDGF$ in fig. 3.7. One edge
CF of the stencil is marked by the points A, A' and N, so that

$$\frac{NA}{NA'} = \sqrt{\frac{V_{n+1}}{V_n}}.$$

A line BP is drawn on the stencil as the perpendicular bisector to
AA' (Bloch, 1936; Jacob, 1938). For ray tracing, the template is
laid on the field plot so that the bisector passes through the point
of incidence P. Moreover, point A is placed on the incident ray El
and N on the normal to the equipotential through P. Then $A'P$

gives the direction of the refracted ray PP_1. This is easily proved from the sine formula trigonometry:

$$\frac{NA}{\sin NPA} = \frac{NP}{\sin PAA'} = \frac{NP}{\sin PA'N} = \frac{NA'}{\sin NPA'}. \quad (3.21)$$

The sines in the second and third item are equal, since, according to construction, $\measuredangle\, PAA' = \measuredangle\, PA'A = 180° - PA'N$. Thus

$$\frac{NA}{\sin \alpha} = \frac{NA'}{\sin \alpha'}$$

follows from the first and last item.

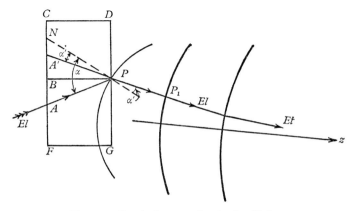

Fig. 3.7. Grapical ray tracing by Snell's law.

The main source of error in the graphical method of tracing with Snell's law is the difficulty in finding the exact directions of the normals of the equipotentials which are needed to plot the angles α and α'. The errors are cumulative, so that after a number of steps the result of tracing might deviate considerably from the true electron path. However, the method is useful in practice, yielding preliminary information quickly.

A certain ambiguity is experienced when the path of an electron has to be traced near the cathode from which it has been emitted with negligible velocity. It is known that the electron travels initially along a field line, i.e. at right angles to the cathode surface and to other equipotentials which are very close to the cathode.

However, as soon as the electron gathers momentum, the radius of curvature r_e becomes greater than the radius r_E of the field line. According to a practical rule by Rajchman (1938) the initial path may be plotted with $r_e \approx 3r_E$ until ray tracing can be started by plotting refractions according to Snell's law.

§3.5. Equations of motion of the electron

A great variety of methods for ray tracing has been proposed which are all based on various ways of integrating the equations of motion of the electron:

$$m \frac{\mathrm{d}^2 z}{\mathrm{d}t^2} = e \frac{\partial V}{\partial z}, \left.\begin{matrix} \\ \\ \\ \\ \end{matrix}\right\}$$
$$m \frac{\mathrm{d}^2 r}{\mathrm{d}t^2} = e \frac{\partial V}{\partial r}. \qquad (3.22)$$

A differential equation for the electron trajectory can be derived by elimination of t from (3.22). In the paraxial region

$$(\mathrm{d}r/\mathrm{d}z)^2 \ll 1,$$

moreover, the first two terms of the power series (3.6) can be substituted for $V(r, z)$. This leads to the 'electrostatic paraxial ray equation':

$$\frac{\mathrm{d}^2 r}{\mathrm{d}z^2} + \frac{1}{2V_0(z)} \frac{\mathrm{d}V_0(z)}{\mathrm{d}z} \frac{\mathrm{d}r}{\mathrm{d}z} + \frac{1}{4V_0(z)} \left(\frac{\mathrm{d}^2 V_0(z)}{\mathrm{d}z^2} \right) r = 0. \qquad (3.23)$$

The focal length of a lens can be obtained from (3.23) which may be written

$$-\frac{\mathrm{d}^2 V_0}{\mathrm{d}z^2} \times \frac{r}{4} = \frac{1}{2} \frac{\mathrm{d}V_0}{\mathrm{d}z} \frac{\mathrm{d}r}{\mathrm{d}z} + V_0 \frac{\mathrm{d}^2 r}{\mathrm{d}z^2} = V_0^{\frac{1}{2}} \frac{\mathrm{d}}{\mathrm{d}z} \left(\frac{\mathrm{d}r}{\mathrm{d}z} V_0^{\frac{1}{2}} \right); \qquad (3.24)$$

integration gives

$$\left[\frac{\mathrm{d}r}{\mathrm{d}z} V_0^{\frac{1}{2}} \right]_{z_1}^{z_2} = -\frac{1}{4} \int_{z_1}^{z_2} \frac{r}{V_0^{\frac{1}{2}}} \frac{\mathrm{d}^2 V_0}{\mathrm{d}z^2} \, \mathrm{d}z, \qquad (3.25)$$

taking parallel incident rays, $(\mathrm{d}r/\mathrm{d}z)_{z_1} = \tan \Theta = 0$,

$$\left(\frac{\mathrm{d}r}{\mathrm{d}z} \right)_{z_2} = -\frac{1}{4(V_0^{\frac{1}{2}})_{z_2}} \int_{z_1}^{z_2} \frac{r}{V^{\frac{1}{2}}} \frac{\mathrm{d}^2 V_0}{\mathrm{d}z^2} \, \mathrm{d}z. \qquad (3.26)$$

This equation can be solved by successive approximation only. For a thin lens, however, where both principal planes coincide with the midplane, $r \approx$ const. for the path of the electron within the lens and

$$-\left(\frac{\mathrm{d}r}{\mathrm{d}z}\right)_{z_2} = \tan \Theta' = \frac{r}{f'},$$

where f' is the focal length and r the aperture of the incident ray.

The power of the thin lens is then

$$\frac{1}{f'} = \frac{1}{4\sqrt{V_0'}} \int_{-\infty}^{+\infty} \frac{\mathrm{d}^2 V_0}{\mathrm{d}z^2} \frac{\mathrm{d}z}{\sqrt{V_0}}, \qquad (3.27)$$

where V_0' is the axial potential in the image space.

In most real lens fields, $\mathrm{d}^2 V_0/\mathrm{d}z^2$ changes sign at one or more points. Thus, in order to evaluate (3.27), it is necessary to split up the range of integration. This is not a convenient procedure and for practical purposes a more useful 'thin lens formula' can be obtained from the 'reduced ray equation'.

Reduced ray equation (cf. Picht, 1939). For computational purposes, a more convenient form of the paraxial ray equation is obtained by substituting in (3.23)

$$R(z) = rV_0^{\frac{1}{4}}; \qquad (3.28)$$

$R(z)$ is called the 'reduced' ray and the equation it obeys is the 'reduced ray equation', viz.:

$$\frac{\mathrm{d}^2 R}{\mathrm{d}z^2} + \frac{3}{16} \left(\frac{\mathrm{d}V_0}{\mathrm{d}z} \bigg/ V_0\right)^2 R = 0. \qquad (3.29)$$

$\left(\dfrac{\mathrm{d}V_0}{\mathrm{d}z} \bigg/ V_0\right)^2$ is a more suitable function than $\dfrac{\mathrm{d}^2 V_0}{\mathrm{d}z^2}$ since it is difficult to obtain accurate values of a second differential coefficient by analytical procedures.

Taking the ratio of the slope to the radial distance leads to the focal length, and using (3.28) we obtain immediately

$$\frac{1}{f'} = -\frac{\left(\dfrac{\mathrm{d}r}{\mathrm{d}z}\right)_{z_2}}{r(z_1)} = -\left(\frac{V_0}{V_0'}\right)^{\frac{1}{4}} \frac{\left(\dfrac{\mathrm{d}R}{\mathrm{d}z}\right)_{z_2}}{R(z_1)},$$

where V_0 and V_0' are the (constant) potentials in the object and image space respectively. Thus,

$$\frac{1}{f'} = \frac{3}{16} \left(\frac{V_0}{V_0'}\right)^{\frac{1}{4}} \int_{-\infty}^{\infty} \left[\frac{\left[\frac{dV_0}{dz}\right]}{V_0}\right]^2 dz. \qquad (3.30)$$

This is the 'electrostatic thin lens formula'. Its derivation is discussed in detail by Sturrock (1954) and its application by Felici (1959). It will be noted that $\left[\frac{dV_0}{dz} / V_0\right]^2$ is essentially a positive quantity.

§3.6. Determination of electron trajectories

There are various methods of obtaining the electron trajectories from (3.22), (3.23) or (3.29), the most powerful of which is the use of the electronic digital computer. However, the method chosen will depend to some extent on the complexity of the problem, the resources available, and sometimes, on the method of determining the field values. It is generally not possible to give analytical solutions for the complicated fields of actual electron lenses; hence very often the equations have to be integrated by computational or numerical methods. It is, however, possible, in some cases to approximate the axial potential by a convenient analytical function for which the paraxial ray equation can be integrated in terms of elementary functions. An example of this technique as applied to the electrostatic two-tube lens is given by Grivet and Bernard (1954) where dV_0/dz is approximated by a 'bell-shaped' function (cf. (3.5)).

Numerical ray tracing. Maloff and Epstein (1934, 1938) divided the lens field into intervals (Δz) in which

$$\frac{1}{2V_0}\frac{dV_0}{dz} \quad \text{and} \quad \frac{1}{4V_0}\frac{d^2V_0}{dz^2}$$

were assumed to be constants. These constants could be substituted into the paraxial ray equation (3.23) leading to an expression of r as a function of z which could be calculated for every step.

Gans (1937) replaced the $V_0(z)$ curve, representing the potential along the axis of the lens, by a polygon of straight lines, so that d^2V_0/dz^2 vanished for each section. In this way the paraxial ray equation (3.23) was reduced to the first two terms only which again could be integrated to give $r = f(z)$ for each section. The method gives only approximate results, but it could be much improved by approximating the $V_0(z)$ curve rather by a sequence of parabolic arcs, and Recknagel (1937) has worked out a step-by-step integration of (3.23) on this principle. This method has subsequently found considerable application, since it leads to a particularly simple form for the ray equations and enables moderately accurate first order properties to be deduced using only a few subdivisions of the axial range. Approximation of $V_0(z)$ by a parabola implies that d^2V_0/dz^2 is a constant, and hence, from (3.7), that the radial field E_r is proportional to r. Thus the radial equation of motion in (3.22) can be solved in terms of simple trigonometric or hyperbolic functions. The method has been applied to the determination of the cardinal points of einzel lenses (cf. §4.6) by Regenstreif (1951), to the design of television display tubes and camera tubes by Schlesinger (1961), and more recently to the analysis of point cathode electron guns (cf. §9.9) by Everhart (1967).

The accuracy of all these methods depends on the size of the integration interval Δz. However, even by choosing very small Δz, one cannot expect to obtain information beyond the range of the first-order theory from methods which are based on the paraxial equation. For tracing lens aberrations, one has to use methods for numerical integration of the general equations of motion (3.22). For this purpose, the experimental values of $V(r, z)$ and its derivatives $\partial V/\partial z$ and $\partial V/\partial r$ have to be tabulated for suitable intervals of Δz and Δr for the region through which the ray passes. Different methods of zonal ray tracing are distinguished again by the ways in which the integration of the equations of motion is carried out. We refer in this connexion to appendix D where details are given of some suitable numerical integration processes, and to the papers by Goddard (1944), Goddard and Klemperer (1944), Motz and Klanfer (1946) and Liebmann (1949b).

Ray tracing by digital computer (see also appendix D). The accuracy of a manual method can be increased by choosing a smaller integration interval, but at the expense of greater labour. Where high accuracy is needed, it is clearly advantageous to adapt numerical ray tracing methods to the digital computer. It may be particularly convenient to do this where the field data must also be obtained by digital computing (cf. §3.2). A great economy can then be achieved by extending the field plotting programme to obtain the trajectories, without the necessity of printing out the field data (cf. Kulsrud, 1967).

The simplest approach to the problem of computing the trajectories is step-by-step numerical integration of the Newtonian equations of motion (3.22), using time as the independent variable. The most important consideration is the choice of the time interval for each step. This must be short enough to keep the error in the location of the trajectory within acceptable limits. Thus, at each step, it is desirable to have some indication of the accuracy of the numerical integration. A possible approach to this is to compute the total energy of the electron at the end of each step, and check that it is remaining constant to better than a prescribed accuracy (cf. Goddard and Klemperer, 1944). If this test indicates that the limit of error is exceeded, the time interval is automatically reduced.

A more sophisticated method of controlling the accuracy of the process is described by Carré and Wreathall (1964), and summarized by Wreathall (1966). These authors use the 'predictor-corrector' method of integrating the equations of motion. The radial and axial components of acceleration of the electron at any instantaneous position are found by interpolation from the field values at surrounding mesh points. From the instantaneous accelerations, the position of the electron at the end of the next time interval can be predicted. The acceleration components are redetermined at this position; in general they will differ from those at the beginning of the interval, the extent of the difference depending on how fast the field is varying in that region. Hence a more accurate estimate of the position of the electron can be obtained by recalculating the new position using a value of acceleration intermediate between those found for the starting position and the predicted position. This new position is known as the 'corrected' position. The difference between the predicted and the corrected position can be used to estimate the error in the integration. If this error would lead to an excessively large accumulation of error over the whole trajectory, the time interval is halved. If, on the other hand, over a number of successive steps, the error is small enough, the time interval is automatically doubled, so that the trajectory can be computed in the minimum time compatible with the required accuracy. The error tolerances are pre-set before starting.

An alternative numerical integration procedure, using a different error controlling routine is described by Weber (1967). Mention should also be made of a paper by Vine (1959) which describes a programmed version of the method of Liebmann (1949b).

Ray tracing using analogue field measurements. There are a certain number of methods of trajectory determination which have been devised for use directly with analogue field plotters. One obvious possibility is to use the voltages derived from a resistance network analogue as input data for an analogue computer set up to integrate the equations of motion. This type of system has been realized by De Beer, Groendijk and Verster (1961/2). The axial potentials obtained from the network were used in order to obtain solutions of the paraxial 'reduced ray equation' (3.29).

Automatic ray-tracing machines are another class of system which are based on the electrolytic trough. These systems are complicated, require skill in operation and careful maintenance, and cannot be produced quickly without much previous experience. There are very few in existence, and their use is likely to diminish rather than increase. For this reason, little space can be given to them.

One type which automatically carries out the 'circle method' of graphical trajectory tracing is described by Gabor (1937), Langmuir (1937a, 1950) and Hollway (1955). A more accurate method, based on the 'inverted trough' of Sander and Yates (1956), is described by Pizer, Yates and Sander (1956). Information about the field is obtained from the electrolytic tank measurements, and the equations of motion (3.22) are integrated by a small digital computer. The results of the computation are converted to a voltage which drives a servo-motor, which in turn causes the measuring probes automatically to trace the trajectory.

In the simplest case of automatic ray tracing, both potential and potential gradient are measured simultaneously by probes dipping into an electrolytic trough and rigidly connected to a small carriage which guides a pencil drawing the electron path. The carriage is

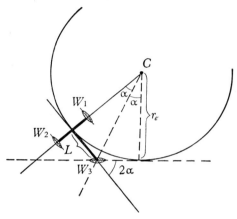

Fig. 3.8. Automatic ray tracing apparatus.

automatically steered by an electric motor. The steering angle is adjusted in the following way. The probe is split up into two wires arranged so that the line joining them is perpendicular to the electron path. Thus it measures not only the potential difference V of the particular point relative to the origin of the electron (for instance, the cathode), but also the field component $E_\perp$ in the direction normal to the trajectory, namely, $E_\perp = -\Delta V/\Delta r$, where ΔV and Δr are the voltage difference and the distance respectively of the two wires of the split probe which is always kept at right angles to the path. Now the steering angle ($= 2\alpha$), i.e. the angle included by the two axles of the carriage, has to be adjusted at each point in order to produce the path with radius of curvature

$$r_e = \frac{-2V}{E_\perp} \tag{3.31}$$

which follows from the centripetal force $mu^2/r_e = -eE$ and the energy equation $mu^2/2 = eV$. As can be seen in fig. 3.8, this path will be followed, if the two planes which are defined by the rims of the wheels W_1, W_2 and W_3 are kept parallel to the tangents of the path. In this case the locus for the centre C of the circle with the radius r_e is the intersection of: (1) the prolongation of the axle of the wheels W_1 and W_2 which is rigidly connected with the carriage and (2) the bisecting line of the angle between the planes parallel to the wheels which passes through the pivot of the wheel W_3 which is steered. Under these conditions it follows immediately that $r_e = L \cot \alpha$ if L is the distance between the pivot and the fixed axle. The necessary control of steering is achieved by applying both V and ΔV to a special circuit, the so-called 'tangential bridge', which is arranged to show balance when the ratio of the two voltages is given by

$$\frac{V}{\Delta V} = \frac{L}{2\Delta r} \cot \alpha. \tag{3.32}$$

CHAPTER 4

SOME ELECTROSTATIC ELECTRON LENSES

§4.1. General properties

According to the considerations of chapter 3 it is clear that the optical properties of an electrostatic lens are fixed by voltage ratios, the absolute value of the potentials at the electrodes being unimportant. Usually, all electrode potentials are measured with respect to a zero level at which the electron ray considered would have zero velocity; the ratio of these electrode potentials alone fixes the position of the cardinal points. For instance, in fig. 3.5 the ratio of refractive indices which is responsible for the angle of refraction at each equipotential is given by V_n/V_0, so that, for example, the cardinal points drawn in fig. 2.2 would be in exactly the same position whether the potentials at the two tubes were 0·1 and 0·5 V or 10 and 50 kV respectively.

It is convenient to measure all geometrical lengths of electrostatic lenses in terms of the radius of one of the electrodes. For instance, the symmetrical two-tube lens of fig. 2.1 with the voltage ratio $V'/V = 5$ has two mid-focal lengths:

$$MF_1 = 5 \cdot 4R \quad \text{and} \quad MF_2 = 4 \cdot 7R.$$

Thus, if the radius of the tube were 2 cm, these mid-focal lengths would be 10·6 and 9·4 cm respectively, or with tubes twice this radius, the mid-focal lengths would be twice as large.

The principal planes of the two-tube lens in fig. 2.2 are crossed over with respect to their corresponding foci. It can be shown from the paraxial ray equation (3.23) that the principal planes are crossed over in every electrostatic lens. The foci of all electrostatic lenses are moved towards the mid-plane if the voltage ratio is increased. A lens may be strong enough to make initially parallel rays cross the axis inside the lens field, so that a real focus is formed before the rays leave the lens. If after passing this focus, the rays are still exposed to the lens field, it is quite possible that

[65]

they are made convergent again to reach a second or even a third focus. Alternatively, after crossing over, the rays would leave the lens as a divergent bundle. However, it can be shown (for instance, from the paraxial ray equation (3.23)) that there exist no proper diverging electron lenses with negative focal length, provided that the lens electrodes are arranged outside the beam and that the lens field is bounded by a field free space at the entrance side and at the exit side of the beam.

§4.2. Symmetrical two-tube lenses*

These lenses are produced by two equal and co-axial metal cylinders at different voltages as shown in figs. 2.1 and 3.5. The positions of their cardinal points are given by the data in table 4.1 for some selected voltage ratios. These data are taken as far as available from computed results by Goddard (1946a) and by Firestein and Vine (1963). In Goddard's investigation, the electron path is obtained with the help of Picht's equation (3.28) operated on an analytical expression by Bertram (1940) for the paraxial potential distribution of the lens. Firestein and Vine, on the other hand, trace the electron path by computation through a field distribution obtained by means of a resistance network (cf. §3.3). The latter investigation covers a great range of voltage ratios; however, the results are presented in graphs which do not permit accurate readings. Experimental methods (cf. §§2.1 and 2.4) used by Spangenberg (1948) and by Klemperer (1953) have supplied supplementary results. These were obtained with smallest possible currents since mutual repulsion between the electrons always tends to increase the focal distances (cf. §8.3).

Since the two-tube lens always makes the electron beam more convergent, it is a positive lens, and this is true whether it accelerates or decelerates the electrons. Ray-tracing results show that the two-tube lens can be considered as consisting of two semi-lenses, a positive one, having equipotentials convex towards the lower potential side, and a negative one, making the beam more divergent, having the equipotentials concave towards the lower

* The term 'two-cylinder lens' used by many authors is not employed here, in order to avoid confusion with the glass-optical usage of 'cylinder lens'.

TABLE 4.1. *Symmetrical two-tube lens: voltage ratio V_2/V_1 and focal length f. Mid-focal distances MF and mid-principal plane distances MP. All absolute values measured in terms of tube radii R*

V_2/V_1	2	3	4	5	6	8	10	
MF_2	26	10·6	6·4	4·7	3·9	3·1	2·6	accelerating
MP_2	5	3·3	3·2	3·0	2·8	2·5	2·4	
f_2	30·6	13·9	9·6	7·7	6·7	5·6	5·0	
MF_1	25	11	7	5·3	4·5	3·5	3·0	decelerating
MP_1	2·9	2·6	2·0	1·9	1·8	1·6	1·4	
f_1	21·6	8·0	4·8	3·4	2·7	1·9	1·6	

$V_1 < V_2$

V_2/V_1	20	30	50	100	
MF_2	1·0	0·7	0·4	0·2	accelerating
MP_2	2·6	2·7	2·8	3·0	
f_2	4·1	4·0	3·9	0·5	
MF_1	2·2	1·8	1·6	1·5	decelerating
MP_1	1·2	1·1	1·1	1·1	
f_1	1·0	0·7	0·5	0·4	

potential side. As can be seen from fig. 3.5, the mid-plane separates the two semi-lenses. The power of the positive semi-lens is always greater than that of the negative semi-lens, so that its effect predominates in the resulting full lens. In fig. 4.3(b) there are shown the paths of the two characteristic parallel beams decelerated and accelerated. The glass-optical analogies of the semi-lenses are also shown in fig. 4.3(a) as convex and concave lenses. The voltage ratios of the two semi-lenses could be defined as the voltage ratio between the mid-plane and the first of the tubes, and as the voltage ratio between the other tube and the mid-plane. The values of these voltage ratios are

$$\left(\frac{V_2 - V_1}{2} + V_1\right) \bigg/ V_1 = \frac{V_2 + V_1}{2V_1} \tag{4.1}$$

for the positive semi-lens, and

$$\frac{2V_2}{V_2 + V_1} \tag{4.2}$$

for the negative one. Thus the voltage ratio of the positive semi-lens increases without limit as a linear function of the total voltage ratio V_2/V_1, but the voltage ratio of the negative semi-lens, being always smaller than that of the positive one, approaches the value 2 as an upper limit as the total voltage ratio is increased indefinitely. If the electrons are accelerated and sufficiently high-voltage ratios are applied, the electrons cross over very close to the mid-plane, and the beam leaves the lens more divergent than it entered. This has been demonstrated by Hamisch and Oldenburg (1964) who made measurements for voltage ratios up to 2000. For voltage ratios $V_2/V_1 > 10^2$ and $< 10^{-2}$ focal lengths and cross-overs of rays have been computed by Typke (1967).

The positions of the nodal points of the electron lens, which are important for a discussion of the magnification, can be calculated from the data of table 4.1. by means of (2.1). In this way it is found that both the nodal points usually lie on the high-voltage side of the mid-plane and are crossed over with respect to their corresponding focal points. The position of the nodal points changes only slightly with the voltage ratio.

The fact that the two nodal points are on the high-voltage side suggests that the magnification y'/y of an image y' produced on the high-voltage side would be smaller than a magnification calculated as the ratio q/p of mid-image to mid-object distance. As these two distances are always easy to measure, the diagram in fig. 4.1, which is due to Spangenberg and Field (1943) is of much practical use. If any mid-object distance is given, the corresponding mid-image distance and the magnification are easily read from the diagram. It can also be seen that if certain values of p and q are required, say by some given dimensions of an electronic tube, the voltage ratio V_2/V_1 applied will be the higher, the larger the radius R of the tubes chosen. Two families of curves are shown which are characterized by given parameters V_2/V_1 and $M = y_2/y_1$ (magnification) respectively.

In the two-tube lenses discussed so far it was understood that the gap between the tubes was to be small in comparison with the tube radius. In the field plot of fig. 3.5, for instance, the gap was chosen to be $2g = 0\cdot2R$, the tube ends being $g = 0\cdot1R$ apart from the

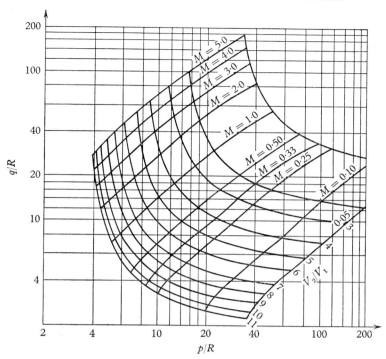

Fig. 4.1. Mid-object against mid-image distances for symmetrical two-tube lens.

mid-plane. Due to the fact that the potential $V(z)$ near the mid-plane is approximately a linear function of the axial coordinate z, the gap width has practically no influence on the lens characteristics as long as the gap is small. The effect of large gap width is demonstrated by table 4.2. There the influence of gap width $2g$ is shown for three different electrode spacings. The voltage ratios V_2/V_1 are set to obtain a given focal length P_2F_2, which in one example is $6R$, in the other example $12R$. The increase in voltage ratio with gap width is plausible since increase of gap width produces an extension of a quasi-homogeneous field inserted between the two semi-lenses. Moreover, it is of interest that with increasing gap width, the mid-plane-nodal point distances and thus the magnifications appear to decrease slightly. This can be seen by substituting the figures of table 4.2 into $MK = MF - P'F'$ and the corresponding equation for MK' which follow from (2.1) and (2.9). Tube lenses with increased gap width have to be shielded carefully to avoid disturbances through the large gap by spurious external electric fields. However, these lenses are of practical importance, because the application of large voltage differences often requires large gaps in order to avoid a breakdown by spark discharges.

TABLE 4.2. *Gap width, voltage ratio, and mid-focal lengths of symmetrical two-tube lens* (Spangenberg and Field, 1943)

$2g$	$P_2F_2 = 6R$			$P_2F_2 = 12R$		
	V_2/V_1	MF_2/R	MF_1/R	V_2/V_1	MF_2/R	MF_1/R
0·2R	8·0	3·0	4·4	3·5	7·4	8·8
1·0R	9·0	3·0	4·0	3·9	7·5	8·2
2·0R	12·0	3·3	3·1	6·4	9·0	6·9

For very high voltages at the electrodes, i.e. high energies of the electron rays, the focal length of the two-tube lens is modified owing to a relativistic increase in electron mass. The relativistic modification of the Picht equation (3.29) is

$$R^{*\prime\prime}(z) + G\,\frac{3}{16}\left(\frac{V'}{V}\right)^2 R^*(z) = 0 \qquad (4.3)$$

where
$$R^* = rV_r^{\frac{1}{4}},$$

V_r being the relativistically corrected voltage (cf. 1.9), and

$$G = \frac{1 + \dfrac{2eV}{3mc^2}\left(1 + \dfrac{eV}{2mc^2}\right)}{\left(1 + \dfrac{eV}{2mc^2}\right)^2}.$$

For the non-relativistic case, we have $G = 1$.

Bas and Preuss (1959, 1963), have used (4.3) to trace rays through a two-tube lens for which $g = 0.4R$, the axial potential distribution having been determined by resistance network analogue. Fig. 4.2 shows the ratio f_r'/f_0' of the relativistic to the non-relativistic image side focal lengths as a function of V_2, for values of V_2 up to 2 MeV. Each curve corresponds to a different voltage ratio V_2/V_1. Somewhat unexpectedly, the curves show maxima and indicate that at very high energies $f'_r < f'_0$. This behaviour is due to the form of the function G in (4.3). The weaker the lens, the smaller is the range in which $f'_r > f'_0$. The

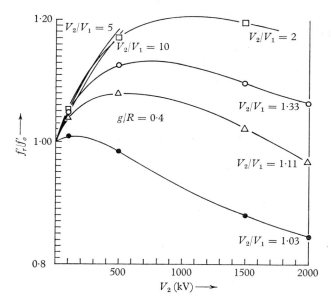

Fig. 4.2. Relativistic variation of focal length for a two-tube lens.

data of Bas and Preuss are complete up to 2 MeV for weak lenses but for higher voltage ratios they only extend to $V_2 = 500$ keV. For stronger lenses reference can be made to the work of Kessler (1961) who calculated cardinal points of tube lenses by integrating a relativistic ray equation using a linear approximation for the axial potential distribution of the lens, and who has presented his results for electrons up to 10 MeV. The magnitude of the relativistic change can be seen, for example, from the result that for voltage ratios $V_2/V_1 = 5$ and 10 respectively, the focal length increases by a factor of $f_r/f_0 = 1.03$ when $V_2 = 10^5$ volts for both voltage ratios. This factor becomes (again for both voltage ratios), $f_r/f_0 = 1.23$ when $V_2 = 10^6$ volts and it becomes $f_r/f_0 = 1.3$ for $V_2/V_1 = 5$ and 1.5 for $V_2/V_1 = 10$ at $V_2 = 10^7$ volts. The cardinal points of lenses in the relativistic region are of practical importance for the design of electron accelerators which are used in high energy nuclear physics (cf. §4.5).

§4.3. Various two-electrode lenses

The position of the cardinal points can be slightly varied if the shape of the electrodes of the electron lens is changed. There exists a very large variety of shapes, and we have to confine ourselves to mentioning just a few simple types in order to explain the principal characteristics and possibilities. In fig. 4.3(b)–(g) are shown six

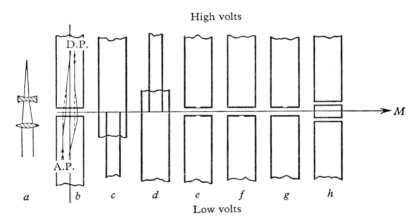

Fig. 4.3. Some simple electrostatic lenses.

examples of two-electrode lenses produced by combinations of cylindrical tubes and straight diaphragms. The straight line M, drawn through all the lenses, coincides in each case with the mid-plane. Only in lenses (b) and (e) does the mid-plane represent a plane of symmetry; in the other cases it is a rather arbitrary plane of reference. It is always practically convenient to take the radius R of the *largest* electrode as a unit of length, since this radius is limited by the size of the evacuated glass tube containing the electron lens. In table 4.3 are given the data concerning mid-focal lengths and the distances between the principal points and the mid-point, the plain symbols corresponding again to the decelerated parallel beam (D.P.), the dashed ones to the accelerated parallel beam (A.P.). It can be seen that the given values are markedly different, although the same voltage ratio ($V_2/V_1 = 5$) was taken in all cases.

TABLE 4.3. *Experimental data concerning the location of foci F and F' and principal points P and P'*

Lens (fig. 4.3)	Electrodes, radius		$V_2/V_1 = 5$				Potential at M (per cent)
	At low voltage	At high voltage	MF	MF'	MP	MP'	
b	Tube, R	Tube, R	5·3	4·7	3·0	1·9	50
c	Tube, $\frac{1}{2}R$	Tube, R'	2·6	3·2	0·1	1·0	31
d	Tube, R	Tube, $\frac{1}{2}R$	4·3	3·4	1·5	2·3	69
e	Diaphragm, $\frac{1}{4}R$	Diaphragm, $\frac{1}{4}R$	1·5	1·4	0·4	0·9	50
f	Diaphragm, $\frac{1}{4}R'$	Tube, R'	2·4	2·9	1·3	0·5	13
g	Tube, R	Diaphragm, $\frac{1}{4}R$	4·4	3·2	1·9	2·6	87

In order to understand the results, we may consider again that any one of these electron lenses can be thought of as consisting of two semi-lenses which are separated by the mid-plane. A large decrease in focal length would then be expected if the smaller electrode is on the low-voltage side. The smaller electrode would then correspond to the positive semi-lens (fig. 4.3(a)), which always overpowers the negative semi-lens, as was pointed out at the beginning of this chapter. The results obtained with lenses 4.3(c) and 4.3(f) support this expectation qualitatively, although the decrease in focal length is rather less than proportional to the decrease in electrode radius. Moreover, if the smaller electrode is at the high-voltage side, it corresponds to a negative semi-lens, which, according to the above considerations should not greatly affect the focal length of the complete lens. The results show, however, that even here the focal length of the complete lens is appreciably decreased in comparison with the symmetrical two-tube lens. The reason for this reduction of focal length with diminishing dimensions of the high-voltage electrode becomes obvious when a field plot of an unsymmetrical arrangement* is inspected. The mid-plane then no longer coincides with any equipotential. The 50 per cent equipotential has become concave

---

* Equipotentials of an unsymmetrical two-tube lens are drawn in fig. 9.4.

towards the smaller electrode and has penetrated appreciably into the larger electrode.

The axial coordinate at which the curvature of the equipotentials changes its sign is generally not far away from the mid-point. The potential at the mid-point is more removed from 50 per cent the greater the difference between the sizes of the two electrodes. In the last column of table 4.3 are noted the actual potential values at the mid-point of the lens. It can be seen that in all cases the voltage ratio of the semi-lens in the smaller tube is decreased and the voltage ratio of the semi-lens in the larger tube is correspondingly increased with respect to the voltage ratios of the semi-lenses in the symmetrical two-tube lens (see (4.1) and (4.2)). Thus, if the dimensions of the electrode at the high-voltage side are reduced, the decrease of focal length and the shifting of the principal points of the complete lens can be understood to be caused by the increased voltage ratio of the positive semi-lens which is contained in the larger tube. From the point of view of magnification it may be expected that an image at the high-voltage side will generally be more magnified if the smaller electrode is on that side.

It must be admitted that there is not much to choose between the various lenses shown in fig. 4.3, for the focal length of any of these lenses can be adjusted at will by simply changing the voltage ratio at the electrodes. However, from the practical point of view, the good electric shielding of some lenses as for instance those in fig. 4.3(c) and (d) with a larger tube overlapping a smaller one, offer some advantage (cf. §6.2). In particular, the lens of fig. 4.3(c) has been frequently employed as a focusing lens in television tubes, the wider tube being formed by the metallized glass neck into which the smaller tube protrudes as the end of the electron gun (cf. fig. 9.4). By using a greater diameter for the second tube, fouling of the beam is avoided when it is scanned by deflecting fields.

Numerical data for two examples of unsymmetrical two-tube lenses are shown in table 4.4. There, the ratio of tube radii is 2/3 and 3/2 respectively, one lens having the larger tube at the high-voltage side ($R'/R = 1\cdot5$), the other lens having the larger tube at the low side ($R'/R = 1/1\cdot5$); in both cases focal lengths and mid-focal distances are measured in units of the large tube radius.

TABLE 4.4. *Focal lengths and mid-focal distances of unsymmetrical two-tube lenses* (Spangenberg and Field, 1943)

V_2/V_1	Lens of fig. 4.3 (c) $R'/R = 1 \cdot 5$			Lens of fig. 4.3 (d) $R'/R = 1/1 \cdot 5$		
	$P'F'/R'$	MF'/R'	MF/R'	$P'F'/R$	MF'/R	MF/R
2	35	33	-24	16·5	12·3	$-19\cdot4$
3	11·2	9·5	$-7\cdot9$	11·8	8·0	$-9\cdot4$
5	6·5	3·9	$-4\cdot3$	8·2	5·0	$-6\cdot6$
7	4·9	2·4	$-3\cdot1$	6·3	3·3	$-5\cdot2$
10	3·8	1·6	$-2\cdot4$	4·8	2·2	$-4\cdot2$
13	3·3	1·3	$-2\cdot1$	4·3	1·7	$-3\cdot6$

TABLE 4.5. *Focal and mid-focal lengths of two-diaphragm lenses*

	V_2/V_1	2	4	6	10	20	40
spacing $S = R$	f_1/R	28·6	6·4	3·7	2·2	1·3	0·86
	f_2/R	40	13	9·0	6·8	5·7	5·4
	MF_1/R	34	9·3	6·1	4·3	3·2	2·8
	MF_2/R	34	8·8	5·3	3·3	1·9	1·2
$S = 4R$	f_1/R	49	10·6	5·8	3·2	1·7	1·1
	f_2/R	70	21	14·3	10·2	7·8	6·8
	MF_1/R	59	16	10	7	5	4·1
	MF_2/R	58	14	8·1	4·5	2·1	0·78

Numerical data for symmetrical two-diaphragm lenses, as shown in fig. 4.2(e), have been calculated by Read (1969a). For the potential distribution, the solution of Laplace's equation in terms of a series expansion of Bessel functions was used. Trajectories were obtained by numerical integration of the Picht equation (3.29) using a digital computer. Two examples are given in table 4.5. The two lenses are distinguished by the distances between the diaphragms, which are taken as one aperture radius and four aperture radii respectively. Focal lengths and mid-focal distances are also measured in units of aperture radii. The tubes which are terminated by the diaphragms are assumed to be large in comparison with the aperture. In this case, the value of the tube radius hardly influences the electron-optical characteristics of the lens. The real function of the tubes consists in providing support for the diaphragms and in the electrostatic shielding of the lens from spurious fields. If the diaphragms are separated by a sufficiently large distance, the power of the lens can be calculated to some approximation as the resultant of a combination of two single apertures (cf. Cosslett, 1946). Focal positions and principal points of diaphragm-tube lenses, similar to those in fig. 4.3(f) and (g), have been measured by Hamisch and Oldenburg (1964).

§4.4. The single aperture

A plane diaphragm may be placed between two plane conductors A and B of different potentials V_A and V_B. Two electric fields will be produced at the two sides of the diaphragm, namely,

$$E = \frac{V_A - V_C}{a} \quad \text{and} \quad E' = \frac{V_C - V_B}{b},$$

where V_C is the potential of the central diaphragm and a and b are its two distances from the external conductors. Now, a small circular aperture may be made in the central diaphragm. Equipotentials will bulge through the aperture, with the stronger field penetrating into the region of the weaker field. This field penetration through an aperture has been studied by Glaser and Henneberg (1935) theoretically and in the field plotting trough. Detailed computations of field distributions about apertured diaphragms of finite thickness have been published by Liebmann (1948).

Two essentially different potential distributions can be distinguished according to the senses of the two fields E and E', which may be in the same or in opposite directions. Examples of such potential distributions are shown in fig. 4.4(a) and (b). If the fields are in the same direction the potential distribution will have no maxima or minima. The equipotentials may penetrate through the aperture to a considerable extent but none of them cross, just one singular equipotential being observed reaching the diaphragm C at right angles. On the other hand, if the fields E and E' are in opposite directions, the potential distribution no longer slopes monotonically but a potential saddle is formed. The saddle-shaped field is like a pass over a mountain, the greatest height of the pass being the saddle point which lies on the axis. But even this saddle point may be flanked on both sides by still greater heights.

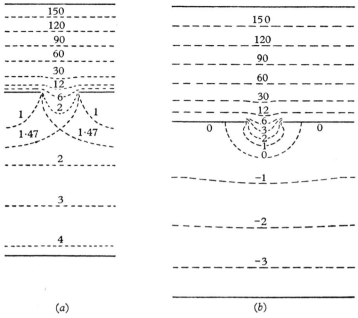

Fig. 4.4. Penetration through an aperture: (a) fields in opposite directions; (b) fields in the same direction.

The saddle point always lies at the side of the weaker field; let its distance s from the centre of the aperture be measured in terms of the aperture radius R and tabulated as a function of the ratio of the weaker field to the stronger field. Taking $E' > E$ the following five values may be given as an example:

s/R	0	0·2	0·5	1	3
E/E'	1	0·6	0·29	0·16	0·01

Here E and E' again correspond to the field strengths on either side of the diaphragm in the case when the aperture is closed.

The potential distribution at a single aperture cannot be considered to be a complete electron lens because it is not bounded by spaces of constant refractive index. Thus lens formulae like (2.8) cannot be applied directly. However, the study of the focusing properties of the aperture is sufficiently interesting for the understanding of various aperture lenses and of emission systems (cf. chapter 9) in which apertured diaphragms are incorporated as elementary units. Moreover, the apertured thin diaphragm is of particular historical interest. Its field was first recognized as having properties of an electrostatic lens, when Davisson and Calbick (1931) investigated experimentally its focusing properties. There, a diaphragm with a small aperture was placed between a thermionic cathode and a fluorescent screen. If the fields at either side of the aperture are small in comparison with the ratio V_C/R (= aperture potential/aperture radius) a formula for the focal length can easily be derived. The axial velocity component near the diaphragm is given by the energy equation

$$u_z^2 = 2\frac{e}{m}V_C. \tag{4.4}$$

Near the aperture, however, the field is inhomogeneous and contains radial components E_r; these produce radial components u_r ($\perp u_z$) of the electron velocity which, according to Newton's law are given by

$$\frac{\mathrm{d}}{\mathrm{d}t}(mu_r) = -eE_r$$

or by integration

$$u_r = -\frac{e}{m}\int_{-t_0}^{+t_0} E_r\,\mathrm{d}t. \tag{4.5}$$

Here we have assumed that the field inhomogeneity extends over a small distance from the aperture only, namely from $-z_0$ to $+z_0$ and these coordinates are reached by the electron at the times $-t_0$ to $+t_0$ respectively. Since $u_z = \mathrm{d}z/\mathrm{d}t = \text{const.}$, we can substitute from this relation for the variable t in (4.5) and we obtain

$$u_r = -\frac{e}{mu_z}\int_{-z_0}^{+z_0} E_r\,\mathrm{d}z. \tag{4.6}$$

A value for the unknown integral in (4.6) can be obtained with the help of Gauss's law according to which the surface integral over the normal component of flux density ($\epsilon_0 E$) is equal to the enclosed charge. Consider a cylinder, symmetrical and co-axial with the aperture, extending from $-z_0$ to $+z_0$. Let the radius of the cylinder be equal to the distance r of the ray from the axis. Since the cylinder contains no charge, we have

$$2\pi r\int_{-z_0}^{+z_0} \epsilon_0 E_r\,\mathrm{d}z + \pi r^2\epsilon_0(E - E') = 0,$$

where the first term covers the curved cylinder surface and the second covers the plane ends, the outward flux being considered positive. Hence the latter equation yields

$$- \int_{-z}^{+z} E_r \, dz = \frac{r}{2} (E - E'). \tag{4.7}$$

We now divide (4.6) by (4.4) and substitute for the integral from (4.7). This leads to:

$$\frac{r}{4} \frac{E - E'}{V_c} = \frac{u_r}{u_z} = \tan \Theta.$$

The tangent of the angle Θ at which the beam emerges, is seen to be proportional to the initial axial distance r from the axis. Hence, rays of different r converge to (or diverge from) a common focus, and the field under discussion possesses the properties of an optical lens. The distance of the focus from the aperture is the focal length f which is given by

$$\frac{r}{\tan \Theta} = f = \frac{4V_c}{E - E'} \tag{4.8}$$

which is the aperture-lens formula of Davisson and Calbick. It should be noticed that the result is an approximation only, applying to paraxial rays ($u_r \ll u_z$) in a thin lens ($z_0 \ll R$). Also, R should be very small in comparison with the dimensions of the arrangement of electrodes which produce E and E'. If this is not the case, the potential V_z in the centre of the aperture differs substantially from V_c and must be substituted for V_c in (4.8) in order to obtain reasonably accurate values for f (Hoeft, 1959); V_z can be found easily from (3.4).

Both focal lengths of the aperture are equal. In the special case that $E > E'$ shown in fig. 4.5, the aperture diverges the rays as the focal length f is negative. This follows from (4.8) since both fields E and E' accelerate the electron in the z-direction, i.e. they both are negative quantities. The diverging effect of the lens also follows from the fact that the stronger field (E) always penetrates into the region of the weaker field (E'), i.e. the equipotentials are bulging through the aperture and form concave surfaces (cf. fig. 4.4b) towards the accelerated electron. Hence according to

Snell's law, the rays are refracted away from the axis. As a further
check on the diverging effect of the aperture, it should be noted that
d^2V_0/dz^2, the second derivative of the axial potential is negative in
the region of interest, hence according to (3.7), an electron beam is
diverged as it passes through the aperture.

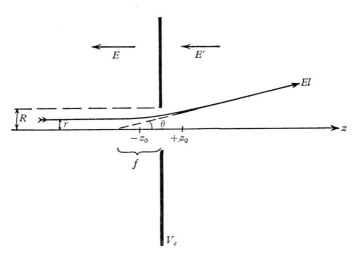

Fig. 4.5. Derivation of Davisson and Calbick's formula.

So far, we have discussed the case of a circular aperture. If we
replace this by a slit aperture, a greatly increased focal length should
be expected since the field which penetrates through a long slot,
decays much more slowly than that through a round hole (cf.
§2.5). Potential distributions both for round holes and for infinitely
long slots have been calculated by Fry (1932b) and by Strashkevitch (1940) and measured by Glaser and Henneberg (1935).
The focal length can be calculated for both apertures in similar
ways, but the result for the slot is just twice as small as that for the
round aperture (cf. (4.8)). The focal length of the slotted diaphragm
is given by

$$f = \frac{2V_c}{E-E'}. \tag{4.9}$$

§4.5. Multielectrode lenses

The electron lenses which have been discussed in §§4.2 and 4.3 are two-electrode lenses. This implies that the potential of the object space is different from that of the image space. The two-electrode lenses are analogous therefore to glass-optical immersion lenses. Multielectrode lenses belong to the same class if the potential of any intermediate electrode ranges between the potentials of the two adjacent electrodes. For example, in fig. 4.3 (h) is shown a three-tube lens in which the potential of the intermediate tube V_2 is variable and lies between the potentials of the first and of the last tube ($V_1 < V_2 < V_3$). It will be observed that the focal lengths of such three-tube lenses are somewhat larger than the focal lengths of the corresponding two-tube lens with the voltage ratio V_3/V_1. As a development from the three-tube lens, still more small intermediate tubes can be inserted between the two external tubes. In this way the 'lens chain' is obtained, by which electrons may be accelerated or decelerated with a relatively small focusing action.

The focal properties of a 1·5 MeV electrostatic acceleratator consisting of eleven identical tubes of radius 5 cm, length 41 cm and spacing 4 cm have been computed by Bas and Preuss (1959) using data already presented in §4.2. This accelerator is a chain of ten two-tube lenses of progressively decreasing voltage ratio. The mid-image distance Z_q (i.e. the distance from the middle of the gap between the first and the second tube to the image plane) is fixed at 605 cm, and, except for the first stage, the acceleration per stage is 150 kV. The voltage V_1 on the first tube then determines the mid-object distance Z_p as indicated in table 4.6.

TABLE 4.6. *Voltage at the first electrode of a lens chain and the required mid-object distance*

V_1	30	60	90	120	kV
Z_p	22	36	52	63	cm

Thus the focal properties of the accelerator can be adjusted by altering the voltage of the first tube, V_1, which is accordingly used

4

as focusing control. The extreme case of a lens chain is an insulating tube sprayed with a very thin metal layer. The resistance of this layer is large enough to maintain a linear potential drop from one end of the tube to the other. Such an arrangement produces a very nearly homogeneous field; it has actually been used to accelerate electrons without focusing them (Farnsworth, 1934), so that the focal length has in this case become infinite.

Also of practical importance is the problem of channelling a long electron beam, by preventing it from spreading with the help of a lens chain; this has been discussed in §2.3. Apart from the application of such lens chains in high energy particle accelerators, the net focusing action of alternatively positive and negative lenses can be used to balance space-charge repulsion in microwave beam tubes (cf. §8.7).

§4.6. Saddle-field lenses

A completely new situation arises if the voltage of the intermediate electrode of a three-electrode lens is adjusted to be outside the range V_1 to V_3. Lenses belonging to this class of multielectrode lenses are called 'saddle-field lenses', because the potential distribution becomes saddle-shaped as shown in the field plot of fig. 4.4(a).

The focal length of a saddle-field lens is shortened gradually as the voltage V_2 of the intermediate electrode either increases beyond V_3 or decreases below V_1. Thus the focal length of the lens can be controlled within a very large range by varying V_2 for any constant value of V_3/V_1. A case of practical interest arises if $V_3 = V_1$. The equivalent electron lens is then surrounded by media of the same refractive indices at both sides, and corresponds to an isolated glass lens, so that in the literature it is often called an 'einzel' lens (i.e. single lens). Since the voltages at both sides of the lens are equal ($V = V'$) the two focal lengths f and f' of the lens must be equal too (cf. (2.8)), and the two nodal points K and K' must coincide with the two principal points P and P' (cf. (2.1)). The electron velocity outside the einzel lens is given by the potential V_o of the two outer electrodes. Inside the lens, the electron is initially accelerated or decelerated according to the choice of the potential

V_i of the inner electrode, but in both cases a converging lens is obtained. The electron while passing through the lens, however, travels much nearer to the axis when $(V_i - V_o) > 0$, than when $(V_i - V_o) < 0$. On the other hand, it is easier to obtain short focal lengths by applying negative voltage $(V_i - V_o) < 0$ than by applying positive voltage $(V_i - V_o) > 0$ to the einzel lens.

The particular case of an einzel lens with $V_i = 0$ is of practical importance. There, the inner lens electrode is simply brought up to cathode potential, while the outer electrodes may be connected to the anode of the system from which the electrons are emitted. Thus only one voltage is applied to such a lens and it is frequently called a 'univoltage' or 'unipotential' lens. Moreover, as the voltage ratio V_o/V_i of the univoltage lens is always infinite, the positions of its cardinal points will be independent of the applied voltage provided the electron velocities are not excessive so that the relativistic increase of the electron mass is negligible.

The fact that the focal length of the univoltage lens does not depend on the applied voltage is of considerable technical interest. Fluctuations of the voltage source are not disturbing. It is even possible to project sharp images with a univoltage lens that is operated with an alternating voltage. On the other hand, the focal length depends largely on the dimensions of the intermediate electrode. If a univoltage lens is used in an apparatus where a given focal length is required, the dimensions of the electrodes must be adjusted with sufficient precision.

The two basic types of the einzel lens are the three-tube einzel lens (fig. 4.3(h)) and the three-diaphragm einzel lens consisting of three plane, apertured diaphragms (fig. 4.6).

The properties of the latter type depend only on the aperture radii R_o and R_i, the spacings S and the thickness T_o and T_i of the diaphragms (see fig. 4.6). Such lenses have been studied originally by Brüche and his collaborators (cf. Brüche and Scherzer, 1934). Chancon (1947) investigated these lenses by ray tracing and found some useful empirical relations for the location of the foci and principal points.

Optical properties of univoltage lenses as a function of the thickness T_i ($T_o \approx 0$) and the spacing S have been determined by Vine (1960). He computed his values from potential distributions

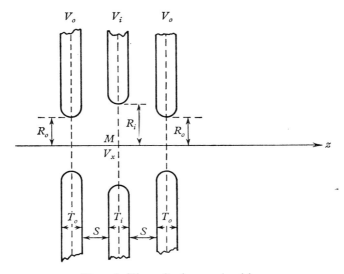

Fig. 4.6. Three-diaphragm einzel lens.

obtained with a resistance network analogue. In table 4.7 we present some of his results read from his graphs.

TABLE 4.7. *Focal lengths and principal plane distances of univoltage lenses*

T_i/R	o	o·4	o·8	1·2	1·6
$S = R\begin{cases} f/R \\ MP/R \end{cases}$	10 o·2	5 o·3	3 o·5	2·1 o·8	2·0 1·5
$S = 2R\begin{cases} f/R \\ MP/R \end{cases}$	4·5 o·05	3 o·1	2·5 o·3	2·1 o·35	2·2 o·6

Regenstreif (1951) derived an analytic function for the potential distribution of einzel lenses by superimposing the potential distributions of three single diaphragms each of which being represented by an equation of the form

$$V(z) = a + bz + cz \tan^{-1}(z/R), \qquad (4.10)$$

where a, b and c are arbitrary constants.

The resulting distribution could be used for an integration of the paraxial ray equation.

On the other hand, Glaser (1956) approximated the potential distribution of the einzel lens by a 'bell-shaped' distribution of the form

$$V(z) = V_o + \frac{\overline{V}_i}{1 + (z/w)^2},$$ (4.11)

where V_o is the potential of the outer electrodes and $\overline{V}_i$ is the potential at the saddle point, i.e. at the centre of the aperture in the inner electrode, $2w$ is the half width of the bell-shaped distribution curve. Equation (4.11) appears to approximate the actual potential distribution very closely and the optical properties of einzel lenses which Glaser derived from it, agree well with experimental data. Glaser's approach has been elaborated by Kanaya *et al.* (1966) and we reproduce here from their paper two diagrams (fig. 4.7, and fig. 6.12 in §6.4) which allow us to read quickly all optical data for

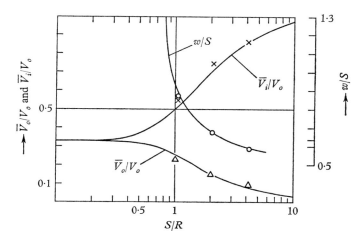

Fig. 4.7. Axial potentials $\overline{V}_i$ and $\overline{V}_o$ and half width $2w$ of bell-shaped potential distribution as a function of diaphragm spacings S.

einzel lenses consisting of three thin diaphragms ($T \lesssim \text{o·1} S$) with equal apertures ($R_1 = R_2 = R_3$). As an example, let us determine from these graphs the properties of an objective with the ratio of diaphragm spacing S to aperture radius R given as $S/R = 2\cdot5$. We read immediately from fig. 4.6 that $\overline{V}_i/V_o = \text{o·75}$, $\overline{V}_o/V_o = \text{o·12}$ and $w/S = \text{o·62}$. Moreover, the abscissa in fig. 6.12 of §6.4 is the potential ratio $\bar{V_i}/V_o$ and for the value $\bar{V_i}/V_o = 0.75$ just obtained we can take the values for

    (i) the focal length $f/2w = 2.9$,

    (ii) the mid-focal length $z_f/2w = 2.8$,

    (iii) the principal plane distance from the mid-plane $z_p/2w = 0.09$

and

    (iv) all data on lens aberrations which we shall discuss in §6.4. For a numerical example, take $R = 0.2$ cm and $S = 0.5$ cm, then $f = 0.93$ cm, etc.

Exact values for focal lengths and mid-focal lengths of einzel lenses have been obtained by Read (1969b) by numerical integration of the ray equations using a potential distribution calculated by a least square collocation method. We present in table 4.8 some of his results for einzel lenses with three equal apertures in plane thin diaphragms.

TABLE 4.8. *Focal lengths f and mid-focal lengths MF of einzel lenses with equal aperture radii R and with spacings S*

$\dfrac{V_i - V_o}{V_o}$	$S = R$		$S = 2R$	
	f/R	MF/R	f/R	MF/R
11	15.3	15.3	7.35	7.25
4	21.1	21.0	10.6	10.5
1	67	67	38	38
0	∞		∞	
−0.5	37	37	27	27
−0.8	6.23	6.10	5.70	5.33
−0.9	3.45	3.05	3.84	2.66
−0.95	2.80	1.77	4.33	0.60
−0.98	3.69	−0.04	−86	90
−0.99	9.23	−5.86	−3.8	7.7
−2.0	0.84	0.07	—	—
−3.0	3.05	2.95	—	—
−6.0	6.63	6.57	2.52	2.20
−11.0	8.66	8.62	3.61	3.40

When the voltage difference $(V_i - V_o)$ between the diaphragms is gradually decreased from large positive values down to zero, the focal length f of the lens increases from small values to infinity. With further decreasing $(V_i - V_o)$ to negative values, the focal length decreases again. An eventual increase of f seen in the last few tabulated values is due to the lens turning into a mirror; this will be discussed in §4.8. As a practical example of a weak einzel lens for relatively slow electrons, we show in fig. 4.8(a) such a lens, the outer electrodes of which have been shaped to improve its optical quality. This lens has been investigated in great detail by Brüche and Scherzer (1934) and their collaborators. The glass

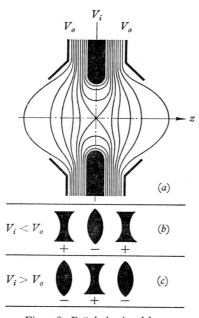

Fig. 4.8. Brüche's einzel lens.

optical analogue which applies to every einzel lens with an initially decelerating potential distribution $(V_i < V_o)$ is indicated in fig. 4.8(b). There the outer fields diverge, and the central field converges the electron beam. On the other hand, for an initially accelerating potential distribution $(V_i > V_o)$ the outer fields converge the electrons while the central field diverges the beam like a concave glass lens shown in fig. 4.8(c). In both cases, however, the overall effect is always a converging action; and this applies for every electrode geometry.

The potential distribution shown in fig. 4.8(a) is symmetrical about the mid-plane (M in fig. 4.6) and about the z-axis of rotational symmetry. Near the intersection of the mid-plane with the z-axis, the equipotentials are symmetrical hyperbolae. The asymptotes are two singular equipotentials which cross on the axis, they represent cones which are co-axial with the lens axis having a semivertical angle to this axis of $54 \cdot 6°$. It is remarkable that this angle is quite independent of the electrode geometry.

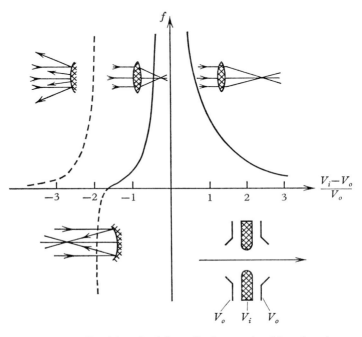

Fig. 4.9. Focal length of three-diaphragm einzel lens (———)
and of electron mirror (– – – –).

Curves for the focal lengths of the particular lens of fig. 4.8 as a function of $(V_i - V_o)/V_o$ are plotted in fig. 4.9. The solid curves belong to a positive focal length (f) for $V_i > V_o$ as well as for $V_i < V_o$; however, strong lenses (small f) can be obtained more readily by making the central electrode negative with respect to the outer electrodes than by making it positive. The focal length for very negative central electrodes is shown by the broken curves in fig. 4.9. There, however, the einzel lens field acts as an electron mirror; this will be discussed in §4.7.

As a practical example of a three-diaphragm einzel lens for relatively faster electrons, we show in fig. 4.10 an electrostatic electron microscope objective. The two external electrodes A and B are on earth potential while the central diaphragm C which is supported by a strong insulator D is at the potential of the cathode which emits the electrons, and which may be at a high negative

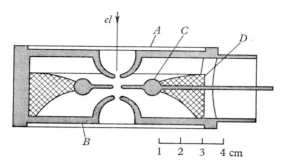

Fig. 4.10. Electrostatic electron microscope objective.

potential (e.g. -40 kV). All three diaphragms have apertures of about a millimetre radius and these apertures are spaced at distances of the order of a millimetre. This very powerful lens has a focal length of a few millimetres only which is independent of the voltage applied to C.

Our general discussion of einzel lenses has so far been confined to such lenses with very thin diaphragms ($T \ll S$ in fig. 4.6). The diaphragms of the practical lenses, however, are frequently of substantial thickness as, for instance, in the examples we have shown in figs. 4.9 and 4.10.

An increase in thickness (T_i) of the central electrode will increase the power ($1/f$) of the lens, the rate of increase in $1/f$ depending on the spacing S between the electrodes. For example, let all electrodes of a univoltage lens have the thickness $T = 0\cdot1R$, where R is the aperture radius. In order to double the power of the lens with $S = R$, T_i must be increased to $0\cdot6R$; to double the power of the lens with $S = 2R$, however, T_i must be increased up to $1\cdot2R$ (Vine, 1960). Detailed nomograms on the influence on the central electrode geometry on the converging power of einzel lenses can be found in the papers by Lippert and Pohlit (1952, 1953, 1954).

The extreme case of very thick electrodes leads to the three-tube einzel lenses (cf. fig. 4.3(*h*)). The great influence of the intermediate tube length L_i on the power of such lenses is demonstrated by fig. 4.11. There, three curves for the mid-focal lengths (*MF*)

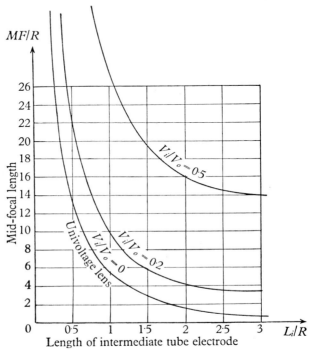

Fig. 4.11. Mid-focal lengths of three-tube einzel lens.

are given as functions of L_i (Klemperer, 1953). The radii (*R*) of the three tubes are equal and the spacings (*S*) between the tubes are small ($S = 0.1R$). Inscriptions to each curve are the voltage ratios $V_i/V_o = 0.5$, 0.2 and 0 respectively, representing constant parameters. In every case, the electrons are initially decelerated inside the lens since $(V_i - V_o) < 0$. The curve $V_i/V_o = 0$ belongs to the series of univoltage lenses.

Exact measurements concerned with the cardinal points of a few selected three-tube univoltage lenses have been published by Liebmann (1949*a*). He found, for instance, the focal lengths for tube lenses with three tubes of equal radius R, and with inter-mediate tube length $L_i = $ o·68R, to be $f = $ 6·20R (and 3·84R) for tube spacings o·26R (and o·76R respectively). Again, for tube spacings equal to one tube-radius, he found focal lengths $f = $ 1·94R (and 1·72R) with $L_i = $ 1·44R (and 2R respectively). The distances between the mid-plane and each of the principal planes were found for all the above lenses to be $MP = $ o·2R within the accuracy of the measurement, with P and F lying on opposite sides of M.

Some investigations on three-tube einzel lenses have been published also in a paper by Gobrecht (1941) which, however, is mainly concerned with various errors of this lens (cf. chapter 6). A paper by Iams (1939) deals with the application of a univoltage lens of three tubes of unequal radius to the focusing of electrons in cathode-ray tubes.

Of some technical importance is the question of finding lenses of shortest possible focal length for use in electron microscopes. Bachman and Ramo (1943) had already noticed a distinct minimum of focal length as a function of either the inner aperture radius R_i or of the voltage ratio V_i/V_o. This minimum has been confirmed by calculations of Bruck and Romani (1944) and it is explained by the fact that the rays will continue to be refracted after passing through the focus (§4.1). In effect, on decreasing R_i or V_i to very small values, the principal point moves towards the electron source more rapidly than the focus. The problem of this 'ultra-focal' refraction will be fully treated in the discussion of the strong magnetic lenses.

The location of foci and principal points for very small R_i/S and for very small or negative values of V_i/V_o has been investigated by various methods theoretically and experimentally. For instance, Regenstreif (1951) has calculated many examples of electron trajectories through einzel lenses. The locations of principal points and foci for very small or negative values of V_i have been measured systematically by Rang (1948) and by Heise and Rang (1949) with particular regard to einzel lenses with large thickness T_i of the inner diaphragm. The large amount of experimental material for f as a function of V_i, S, T_o, T_i, R_o and R_i can be divided into several regions which are classified by the fact that the electron trajectories may cross the lens axis once, twice or even three times respectively. The first region shows characteristic minima and the second region shows characteristic maxima for the focal length as a function of the voltage ratio.

Apart from the simple three-diaphragm lenses with plane, apertured diaphragms, other types of three-diaphragm einzel lenses with specially shaped electrodes have received detailed treatment in the literature. Here, we refer to the papers by Ramberg (1942) and by Liebmann (1949a). The latter authors investigated a lens with an inner diaphragm which was tapered towards its relatively wide aperture. We also mention the thoroughly investigated lens type shown in fig. 4.9 (cf. Johannson and Scherzer, 1933), where the two external diaphragms are extended outwards in the shape of co-axial cones. Relations between the properties of some asymmetrical and similar symmetrical einzel lenses have been discussed and tested experimentally by Hanszen (1956, 1958).

A point of general interest is the behaviour of einzel lenses towards fast electrons in the relativistic region. Laplume (1947) calculated the focal length f of a three-diaphragm univoltage lens ($V_i = $ o) as a function of the energy V_o of the incident electrons. Some of his values are quoted here in table 4.9, where S is again the distance between two consecutive electrodes. It can be seen that the focal length f of the univoltage lens reaches a distinct maximum in the region of 1 MeV.

TABLE 4.9. *Focal lengths of univoltage lenses in the relativistic region*

V_o in MeV	o	0·3	1	2	5	10	∞
f/S	1·28	1·33	1·38	1·36	1·28	1·20	1·06

§4.7. Electron mirrors

If the potential of the intermediate electrode in a saddle-field lens is gradually made more and more negative, the path of the electrons will finally be blocked and all those moving up the potential mountain will be returned. At first the marginal electrons are returned, but when the inner electrode becomes still more negative, the axial electrons are also reflected. In this way the einzel lens can be used to modulate the intensity of an electron beam. The lens behaves as if closed by an iris diaphragm when the negative voltage V_i is increased.*

If a narrow pencil of inhomogeneous electron velocities is transmitted along the axis of a saddle-field lens, the electrons of various velocities can be sorted by control of the inner lens electrode which is kept at negative potential. Only those electrons will be transmitted through the lens whose energy is sufficient to surmount the height of the potential saddle; all slower electrons will be reflected by mirror action. The lens is thus used as an 'electron velocity filter'. The action and construction of these filters will be discussed in detail in §7.3 in connexion with chromatic errors.

If the electrons emitted by any extended object are projected backwards by such a reflecting, axially symmetrical field, the reflected electrons form a real image. This was predicted theoretically by W. Henneberg and A. Recknagel (1935) and was verified later by G. Hottenroth (1936). The arrangement acts as a mirror. By changing the voltage at the negative electrode, different equipotentials can be chosen to act as reflectors. Thus the focal length and the principal plane of the mirror can be varied. In fig. 4.9 two broken curves show the focal length of the electron mirror produced by the electrode arrangement of a three-aperture lens. This focal length at first increases with decreasing negative bias; then suddenly it assumes negative values, which means that the mirror reflects the electrons like a concave glass optical mirror. For strongly negative potentials V_i, of the inner electrode, the mirror becomes periodically 'concave' and 'convex'.

A simple type of electron mirror is produced by the electric field between two tubes. Electron reflexion by such a two-tube mirror is shown in fig. 4.12. At the moment of reflexion, the axial velocity component of the electron is reduced to zero. Thus the reflecting equipotential is very nearly at cathode potential, and it is often called the 'zero equipotential'. In fig. 4.12 the electron is reflected at the mid-plane of the lens. Since the potential of the mid-plane

---

* In Gaussian optics, all electrons should have equal axial velocities independent of their axial distance. Thus either all electrons should be transmitted or reflected simultaneously. The finite slope of any actual modulator characteristic is due to lens errors (cf. §9.5).

is $\frac{1}{2}(V_1 + V_2)$, i.e. half-way between the potential of the electrodes, the cathode potential must have been half-way between the electrode potentials, and $V_2 = -V_1$.

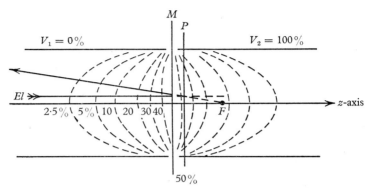

Fig. 4.12. Two-tube electron mirror; measurement of focus F and principal plane P.

For the present purpose it is again suitable to label the equipotentials as in fig. 3.3 in terms of percentages of the total potential difference. If V_1 is taken as a positive voltage and V_2 as a negative voltage with respect to the cathode, the zero equipotential of the electron mirror is equivalent to the equipotential labelled

$$V_C = \frac{100}{(1 - V_1/V_2)} \text{ per cent.}$$

Some of these equipotentials are shown in fig. 4.12 with the proper designation. Focal point F and principal plane P can be measured in the way described in §2.1, by tracing initially parallel electron pencils. The pencils are made visible by a sliding fluorescent target before and after having passed the reflecting field. Such a target, however, has to be traversed by the incident beam. Klemperer (1953) used a target made from a fine metal gauze, the wires of which were coated with a thin layer of fluorescent material. The pencils passing through the meshes grazed the fluorescent wires and thus became visible. The target was then moved along the z-axis and the semi-apertures of the incident and of the emerging pencil were measured as a function of the z-coordinate with a microscope viewing 'through' the mirror, i.e. against the direction of the incident beam. Focus F and principal plane P were traced by finding the extrapolated intersections of the emerging pencil with the axis and with the extrapolated incident pencil respectively. Table 4.10 shows the results. In all cases, the electron mirror acts like an optical mirror in combination with an optical lens, the lens being represented by the field which is passed before and after the reflexion. The mirror converges the electron rays, until the voltage ratio reaches -0.412; for still more negative values of V_2 the mirror diverges the rays. At $V_2/V_1 = -0.412$ the field behaves for paraxial rays like a plane mirror.

TABLE 4.10. *Cardinal points of two-tube electron mirror*

Mirror	Converging			Plane	Diverging	
V_2/V_1	-0.2	-0.25	-0.35	-0.412	-0.50	-1.0
Reflecting equipotential:						
Percent	17	21	27	29·5	34	50
At	$+0.66R$	$+0.56R$	$+0.44R$	$+0.40R$	$+0.30R$	0
MF/R	-0.3	-0.9	-3.2	∞	$+3$	$+0.8$
MP/R	$+1.1$	$+0.9$	$+1.1$	∞	$+0.5$	$+0.2$

For more details of image formation by the two-tube mirror, including magnification values, the reader may refer to the results of Nicoll (1938). In his experimental arrangement which has later been used in some electron microscopes (e.g. Mahl and Pendzich, 1943) the electrons coming from an object are projected through a tube containing a plane diaphragm with a small aperture. At the end of this tube, they are reflected by an electron mirror, and travelling back through the tube they strike a fluorescent screen on the back of the plane diaphragm. There, an image is projected, the focus being controlled by the voltage applied to the mirror. A small part of the image is, of course, lost owing to the aperture in the diaphragm.

A simple form of electron mirror, which is of some practical importance, e.g. in the electron mirror microscope of fig. 4.13 consists of a plate T at cathode potential, at distance d from a diaphragm with an aperture A of radius R and at anode potential. If $R \ll d$, it is easy to show, by application of the aperture-lens

formula (4.8) to both incident and reflected beams, that such a
mirror is a diverging element of focal length $4/3d$, whose principal
plane is $d/3$ behind the plate T. If R is not small compared with d,
it is possible to derive an expression for the cardinal points which
is accurate to a few per cent in the range $0 < R/d < 0.6$ (Barnett,
Bates and England, 1969). The mirror is diverging with focal

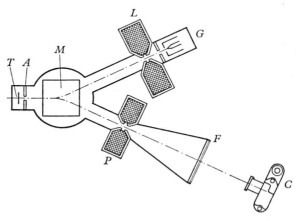

Fig. 4.13. Electron mirror microscope.

length f_m equal to $1.33d(1 - 0.41R/d)$ and its principal plane is $\frac{1}{4}f_m$
behind the plate T.

In order to obtain a distortion-free image projection from an
electron mirror, the electron beam must enter and leave the mirror
along its optical axis. This condition can be fulfilled in a completely axially symmetrical arrangement only by sacrificing the
paraxial part of the image. In another arrangement which was
introduced already by Hottenroth (1936) this condition is satisfied
with the help of a deflecting magnetic field which separates the
incident from the reflected beam. We illustrate Hottenroth's
arrangement by a more recent application of it in the electron
mirror microscope. This was developed in detail by Mayer (1955,
1956); the design we show in fig. 4.13 is due to Bartz, Weissenberg
and Wiskott (1954). There G is an electron gun, the electron rays
are made parallel by the magnetic lens L, they are deflected by the
magnetic field M and they enter through the aperture A into the

retarding electrostatic field produced by the plate T, the potential of which is near to the potential of the cathode of the gun. The electrons reflected by the mirror field in front of T, pass again through M where they are deflected to the other side. The rays are now slightly divergent owing to the field about the aperture A and they are projected by the magnetic projector lens P onto the fluorescent screen F. If the potential of the surface of T is not uniform, transverse fields exist near its surface. These fields alter the transverse velocities of electrons which are reflected very close to T and modulate the intensity of the electron beam, thus creating an image related to the form of the surface potential distribution, which by means of the lens P is projected on to F. There, the fluorescent image can be photographed by the camera C. To ensure that the beam does not strike the specimen, T is held at a volt or so negative with respect to the cathode.

Another electron optically interesting application of the mirror is seen in the energy sorting electron microscope (fig. 4.14) of Castaing and Henry (1962, 1963, 1964). There, the mirror is produced by the electrostatic field of two slits and a backing plate. Electrons are directed into the mirror by a magnetic prism (§ 10.5) and the reflected beam leaves through the same field. In this particular arrangement, incident and reflected beams are not only separated, but are also brought back into line. Here, these beams are aligned with the axis of a highly resolving electron microscope. It will be explained in § 10.8 how the prism disperses the energy components in the beam, so that the arrangement can be used for the study of separate microscopic images which are produced by separate energies owing to energy losses of the electrons in the specimen object.

For sufficiently small beam currents, the definition of the image projected by an electron mirror is sufficient for incorporation in the design of high quality electron optical apparatus. It is for instance possible to use a diverging mirror as a projector lens in a transmission electron microscope (Mahl and Pendzich, 1943), thus halving the column length necessary to obtain a given final magnification. A more recent example of a mirror used as a projector lens in the scanning mirror microscope has been described by Bok, Kramer and Le Poole (1964).

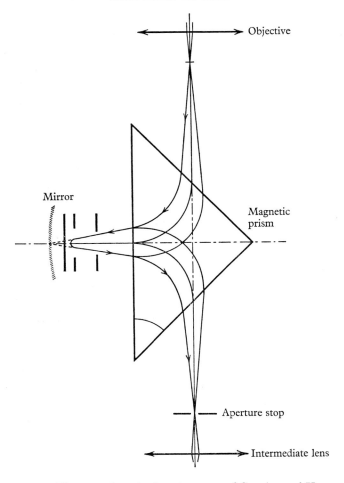

Fig. 4.14. Electron mirror in the microscope of Castaing and Henry.

§4.8. Gauze lenses

In the earliest development of electrostatic electron lenses, shaped
metallic wire gauzes or screens were used to represent the
refracting surfaces. Knoll and Ruska (1932) used a spherical
condenser consisting of two concentric spheres at different potentials
and provided with wire gauze windows. They also discussed the
use of other lenses of similar kind made of two closely spaced

parallel wire screens at different potentials dividing the space into two equipotential regions and thus simulating the glass surfaces of light lenses. In all cases, however, disturbances are produced by a certain amount of field penetration through the meshes of the gauze. Each aperture of the gauze acts as a separate little 'facet lens', the focal length of which could be estimated in the way explained in §4.4. Verster (1963) has made a detailed study of the scattering of the electron beam by these facet lenses. There, the scattering increases with r/p, where r is the radius of the inter-secting gauze wires and p is the pitch, i.e. the side of the square aperture $2r$. Hence a gauze of rather thick wires scatters less than one of thin wires. However, one pays for this advantage by having a smaller fraction of transmitted electrons.* Gauze lenses with relatively small field non-uniformities are produced if gauze screens are made to the shape of an equipotential of an electrode arrangement and inserted in place of this equipotential. For instance, it is practicable to replace the mid-plane of the sym-metrical two-tube lens by a plane wire gauze. In this way the two semi-lenses discussed in §4.2 could be separated by inserting a plane gauze in place of the mid-plane. By application of different potentials to this mid-plane gauze, the focal lengths of the two semi-lenses could then be controlled independently. Moreover, by omitting either of the tubes, each of these semi-lenses could be obtained separately. If the gauze is at the higher potential, the focal length of the tube-gauze lens is appreciably shorter than that of the two-tube lens (cf. table 4.1). This applies particularly to relatively low voltage ratios, as can be seen from the experimental values (Klemperer, 1953) in table 4.11. On the other hand, if the gauze is at the lower potential, a diverging electron lens is obtained. This is a remarkable result, since ordinary electron lenses are always converging lenses.

Einzel lenses with gauze screens have been investigated by Knoll and Weichardt (1938) and by Liebmann (1949a). In a three-aperture einzel lens the inner diaphragm may be replaced by a plane gauze. There, a converging lens is obtained if the potential of the gauze is positive with respect to the outer diaphragms ($V_i > V_o$).

* The finest wire gauzes obtainable have a pitch of 10 μ (made by EMI, Hayes, Middlesex).

TABLE 4.11. *Tube-gauze lens: focal lengths* $(P'F')$ *and mid-plane to principal plane distance* (MP') *expressed in terms of the tube radius* (R) *for various voltage ratios* V'/V

V'/V	1·5	2	3
$P'F'/R$	8·7	5·5	3·7
MP'/R	1·0	1·0	1·0

A diverging lens is obtained if the potential of the gauze is made negative with respect to the outer diaphragms $(V_i < V_o)$. The principal planes of such lenses are not crossed; thus the mid-focal distances are greater than the focal lengths. For small voltage ratios $(V_i - V_o)/V_o$, the focal length f of the gauze lens is much shorter than that of the ordinary einzel lens. For instance, at

$$(V_i - V_o)/V_o = 5,$$

f of the einzel lens (cf. fig. 4.6) with $R_i = R_o/2$ is reduced by a factor 7 when the central aperture is covered by a wire gauze.

Gauze lenses can be made which suffer less from spherical aberration (§6.4) or from distortion (§6.8) than the corresponding ordinary electron lenses. They are, however, not suitable for the production of sharp foci or for the projection of high-quality optical images. Foci and images will always be impaired by the inevitable non-uniformity of the field in the gauze.

However, gauze lenses have successfully been used to improve the performance of some mass spectrographs, electron multipliers, electron guns, velocity filters, particle accelerators, etc. when no sharp focus was required.

§4.9. Electrostatic quadrupole lenses

We have so far treated (in §2.5) an ideal quadrupole field (fig. 2.7) produced between two pairs of opposing electrodes of hyperbolic cross-section and of infinite length. The location of the cardinal lines of these fields was derived as a function of the excitation constant β in (2.39) and of the length L for the weak lens approximation neglecting fringe fields. We show now in fig. 4.15 the position of the cardinal points of the same model for values of βL

up to 2·0 (Septier, 1961). In order to account for the fringing fields
the actual length L_0 may be replaced by an equivalent length

$$L_1 = L_0 + 1 \cdot 1 a, \qquad (4.12)$$

where a is the semi-aperture of the lens (cf. fig. 2.7).

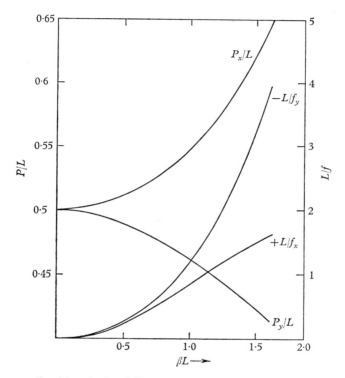

Fig. 4.15. Focal lengths f and distances P of the principal planes from the exit
face in two principal sections xz and yz. These are given in terms of the length L
of the quadrupole and plotted against βL, where β is the excitation constant—see
(2.27).

Practical electrostatic quadrupole lenses are usually not produced
by electrodes of hyperbolic cross-section. We show in fig. 4.16 some
electrodes of other systems with quadrupole symmetry, which are
manufactured more easily. The potential distribution in these
lenses is somewhat similar to that of the lens with hyperbolic

electrodes (fig. 2.7), in particular, near the z-axis. The quadrupole potential distribution is quite generally described by

$$\frac{V}{V_a} = K_1 \left(\frac{x^2 - y^2}{a^2}\right) + K_2 \left(\frac{x^6}{a^6} - \dots\right). \qquad (4.13)$$

For the lens with hyperbolic electrodes (§2.5, (2.20)) the constants K_1 and K_2 are obviously 1 and 0 respectively. The lenses represented by the cross-sectional drawings of fig. 4.16 (a)–(c) have

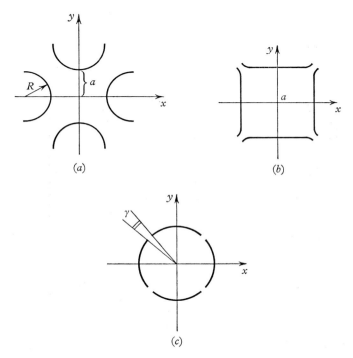

Fig. 4.16. Various electrostatic quadrupole lenses.

been investigated by Bernard (1954). The field between the convex circular electrodes (fig. 4.16(a)) is substantially the same as that between the hyperbolic electrodes. Also with plane electrodes (fig. 4.16(b)) the field is little changed and one finds $K_1 = 1\cdot037$, $K_2 = 0\cdot009$ but for the circular concave electrodes (fig. 4.16(c)) these constants are $K_1 = 1\cdot27\,(\sin\gamma)/\gamma$, $K_2 = 0\cdot042\,(\sin 3\gamma)/\gamma$ where

γ is the central angle measuring the width of the gap between the electrodes A.

Development of the lens with circular concave electrodes has led to the model shown in fig. 4.17(a) which can be easily constructed

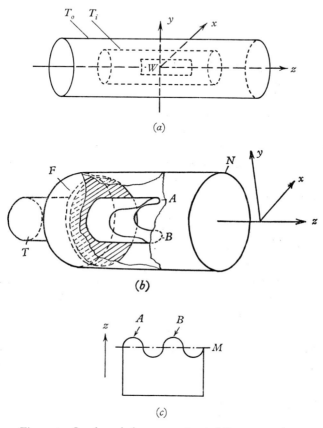

(a)

(b)

(c)

Fig. 4.17. Quadrupole lenses constructed from two tubes.

from two tubes and which provides good shielding against external fields. There, an inner tube T_i is held by insulating rings in a position co-axial to an outer tube T_o. Two windows W are cut opposite to one another in T_i. T_i is held at anode voltage and T_o at cathode voltage, the field from T_o penetrating into T_i through the windows W. Gregory and Sander (1962) have traced the focal length and aberrations of such quadrupole lenses of various geometrical dimensions.

If the quadrupole consists of four spheres of radius R at a distance a from the axis and at voltages $\pm V_a$, its two focal lengths can be calculated (cf. Archard, 1954) to be

$$\mp f_q = \frac{V_0}{V_a} \frac{a^2}{4R}, \qquad (4.14)$$

where V_0 is the accelerating voltage of the electron beam.

Another kind of easily constructed quadrupole lenses is made up from 'lipped' tubes; a great variety of these lenses has been investigated by Klemperer (1953). An example of a lipped tube lens is shown in the perspective drawing of fig. 4.17(b). There, a smaller tube T protrudes into a wider tube N, such as for instance the neck of a cathode ray tube. T is terminated by two 'lips' A and B. The lips are easily produced, e.g. by carving out the straight end of the tube with a half-round file. The shape of the lens electrode T is further illustrated by fig. 4.17(c) which shows the cylinder surface developed into a plane. The lens may be shielded by a flat circular flange F. If the voltage ratio, applied to T and N, accelerates an electron beam moving in the z-direction, this beam is brought to a line focus parallel to the y-axis, as drawn in fig. 4.17(b). Simultaneously the beam is diverged in the y-direction corresponding to a virtual focus parallel to the x-axis. On the other hand, if the beam moving in the z-direction is decelerated, a real line focus $\|x$ and a virtual line focus $\|y$ is produced. In table 4.12 the midfocal length MF of the lens of fig. 4.17(b), measured in terms of R, the radius of the tube T, is given as a function of the accelerating voltage ratio V_2/V_1. The midplane M is situated in the middle of the lips as shown in fig. 4.17(c).

TABLE 4.12. *Midfocal length of quadrupole lens of fig.* 4.17(b) *versus voltage ratio. Ratio of tube radii* $R_2/R_1 = 2$. *Radius of curvature of lips* $r_L = \frac{1}{4}\pi R_1$

V_2/V_1	1·2	1·3	1·4	1·5	1·7
$\dfrac{MF_y}{R_1}$	13	9	7	5·8	4·9

The focal line F_y which corresponds to the 'accelerating' lens is parallel to y. For a reversed voltage ratio, i.e. for V_1/V_2, the 'decelerating' lens projects a focal line F_x parallel to x, and for this particular lens $MF_x \simeq MF_y$ if $V_2/V_1 = V_1/V_2$ (this is not the case for other lipped lenses). At any given voltage ratio, the midfocal length of a lipped tube lens is substantially smaller than that of the plain two-tube lens (cf. §4.3) which can be considered as a lipped lens with $r_L \to \infty$. Quite generally, MF increases with decreasing r_L.

Single electrostatic quadrupoles have found little practical application. We mention here the 'scan magnification' which produces an increase of the deflexion of an electron beam. It is based on the fact that the deflexion of a narrow pencil traversing the quadrupole field is proportional to its distance from the axis. This property of the diverging field has been used for increasing the angle of deflexion in a cathode ray tube by El-Kareh (1961) and by Johnson (1962). The technique uses the diverging quadrupole between the deflector plates of an oscilloscope and the screen. An overall diverging effect in the converging plane can be obtained by making the field so strong that over-convergence occurs.

Detailed considerations for a wide-band oscilloscope with scan magnification have been published by Himmelbauer (1969). The oscilloscope which he constructed contained a triode gun with a quadrupole prefocusing doublet. This was followed by the vertical deflexion plates and a scan magnifier consisting of 5 diaphragms with non-circular apertures in alternating orientations. Finally the beam traversed the horizontal deflexion plates and a post accelera-tion mesh (see §10.2). The sensitivity (see §10.2) was 2–3 times that of a similar tube without the scan magnification.

A quadrupole doublet can be used to produce a point focus. For this purpose, the two quadrupoles should be identical, closely spaced and turned about the z-axis by $\frac{1}{2}\pi$ with respect to each other. The effective excitation constant (2.27) of such a doublet of quadrupoles as in figs. 2.7 or 4.16(a) has been given by Bauer (1966) as $B = \pi/L$, where L is the length of one quadrupole element. The focal length of this doublet in the approximation of (2.34) is calculated to be

$$f = L/\pi^2. \qquad (4.15)$$

The doublet behaves like a strong, thick lens of circular symmetry; however, the foci are located inside the field and at a substantial distance from the end of the electrodes. This precludes the use of the system for many purposes. The use of electrostatic quadrupoles in lenses consisting of three or still more quadrupole elements forming objectives of high performance will be discussed in chapters 6 and 7, as such combinations have become quite promising for constructing lenses free from aberrations.

Of much interest is the application of electrostatic quadrupoles for strong focusing in some travelling wave tubes and in some linear accelerators. There, a large number of identical quadrupoles is used to channel electrons or ions over relatively great distances. In this 'alternating gradient' focusing the beam encounters lenses which are alternately convergent and divergent (cf. §2.3). In linear accelerators, the particle traverses successively drift tubes and accelerating gaps, and between two successive gaps the time of flight must be equal to the period of a high frequency oscillation. This complicated problem which has been studied by many authors cannot be followed up in this book; we refer the reader to the paper by Teng (1954) where the use of quadrupoles in linear accelerators is discussed in detail.

CHAPTER 5

MAGNETIC ELECTRON LENSES

§5.1. Electron path in homogeneous magnetic fields

In a magnetic field the electron is accelerated in a perpendicular direction both to the field vector and to its own velocity. A static magnetic field does not change the kinetic energy of the electron but only changes the curvature of its path. Equating the magnetic force to the centrifugal force of the electron one obtains

$$euB_z = \frac{mu^2}{r_e}, \qquad (5.1)$$

$-e$ and m being charge and mass of the electron respectively, u its linear velocity, B_z the flux density of the magnetic field at right angles to the velocity u and r_e the radius of curvature of the path.

If the electron moves parallel to the direction of a homogeneous magnetic field, it is not influenced and travels along a straight line. If it moves in a direction perpendicular to the homogeneous field, its path is a circle. The time t_c required to traverse a complete circle, i.e. the period of the circular movement, is given by $u = 2\pi r_e/t_c$. Thus the angular frequency of rotation is

$$\frac{2\pi}{t_c} = 2\pi\nu_c = \omega_c = \frac{eB_z}{m}; \qquad (5.2)$$

ν_c is the 'cyclotron frequency', which only depends upon the magnetic flux density but is independent of the electron velocity* and of the radius of the electron path.

If the electron moves through the homogeneous magnetic field in a direction making an angle Θ with the lines of force, the velocity

---

* In the relativistic region, the cyclotron frequency decreases with increasing electron velocity due to the increase in electron mass m. Substituting in (5.2) for m from the relativistic equation

$$mc^2 = e(V_o + V),$$

we obtain $\qquad \omega_c = Bc^2/(V_o + V), \qquad (5.3)$

where V is the kinetic energy of the electron expressed in electronvolts and $V_o = 511 \times 10^3$ eV. is its rest energy.

u can be imagined to be split up into two components. The first component

$$u_z = u \cos \Theta$$

is parallel to the field and is not influenced by the field. The action of the field on the other component

$$u_y = u \sin \Theta,$$

which is perpendicular to the field, will not depend on the existence of the first component. Thus the projection of the electron path on a plane perpendicular to the lines of force must be a circle, and the path of the electron itself must be a spiral lying in the wall of a cylinder. It is useful to note that the electron will follow the spiral path moving forward in the sense of a right-hand corkscrew if u_z and B_z have the same direction.

The pitch of the spiral path may be obtained by considering the distance z_f which the electron moves in longitudinal direction while performing just one circle with the cyclotron frequency:

$$z_f = u_z t_c = (u \cos \Theta) \frac{2\pi}{B_z} \frac{m}{e}. \qquad (5.4)$$

Suppose a number of electrons of given velocity u cross any given line of force at a point J at a number of different inclinations Θ, which may be assumed to be so small that $\cos \Theta \approx 1$. All these electrons will again cross the same line of force at another common point J', because the component perpendicular to the line of force performs just one full revolution within the cyclotron period t_c given by (5.2).

Using in (5.4) $mu/e = (Br)_e$ for the electron momentum,* we obtain

$$JJ' = z_f \approx \frac{2\pi}{B_z} (Br)_e. \qquad (5.5)$$

Thus it appears that a homogeneous magnetic field is able to produce an electron optical picture of unit magnification within a distance z_n from the object, z_n being equal either to z_f as given by (5.5) or to a multiple of z_f. z_n is perfectly defined by the ratio of the electron velocity u and the flux density B_z of the magnetic field.

---

* It is convenient to bracket together the product of B and r with the subscript e. $(Br)_e = p/e$ (p = momentum) is called the 'electron momentum' or 'magnetic rigidity of the electron ray'.

The focal length of the homogeneous field is always infinite; thus optically it does not represent a simple lens but it belongs to the telescopic systems.

It may seem difficult to understand the motion of the electron in a magnetic field from an optical point of view. The difficulty is due to the fact that the refractive index of the beam is not a simple function of its coordinates only but, as follows from the above considerations, it also depends on the direction of the beam. Such a refractive index has already been specified in (1.19). Now Störmer (1933) and Dosse (1936) have shown that the problem of the refraction of electrons in all magnetic fields of circular symmetry can be attacked in a much simplified way by considering two-dimensional motion of the electron in two separate planes. The two planes are:

(1) The 'equatorial' plane at right angles to the axis of circular symmetry, moving with the longitudinal speed u_z of the electron.

(2) The 'meridional' plane rotating with the electron about the line of force that passes through the object point and the image point.

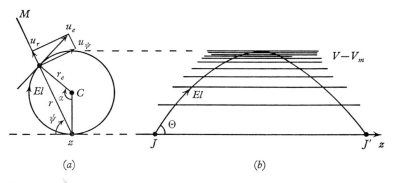

Fig. 5.1. Motion of the electron in a homogenous magnetic field: (a) path in the equatorial plane; (b) path in the rotating meridional plane.

This line may be considered as the axis of the system.

The electron path in the rotating meridional plane and in the equatorial plane is shown in fig. 5.1. In the equatorial plane, the electron El moves along a circle, starting from J. (5.1) and (5.2) describes this circular motion about the centre C in terms of the radius of curvature r_e and of the angular velocity $d\alpha/dt = \omega_c$, i.e. the cyclotron frequency.

We may also describe the electron motion in the homogeneous field with respect to the z-axis through J. The distance of the electron from this axis, i.e. the radius vector r, corresponds to the ordinate of the electron in the meridional plane M. The vector r always lies in this rotating meridional plane including an angle $\psi = \frac{1}{2}\alpha$ with the initial position of the plane. Using (5.2), the angular velocity of the meridional plane is given by

$$\frac{d\psi}{dt} = \frac{e}{m}\frac{B_z}{2}. \tag{5.6}$$

On the other hand, the angular momentum of the electron with respect to the axis is

$$G = mru_\psi, \tag{5.7}$$

where u_ψ is the tangential velocity component. The magnetic force is

$$F_\psi = eu_r B_z, \tag{5.8}$$

where u_r is the radial velocity component. Now the change in angular momentum equals the moment of forces, i.e.

$$\frac{dG}{dt} = rF_\psi. \tag{5.9}$$

(5.9) with (5.7) and (5.8) leads to

$$\int_{(1)}^{(2)} d(mru_\psi) = \int_{(1)}^{(2)} eB_z r\, dr,$$

or
$$(mru_\psi)_1 - (mru_\psi)_2 = \frac{e}{2\pi}[(B_z \pi r^2)_1 - (B_z \pi r^2)_2], \tag{5.10}$$

which can be written

$$G_1 - G_2 = \frac{e}{2\pi}(\phi_1 - \phi_2). \tag{5.11}$$

This is 'Busch's theorem' which says that the angular momentum of an electron proceeding from a position (1) to a position (2) changes proportionally to the magnetic flux which passes between the two co-axial circles drawn through the initial and the final positions respectively. We now use Stokes's equation

$$\int \text{curl } A \cdot da = \oint A \cdot ds, \tag{5.12}$$

A being the vector potential defined by $B = \operatorname{curl} A$. (5.12) shows that the integral over the scalar product of B and the surface element da (i.e. over the product of the normal component B_z with da) equals the line integral over the vector potential by the line elements ds which form the boundary enclosing the surface. Since in the homogeneous field B_z is constant over this surface and perpendicular to it everywhere,

$$B_z \pi r^2 = A_\psi 2\pi r, \qquad (5.13)$$

A_ψ being the only component of A that can exist in the homogeneous field. Substituting (5.13) into (5.10) and dropping the index of A gives

$$r(mu_\psi - eA) = C. \qquad (1.20)$$

This equation represents the conservation of the canonical angular momentum. It has already been introduced in §1.2 and the 'meridional refractive index'

$$N_m = \sqrt{V_m} = \sqrt{\left(V - \frac{e}{2m}A_\psi^2\right)} \qquad (1.24)$$

has been derived there.

To obtain numerical values of N_m in a field of given flux density B_z, the value for A given by (5.13) may be substituted in (1.24). In particular, when dealing with homogeneous fields, one has $B_z = \text{const.}$, and $N_m^2 = c_1 - c_2 r^2$, where c_1 and c_2 are constant. Consequently, in homogeneous fields the lines of constant meridional refractive index are straight lines parallel to the axis with the value of N_m decreasing with the distance from the axis. Such lines of constant N_m are drawn, for example, in fig. 5.1, and an electron passing through the refracting field in the meridional plane is seen to describe a path of the shape of a sine curve. In a gravitational model, the electron moving away from the axis would have to climb up a potential mountain, reaching its maximum 'height' at its greatest distance from the axis.

§5.2. Paraxial rays in inhomogeneous magnetic fields of circular symmetry. The thin lens

The treatment of the electron motion in the rotating meridional plane which has just been given for the case of a homogeneous field also applies to the paraxial region of an inhomogeneous magnetic field of circular symmetry. For relatively small distances r, from the axis, the axial component of the flux density B_z may be considered to be independent of r. In this case also, the speed of rotation of the electron, given by (5.6), is independent of r, and rays of different aperture travelling at any time in a given rotating meridional plane will remain in this plane. Again, the integration of (5.9) leads to (5.10) and (5.13) only as long as B_z is considered to be independent of r.

This may be illustrated by the power series

$$B_z(r, z) = B_z - \frac{r^2}{4}\frac{\mathrm{d}^2 B_z}{\mathrm{d}z^2} + \frac{r^4}{64}\frac{\mathrm{d}^4 B_z}{\mathrm{d}z^4} + \dots, \qquad (5.14)$$

which is derived in appendix A. The series (5.14) shows the above restrictive condition to be fulfilled for small r for which the higher terms can be neglected. Moreover, substituting $B_z(r, z)$ of (5.14) into Stokes's equation (5.13) gives the series

$$A_\psi = \frac{r}{2}B_z - \frac{r^3}{16}\frac{\mathrm{d}^2 B_z}{\mathrm{d}z^2} + \frac{r^5}{384}\frac{\mathrm{d}^4 B_z}{\mathrm{d}z^4}. \qquad (5.15)$$

(5.15) becomes identical with (5.13) if r is small enough to allow higher terms to be neglected. Again, in inhomogeneous fields of circular symmetry A_ψ is the only existing component of A so that the index ψ may be dropped.

The constant C in (1.20) now vanishes not only for all axial rays as in §5.1, but in addition it vanishes also for all electrons which start in a field-free space where $u_\psi = 0$ and $A = 0$. The field-free space is the object space from which a complete paraxial image can be projected by the magnetic lens into the image space. The meridional refractive index N_m is again given by (1.24). Since, however, B_z, now a function of z, reaches a maximum somewhere in the lens, the lines of constant meridional refractive index in the inhomogeneous field are curved.

The focusing action of an inhomogeneous field of circular symmetry is shown in fig. 5.2. There, two different electron paths are drawn starting at the same point J on the axis with different angles Θ. These paths are seen in the equatorial plane (fig. 5.2(a)) and in the rotating meridional plane (fig. 5.2(b)). Any electron starting on the axis at a point J travels along a substantially straight path until it gradually enters the magnetic field. This field extends round an equatorial plane G and decays rapidly in both directions along the z-axis. To give an impression of the shape of the lines of constant meridional refractive index N_m, two such lines are drawn in fig. 5.2(b). The electrons are deflected gradually in the field and then proceed to the substantially field-free space where they travel along straight lines; eventually they intersect the axis at J'.

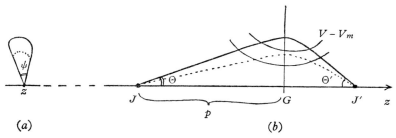

Fig. 5.2. Focusing action of a magnetic lens: (a) electron path in the equatorial plane; (b) electron path in the meridional plane.

For a given mid-object distance $JG = p$, the position of the image does not depend on the slope angle Θ as long as Θ is small. The meridional plane through which the electron travels is of course stationary while the electron moves through the field-free space. While the electron passes the magnetic lens, the meridional plane rotates through the angle ψ which is shown in fig. 5.2(a). This angle of rotation is given by the integral of (5.6)

$$\psi = \int_{z_1}^{z_2} \frac{e}{m} \frac{B_z}{2} \, \mathrm{d}t.$$

Replacing $\mathrm{d}t$ by $\mathrm{d}z/u_z$ and again introducing the energy $V = mu^2/2e$ one obtains

$$\psi = \sqrt{\frac{e}{8mV}} \int_{z_1}^{z_2} B_z \, \mathrm{d}z = \frac{1 \cdot 48 \times 10^5}{\sqrt{V}} \int_{z_1}^{z_2} B_z \, \mathrm{d}z$$

$$= \frac{1}{2(Br)_e} \int_{z_1}^{z_2} B_z \, \mathrm{d}z \ \text{[radians]}. \tag{5.16}$$

In the last part of (5.16), $(Br)_e$ is used again as a convenient measure for the momentum of the electron, and thus this last part is relativistically correct for highest electron speeds. ψ increases clockwise on looking in the direction of B_z, if the electron travels in the direction of B_z. According to (5.16) the rotation ψ between two points z_1, z_2 on the axis is proportional to the magnetic potential difference

$$V_H = \int_{z_1}^{z_2} \frac{B_z}{\mu}\, \mathrm{d}z$$

between these points. If object and image are located at z_1 and z_2 respectively, ψ represents the relative rotation of the image. The sense of rotation is reversed by a change of sign of B_z, but the convergence of the electron beam depending, as it does, upon B_z^2 is unaffected.

The study of the electron motion in the rotating meridional plane allows one to visualize all geometrical properties of the complicated spiral path of an electron through inhomogeneous magnetic fields. The lens properties of such fields which are responsible for image formation and which have already been mentioned in this section can easily be derived from the equations of motion:

$$m\frac{\mathrm{d}^2 r}{\mathrm{d}t^2} = -eB_z u_\psi + \frac{mu_\psi^2}{r}$$

$$= -eB_z r\frac{\mathrm{d}\psi}{\mathrm{d}t} + mr\cdot\left(\frac{\mathrm{d}\psi}{\mathrm{d}t}\right)^2. \tag{5.17}$$

It is understood that (5.17) refers to the radial acceleration (r = radius vector) in the equatorial plane. With respect to the z-axis, the magnetic force ($-eB_z u_\psi$) is no longer balanced by the centrifugal force mu_ψ^2/r. Now (5.6) which, as we pointed out, applies to paraxial rays in any field of circular symmetry, may be substituted into (5.17) giving

$$m\frac{\mathrm{d}^2 r}{\mathrm{d}t^2} = -\frac{e^2 r B_z^2}{4m}; \tag{5.18}$$

or, for constant u_z with $\mathrm{d}t = \mathrm{d}z/u_z$ and $mu_z^2 = 2eV$,

$$\frac{\mathrm{d}^2 r}{\mathrm{d}z^2} = -\frac{e}{8m}\frac{B_z^2 r}{V}. \tag{5.19}$$

This is the 'paraxial magnetic-ray equation' (corresponding to the electrostatic equation (3.23)). The lens properties of the inhomogeneous circular-symmetrical field are shown by (5.18), since the radial acceleration appears to be proportional to the value of the aperture r. Integration of (5.19) gives

$$\left(\frac{dr}{dz}\right)_1 - \left(\frac{dr}{dz}\right)_2 = \frac{e}{8mV} \int_{-\infty}^{+\infty} rB_z^2 \, dz. \qquad (5.20)$$

(5.20) immediately leads to the focal length of the lens, if this lens is so short that the beam aperture r is substantially a constant in the lens field, i.e. if $r = r_0 = $ const., where $B \neq 0$. With this restriction and taking dr/dz as the slope of the ray in the field-free object space (1) and image space (2) respectively, it follows for parallel incident rays that

$$\left(\frac{dr}{dz}\right)_1 = \tan \Theta' = 0,$$

and

$$\left(\frac{dr}{dz}\right)_2 = \tan \Theta = \frac{r}{f}.$$

Therefore

$$\frac{1}{f} = \frac{e}{8mV} \int_{-\infty}^{+\infty} B_z^2 \, dz = \frac{2 \cdot 20 \times 10^{10}}{V} \int_{-\infty}^{+\infty} B_z^2 \, dz$$

$$= \frac{1}{(2Br)_e^2} \int_{-\infty}^{+\infty} B_z^2 \, dz. \qquad (5.21)$$

For great electron velocities V_r of (1.9) must be substituted for V in (5.19)–(5.21) while the last part of (5.21) is relativistically correct anyway. (5.21) is the famous 'short-lens formula' derived by Busch (1926) in his historic first paper on electron optics, §1.1.

§5.3. Focal length and image rotation of thick magnetic lenses

Axially symmetrical magnetic fields can be produced by ordinary electric coils (air coils), by iron-shielded coils or by permanent magnets. The formulae of §5.1 find approximate application for the limiting case of very long coils only. On the other hand §5.2, and in particular Busch's equation (5.21), apply to paraxial rays in very short fields only. There, the geometrical extension of the lens field along the axis should be negligibly small in comparison with the focal length and with object and image distances. Most frequently met with in practice, however, is the intermediate case of the thick lens, for which no simple theoretical formulae can be derived.

For practical purposes, the focal length of the thick lens can be well estimated with the help of an empirical formula:

$$f = G\frac{V_r}{(nI)^2} = \frac{G}{7\cdot20}\frac{(Br)_e^2}{(\mu nI)^2}, \qquad (5.22)$$

where V_r is the relativistically corrected accelerating voltage of the electron ((1.9) of §1.2), nI is the number of ampere-turns of the coil,

$$\frac{nI}{\sqrt{V_r}} = k_f \qquad (5.23)$$

is known as the 'excitation coefficient' of the lens. G is an empirical factor called 'coil-form factor'. The second expression in (5.22) contains $(Br)_e$ as a measure of the electron momentum and $\mu = 1\cdot26\times 10^{-6}$, the permeability of space; it is relativistically correct for highest electron energies.

Various definitions have been given for a coil-form factor by Ruska (1934), Klemperer (1935), Deutsch, Elliot and Evans (1944), Cosslett (1946) and others. The coil-form factor G in (5.22) is a constant of the value G_0 for a given air coil, if practically the whole lens field is contained between object plane and image plane, i.e. if the field does not spread substantially beyond this interval. Empirical values of the form factor G_0 for all kinds of air coils of different shapes are found within the limits

$$97R < G_0 < 110R,$$

where R is the mean radius of the coil. The values of G_0 are found to be the smaller the more concentrated the field towards the mid-plane. This may be explained in the following way: The total magnetic potential difference generated by the coil equals the number of ampere-turns, namely,

$$V_H = nI.$$

A certain fraction of this potential difference extends along the lens axis between object position z_1 and image position z_2 and is given by

$$_{z_1}^{z_2}V_H = n\frac{I}{2}(1-\beta) = \frac{1}{\mu}\int_{z_1}^{z_2} B_z\,\mathrm{d}z, \qquad (5.24)$$

where β is usually a small correction term which accounts for the 'ultrafocal field' on the axis. For, given $^{z_2}_{z_1}V_H$, however, the integral $\int_{z_1}^{z_2} B_z^2\,dz$ in Busch's equation (5.21) will be the larger the narrower the B_z distribution. The narrowest physically possible B_z distribution of all air coils belongs to the single wire loop.* Its G_0 value is calculated from (5.21) and (5.22) to be

$$G_0 = 97R,$$

where R is its radius.

Of some interest is the problem of transition from the thin lens to the 'infinitely' long coil which frequently appears in the design of practical apparatus. Consider first the case where object and image are both outside the field of a coil of the length L. There, the distance between the two principal planes $2MP$, and the focal length f depend relatively little upon the coil radius R. Only for weak excitation, can the coil be treated as a thin lens; this is seen from the example given in table 5.1 where for a coil with

$$0.2 < L/R < 5$$

the measured values of f and MP are compared with the data calculated for a thin lens, namely f_{Busch} (cf. (5.21)) and $MP = 0$ respectively.

TABLE 5.1. *True focal length f and focal length f_{Busch} calculated from the axial field distribution (MP = distances between mid-plane and principal plane, L = length of coil)*

f/L	0·96	1·19	1·43	2·68	5·17
MP/L	0·08	0·065	0·055	0·04	0·03
f/f_{Busch}	0·78	0·86	0·87	0·93	0·97

Still more complicated is the case of the thick lens where object or image (or both) are deeply immersed in the field of the lens, i.e. where β the coefficient accounting for the ultrafocal field (5.24) is large. In this case f is increased substantially by the loss of focusing power owing to the increase of β, and (5.22) ceases to

---

* The large reduction of G effected by iron shielding the coil will be discussed in §5.4.

be appliable, as the power of $(Br)_e/(\mu nI)$ (see 5.32)' is gradually reduced from 2 to 1. When the length of the solenoid is increased to 1·5 times the object to image distance, the paraxial field within this distance is within 1 per cent homogeneous, and z_f can again be calculated from (5.5) (cf. Graham and Klemperer, 1952).

A useful approach for obtaining practical formulae for the focal lengths of thick lenses is based on the possibility of an empirical representation of the experimental field distribution (cf. §5.5) by an analytical formula. The flux-density distribution along the axis (z) of a circular current (single wire loop) can be calculated with Ampère's law to be

$$B_z = B_0 \Big/ \left[1 + \left(\frac{z}{R_0}\right)^2 \right]^{\eta}, \qquad (5.25)$$

where B_0 is the maximum flux density in the centre, R_0 is the radius of the loop and $\eta = \frac{3}{2}$ is a constant.

Glaser (1941 b) proposed to present the field of any magnetic lens by (5.25) taking empirical values for R_0 and for η, and for the half width $2w$ of the experimental field distribution.

For the particular case $\eta = 1$, $R = w$ (5.25) becomes

$$B_z = \frac{B_0}{1 + \left(\dfrac{z}{w}\right)^2}. \qquad (5.26)$$

For this simple distribution Glaser (1941 a, b) derived formulae for all lens properties (see also §6.5). For instance for the focal length he obtained:

$$PF = f = \frac{w}{\sin\left[\pi/\sqrt{(1 + K_H^2)}\right]}, \qquad (5.27)$$

where

$$K_H^2 = \frac{(wB_0)^2}{4(Br)_e^2} = \frac{e(wB_0)^2}{8m_0 V_r}. \qquad (5.28)$$

We shall refer to K_H as 'Glaser's magnetic lens strength parameter'. Comparison with (5.22) shows that K_H is roughly proportional to G/f (cf. Ments and Le Poole, 1947). For low field values, i.e. for $K_H \to 0$, Glaser's formula (5.26) agrees with Busch's thin-lens equation (5.21), both leading to the focal length

$$f \approx 2w/(\pi K_H^2), \qquad (5.29)$$

i.e. in the region of low field values, the power of the lens is proportional to K_H^2.

A frequently used alternative analytical representation of the flux
density in the magnetic lens, namely

$$B_z = B_0 \operatorname{sech}\left(\frac{z}{0\cdot759w}\right), \qquad (5.30)$$

is due to Lenz (1951, 1952) and to Grivet (1952, 1965). Either
(5.25) or (5.30) can be substituted into the paraxial ray equation
(5.18) and transformed into an equation which can be integrated
exactly. The result is complicated, in particular for strong excitation when the beam forms 'immersion foci' as it crosses the axis
several times inside the field. Numerical results have been tabulated
by Grivet (1952, 1965); these are in agreement with the experimental values which we shall report in the next section.

The image rotation by a thick lens is given by (5.16). In the case
of microscope lenses, this rotation can be calculated conveniently
from a formula given by Ments and Le Poole (1947), namely

$$\psi = 0\cdot19nI/\sqrt{\{V(1 + 10^{-6}V)\}}. \qquad (5.31)$$

The angles of rotation generally are somewhere between 0 and 180°
and the limiting cases are: (i) The short lens with $\psi \to 0$, i.e. with
negligible rotation, leading to an inverted image as in the case of
the glass lens. (ii) The solenoid with $\psi = 180°$ yielding an erect
image. The rotation by lenses of given focal length is the larger
the more the field extends along the axis. Thus, the image rotation
is always larger for an air coil than for an iron-shielded coil of the
same focal length.

Magnetic lenses without image rotation can be obtained by
using a pair of similar coils traversed by the same current in
opposite directions. The electrons pass in succession two fields of
opposite directions, and the rotation in the first field is eliminated
by an opposite rotation produced in the second field (Stabenow,
1935; Becker and Wallraff, 1940b). Such rotation-free lenses,
however, are not often met in practice: (i) since they need a
multiple of the number of ampere-turns of the ordinary lens,
(ii) since they are much more sensitive to misalignment, (iii) since
they have very much greater aberrations than the corresponding
single coils.

Of some interest is the possibility of producing magnetic lenses of negative focal length. Graham (1949) and Hubert (1951) have shown that a 'paraxial baffle coil', i.e. a very small coil arranged coaxially with an electron beam so as to intercept paraxial rays, will diverge the zonal and marginal rays of this beam. The diverged rays of the beam remain substantially homocentric; thus the paraxial baffle coil acts as a lens of negative focal length.

§5.4. Practical lens design

Very large air coils are used for focusing very fast electrons. The problem of saving electrical power and weight of copper wire is frequently of some practical importance.

Apparently, a given number of ampere-turns is employed with the best efficiency if the flux along the axis is all concentrated between object and image. Hence, the dimensionless factor

$$\frac{(Br)_e}{\mu nI} \qquad (5.32)$$

has been termed 'focusing efficiency'; under given geometrical conditions, it indicates the number of ampere-turns required to focus rays of a given momentum $(Br)_e$ (cf. (5.22)). Most efficient of all coils is the single wire loop, but its diameter should be chosen as small as possible in comparison with object to image distance.

The focusing efficiency as a function of coil length, coil radius and of object-to-image distance have been investigated by Graham and Klemperer (1952). Very short coils seem to be wasteful, since the required ampere-turns have to be wound in very many layers. Hence the mean radius of a very short coil becomes very large with subsequent large spreading of the lens field outside the object-to-image range. On the other hand, a very long coil will again suffer from the disadvantage of widely spread field regions, particularly when the coil extends beyond the object-to-image distance.

For a powerful air coil placed symmetrically between object and image a minimum weight of copper wire and least consumption of electrical energy are required if the coil length (L) is chosen of the order of half of the distance ($2s$) between object and image. The optimum for L/s is not critical, its exact value depends of course on the required aperture R_i of the coil and on the momentum of the electrons which have to be focused.

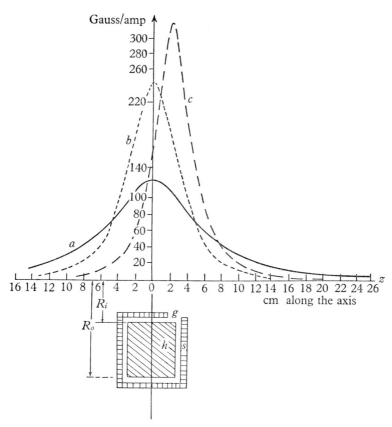

Fig. 5.3. Field distributions along the axis of three magnetic lenses: (a) air coil; (b) iron shielded with wide gap; (c) iron shielded with narrow gap.

Great economy in electrical energy is gained by shrouding the coil with material of high permeability. The scheme of an iron-shielded coil is shown in a meridional section in fig. 5.3. The coil h (produced by rotation of the rectangle h about the axis z) has the inner and outer radii R_i and R_o. The soft iron shield s completely surrounds the coil leaving only a relatively small annular gap g.

The total flux passing through the iron has to cross this gap. It produces an intense stray field by which the electrons are focused.

The axial component of flux density along the axis of the lens is plotted as curve c in fig. 5.3. The relevant figures of flux density have been obtained with a coil of 1550 turns and with an iron shield 5 mm thick leaving a gap of 2 cm. The coil dimensions, length $2L = 15$ cm, inner and outer radii $R_i = 5$ cm and $R_o = 12$ cm, are indicated in the figure. For $I = 1$ amp, B_z reaches a maximum of 320×10^{-4} Wb m^{-2} in the middle of the gap. For the coil of fig. 5.3, the gap width of 2 cm presents an optimum; maximum flux density in the middle of the gap decreases for wide gaps as well as for smaller gaps. If the gap is enlarged to 6 cm (i.e. if the whole *inner* iron cylinder is omitted) the maximum is decreased to 240×10^{-4} Wb m^{-2} A^{-1}, the corresponding axial flux density distribution being shown in curve (b). Curve (a) applies to the 'air coil', the iron shield being omitted completely. The centre of the coil is chosen in each case as origin of the co-ordinates. The width of the flux-density distribution of an air coil is always reduced by the application of an iron shield. For example, in the two cases of curves (a) and (b) the field decays to 10 per cent of its maximum value within distances of about two mean coil radii and within about one mean coil radius respectively.

Axial components of flux density along the radii of the above lenses are shown in fig. 5.4. The curves a, b, c again belong to the air coil, the iron coil with wide gap and to the iron coil with small gap, respectively, as given in fig. 5.3. Curves a and b are measured along a radius from the centre of the coil. For curve c the measurements are taken from the middle of the gap which is displaced from the centre of the coil. All curves show increasing values of flux density with increase in radius. Curve d shows the radial distribution of the iron coil with the large gap at about 5 cm from the centre of the coil. This curve shows the remarkable decrease of field for the peripheral regions.

The focal lengths of the magnetic lenses of fig. 5.3 as a function of the coil current and of the focused electron velocities can be calculated by means of the formula for thick lenses (5.22). The value of the coil-form factor G may be determined by the measurement of one focal length with a known electron velocity at a given

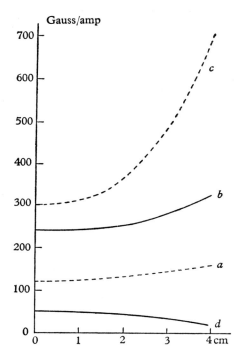

Fig. 5.4. Radial field distributions in magnetic lenses.

coil current provided the number of turns of the coil is known. In this way, for instance, with object and image each at a distance of six mean coil radii from the mid-plane the following values have been obtained: for the air coil of fig. 5.3, $G = 9 = 106R$, where R is the mean coil-radius; for the corresponding iron-shielded coil (2 cm gap) $G = 2\cdot4$ [m.k.s.]. An increase of the latter value due to saturation of the iron shield sets in very gradually and slowly. For instance, when the iron-shielded coil with 2 cm gap reaches a maximum flux density of $0\cdot1$ Wb m^{-2} on the axis, G is increased by 5 per cent in comparison with its value at low-coil currents. On the other hand, for the same shielded coil with 6 cm gap the increase of G is only of the order of 1 per cent at the above flux density.

As an example for practical lens design which has been used in some image converters and in television pick-up tubes, we describe here a relatively weak iron-shielded lens (cf. figs. 5.5 and 5.6, 5.13 and 5.14). The coil has 1 500 turns of fine wire, it has an inner and outer radius of 58 and 75 mm respectively and is shielded by 4 mm iron sheet. The shape of the shield is shown in fig. 5.5. This figure also contains a map of contour lines of the axial component of flux density. The striking similarity of these

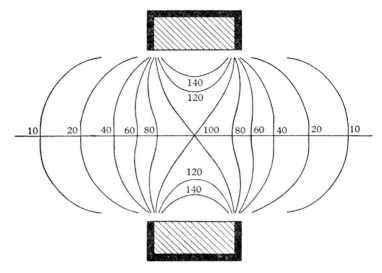

Fig. 5.5. Contour lines of axial components of flux density in iron shielded coil.

contour lines to the equipotential lines in a symmetrical electro-static saddle-field lens is seen on comparison with fig. 4.8. The figures given against the contour lines indicate the axial component of flux density with respect to that at the centre of the lens as 100 per cent. The usual graphical representation of the same flux distribution is given in fig. 5.6. There, the axial components of flux density are plotted against the axial z-coordinate along the axis and along lines parallel to the axis and in various distances from it. The flux density in the centre is again taken as $B_0 = 100$ per cent. The batch of curves of fig. 5.6 covers the whole lens field by a network of flux-density values; it is a very convenient representation of the lens for the purpose of ray tracing, as will be seen in the next section.

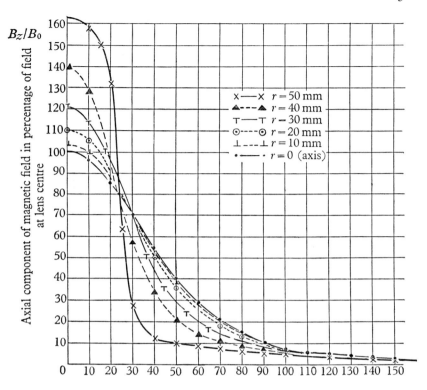

Fig. 5.6. Field distributions in the magnetic lens of fig. 5.5.

Of quite a particular design are the magnetic electron lenses that are used in electron microscopes. The strongest possible fields have to be concentrated within the shortest space along the axis in order to obtain the smallest possible focal lengths. Microscope objectives are produced by powerful coils (order of 5×10^3 ampere-turns) which are shielded by iron excepting for a small gap between pole-pieces. Microscope projector lenses are of similar design. The requirements for the two types, however, are rather different: the ray passes right through the projector lens, whereas in the objective, the rays effectively originate from an object which is near the focal point.

The main problem in the construction of microscope lenses occurs in the design of the magnetic shield; this has been discussed in detail by Liebmann and Grad (1951), Liebmann (1955), Durandeau and Fert (1957) and by Mulvey and Wallington (1969). An iron shield, such as in fig. 5.3(c), when scaled down to small dimensions in order to produce a very short focal length would produce a narrow bell-shaped field distribution at weak excitation only. When the excitation is increased, say, for mild steel above $nI \approx 1000S$ (S = gap width in mm) saturation becomes noticeable, and first, the bell-shaped distribution widens rather than increases in height. With further increase of nI, a parasitic field

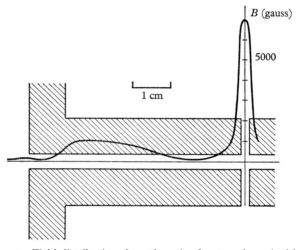

Fig. 5.7. Field distribution along the axis of a strongly excited lens.

appears in the channel. This is demonstrated by the field distribution curve shown in fig. 5.7 representing a 1 cm thick iron shield with 4 mm bore and 2 mm gap, the lens being excited by $nI = 8000$ ampere turns.

By a proper design of the shield as shown in fig. 5.8 parasitic fields can be largely avoided while the field in the gap is maximized. There, the cross-section of the shield increases with increasing distance from the gap, the semi-angle of the conical increase being conveniently 10 to 15°. The exact shape of the pole pieces immediately facing the gap appears to be unimportant; the performance of the lens depends substantially on the width of the gap S and of the bore D only.

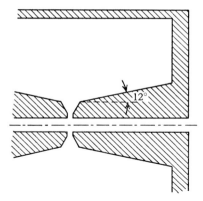

Fig. 5.8. Magnetic shield of an efficient microscope lens.

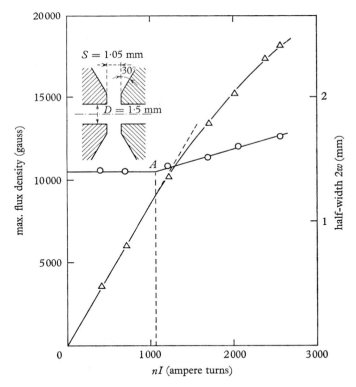

Fig. 5.9. Maximum flux density B_{max} ($\triangle$) and half width $2w$ ($\bigcirc$) of the field distribution along the axis of a microscope lens.

The increase of the maximum field and of the half width of the
field distribution curve of a typical well designed microscope lens
are shown in fig. 5.9. At the point A, saturation effects first become
noticeable.

Microscope lenses are always sufficiently strong to have the rays
cross the axis substantially inside the lens field. Hence it is
important to have the field concentrated within the shortest
possible distance. This is explained by fig. 5.10 which is an exact

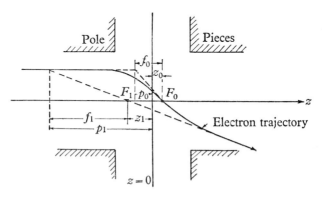

Fig. 5.10. Electron path, cross-over distance f_0 and
optical focal length f_1 of a microscope lens.

ray tracing diagram due to Liebmann and Grad (1951). The
cross-over point F_0 is found at the distance z_0 from the mid-plane
($z = 0$), the principal plane is seen at the distance $-p_0$; we can
define the cross-over distance as $f_0 = p_0 + z_0$. (This cross-over
distance f_0 controls the magnification obtained when using the lens
as a microscope objective, as will be explained later.) After passing
F_0, the trajectories are again converged by the ultrafocal field.
Hence, the optical focal point F_1 which is produced as the intersection of the emerging straight part of the trajectory with the
z-axis is found at the distance $-z_1$ from the mid-plane. The optical
principal plane, i.e. the plane of the intersection of the produced
incident and emerging rays is found at a distance $-p_1$ from the
mid-plane. Hence $f_1 = p_1 - z_1$ is the optical focal length, which,
according to §2.2, controls the magnification when using the lens
as a projector. On the other hand, when using the lens as an

objective, the object is placed at the cross-over point F_0. In fig. 5.10 the rays originating from F_0 would travel through the objective from right to left and emerge nearly parallel to the axis. Thus the image distance $s' \gg f_0$ and the magnification s'/f_0 can be made very large.

With increasing coil current, the distance z_0 gradually decreases while the ultra focal field increases in intensity and extension. Hence, the optical focal length f_1 first decreases, then passes through a minimum at a given optimum number of ampere turns. This is shown in fig. 5.11 for the example of a well-designed microscope lens where the optical data are plotted against the excitation coefficient (5.22). The minimum of the optical focal length (f_1) occurs at the point P.

Fig. 5.11 applies to a bore-to-gap ratio $D/S = 1$. Experiments with increased D/S ratio have shown that in the region

$$0.8 < D/S < 5,$$

the minimum $(f_1/D+S)$ value remains almost constant. In particular, Mulvey and Wallington (1969) have shown that in this region of D/S the minimum of f_1 is given by

$$(f_1)_{min} = 0.55(S^2 + 0.56D^2)^{\frac{1}{2}} \qquad (5.33)$$

and that the excitation coefficient $nI_A/\sqrt{V_r}$ for producing $(f_1)_{min}$ increases from 9.25 to 10.00 to 10.85 when D/S increases from 0.2 to 0.5 to 2, thus, for practical purposes, optimum nI_A is given by $10\sqrt{V_r}$. The minimum cross-over distance $(f_0)_{min}$ which controls the magnification of the objective, occurs at $1.15nI_A$, namely at slightly larger excitation than that of the projector focal length.

As an example for practical microscope lens construction we show in fig. 5.12 a type of lens which has been used by Durandeau and Fert (1957) for electrons of 50–75 kV. $z'...z$ is the axis of the lens, P are the pole pieces, C is the magnetic yoke and B are the windings for an excitation of up to 4000 ampere turns. Moreover, g is a socket for a joint, e.g. a rubber ring to prevent vacuum leaks and c is a channel to give access to the gap. The same design can be employed for the construction of an objective, a projector or a condenser lens with the exception of the pole pieces which must be different for the different applications. For an objective the authors

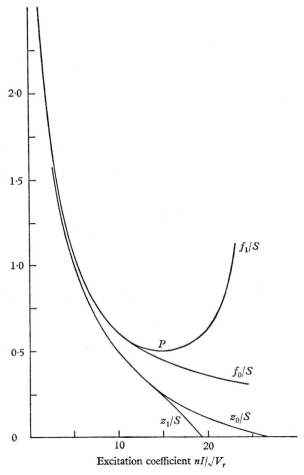

Fig. 5.11. Optical focal length f_1, cross-over distance f_0 and mid-plane to F-distances z_0 and z_1 for magnetic shield dimensions $D/S = 1$.

recommend the dimensions $S = 3$ and $D = 4$ yielding a focal length $f_0 = 2 \cdot 1$ mm while the beam passes through the lens in the direction from z' down to z. For a projector lens, on the other hand, pole pieces with $S = D = 1 \cdot 7$ mm are recommended yielding a minimum $f_1 = 1 \cdot 2$ mm at $nI = 3900$ a.t. and $V = 75$ kV, but here the beam must pass through the lens in the direction from z to z'.

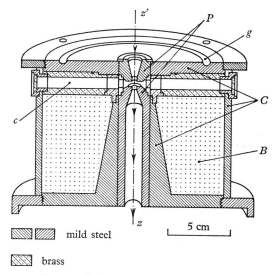

mild steel		
brass		

Fig. 5.12. Magnetic microscope objective.

§5.5. Permanent magnets. Superconductors

Magnetic steels of great permanence (e.g. $1-1\cdot5$ Wb m^{-2}) are now available for the construction of magnetic lenses (cf. Wohlfarth, 1959; Parker and Studders, 1962). The simplest permanent lenses are set up by tubular, cylindrical pieces of magnetic material through which the electrons travel.

The field set up by a tubular permanent magnet is qualitatively different from the field of a solenoid. The permanent magnet of fig. 5.13, with z being its axis of circular symmetry, has the circular poles N and S to which all the lines of force converge. On the other hand, the solenoid coil C of fig. 5.14 sets up closed lines of force which encircle the coil. In fig. 5.15 are shown distributions of flux densities along the axes of the two lenses, both being reduced to the same maximum B_0. While the flux density along the axis of the coil gradually decays from the maximum to infinity, the flux density along the axis of the permanent magnet decays more rapidly and becomes zero at the two points P_1 and P_2, where it reverses its direction. Electron optically, this implies a reversal of the sense of rotation of the rotating meridional plane at P_1 and P_2,

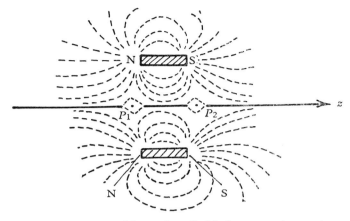

Fig. 5.13. Lines of force of a cylindrical permanent magnet.

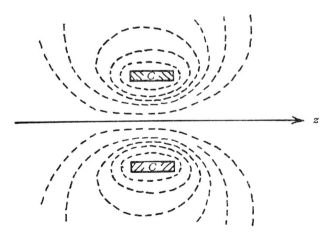

Fig. 5.14. Lines of force of a current coil.

resulting in a relatively small image rotation produced by the permanent lens. The points P_1 and P_2 are clearly seen in the lines-of-force picture of fig. 5.13.

Fields set up by the modern permanent magnetic lenses are surprisingly strong. For instance, a cylinder 40 mm long of 40 mm diameter and 8 mm wall thickness may easily retain a magneto-motive force of 10^3 ampere-turns between its end-faces,

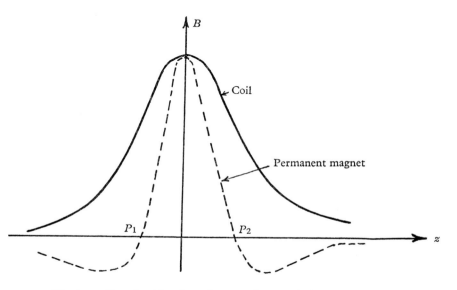

Fig. 5.15. Flux densities along the axes of magnetic lenses produced by a current coil and by a permanent magnet.

producing a maximum flux density of 4×10^{-2} Wb m^{-2} in the centre of the cylinder.

If permanent magnetic lenses are used for the sharp focusing of electron images, a control of focal length is desirable. Such control has been obtained with the help of magnetic shunts. In fig. 5.16, for instance, the flat circular discs D_1 and D_2 made from mild steel, are attached to the permanent magnetic cylinders A. If the shunt is used outside the magnet as in fig. 5.16, these discs D_1 and D_2 extend outside the cylinder A. In this case, the shunt is a cylinder C made from mild steel with a screw thread on its inside so that it may be screwed over the disc D_1 in order to close up the magnetic gap G towards D_2.

Some permanent magnetic lenses for television tubes have been described by Hadfield (1950) and by Jonker (1953). The application of permanent magnetic lenses to electron microscopes has led to relatively cheap, small instruments. Borries *et al.* (1940) have described an experimental instrument for 50 keV in which objective coil and projector coil were replaced by about a dozen bar magnets each. 7·5 and 2·7 mm focal lengths respectively were

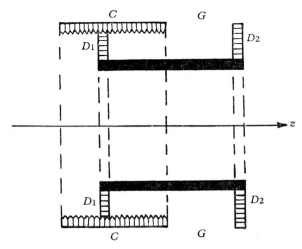

Fig. 5.16. Permanent magnetic lens of variable focal length.

obtained in these lenses. Reisner and Dornfield (1950) have described a small commercial RCA microscope, and Reisner (1951) has given details about its lens system, the cross-section of which is shown in fig. 5.17. Four rectangular parallelepipeds M (two are visible in the figure), made from 'Alnico V' and ground to shape, are placed symmetrically about the axis and attached to mild steel pole-pieces of circular symmetry. There are two magnetic gaps filled with brass spacers. The fields of the two electron lenses formed by these gaps are in opposition, hence the image rotation is small. An outside shell of mild steel which reduces stray fields to a minimum is sensibly a surface of constant magnetic potential. Potential differences of 1000 and 700 ampere-turns can be measured across the objective lens (gap 1) and across the projector lens (gap 2) respectively. The stability of Alnico magnets is great, and lens assemblies have been in use for years without apparent loss in field strength.

Langner (1955) constructed a permanent magnetic microscope

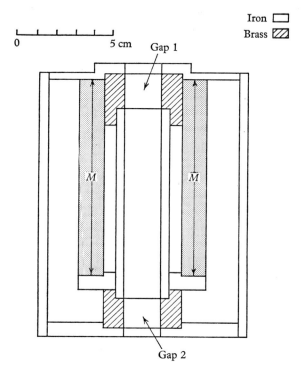

Fig. 5.17. Permanent magnetic microscope objective and projector lens.

for 60 kV. The fields in both the condenser and the objective could be finely adjusted by a movable shunt, the objective having a magnetic potential difference (see §5.6) of 3250 ampere-turns and a minimum focal length of 1·6 mm. The projector lens had a maximum induction of over 20 kgauss, the magnification being regulated by sliding the fixed pair of pole pieces over the gap. Though permanent steel magnets appear to be quite suitable for the construction of microscope lenses, they have found little application in the lenses of commercial instruments.

The situation is similar in respect of the application of permanent lenses to cathode-ray tubes. There, the use of ferrites of high coercitivity (cf. Standley, 1962) is of some interest. These materials are ceramics of small conductance so that if alternating fields are superimposed over the lenses, e.g. scanning magnetic fields which frequently overlap the focusing field in the television tube, eddy current formation is avoided. Verhof (1954) has described the use of ferrite magnets for focusing of television picture tubes. His lens consists of two flat rings made of 'Ferroxdure', co-axially mounted to the tube and axially magnetized in opposite directions. Stray fields are considerably less than those of permanent magnetic lenses of the conventional type. The strength of the lens is varied by the mutual separation of the two rings. Centring of the picture is achieved by moving a soft iron centring ring.

Unique advantage with regard to both instrumentation and specimen preservation can be derived from the use of superconducting magnetic lenses operating at liquid helium temperature. There, superconducting coils, or discs, or cylinders have been used; these have been made from the Niobium alloys: NbZr ($B_c \approx 6$, $T_c \approx 10$), NbTi ($B_c \approx 12$, $T_c \approx 9$), Nb$_3$Sn ($B_c \approx 30$, $T_c \approx 18$) where B_c and T_c are the critical fields and temperatures measured in Wb m^{-2} and °K respectively (cf. Kittel, 1966). With such magnetic lenses, extremely stable and noise-free fields can be produced which are homogeneous to one part in 10^6 to 10^7. Persistent currents up to 10^3 amp can be induced by external fields in these superconducting lenses and current densities of 90 amp mm^{-2} can be tolerated which compare with 1 amp mm^{-2} in conventional magnetic microscope lenses. Fernandez-Moran (1965, 1966) has described the use of these lenses for electron microscopy; he used superconducting solenoids in conjunction with stigmator coils (§6.7) and iron pole pieces. For fine focusing purposes the current could be adjusted to a precision of 10^{-9} by the use of vernier control circuits.

Boersch, Bostanjoglo and Lischke (1966) have used as a microscope objective a hollow superconducting cylinder arranged inside the gap of the iron shield of a strong magnetic coil. This cylinder was mounted on a heat conducting copper tube which was cooled in a liquid helium cryostat. The current in the superconducting cylinder was induced when the current of the lens coil was switched off. The trapped magnetic field was of greater constancy and of smaller dimensions than that of conventional magnetic lenses.

Kitamura, Schulhof and Siegel (1966) have measured in detail the performance of a variety of superconductive lenses for 100 keV and 1 MeV microscopes. The lenses were made from niobium–zirconium coils of 5 mm inner diameter or from thin discs of platinum substrate on which Nb_3Sn vapour was deposited. These were kept at liquid helium temperature and a current of $\sim$ 80 amp was induced, producing a maximum field of $2 \cdot 8$ Wb m^{-2} with $1 \cdot 3$ mm focal length for 100 kV electrons. The field was shaped with ferromagnetic pole pieces or with diamagnetic shields of other superconductors, to minimize aberrations.

§5.6. Determination of magnetic field distributions

An exact calculation of magnetic lens fields by integration over the field components of the current elements is possible for air coils only. For every lens field, however, there can be used computational methods. For example, Hesse (1950) has shown how field distributions in magnetic lenses can be obtained by a relaxation method.

Most important are direct determinations of the flux-density distribution, and much effort has been expended in developing suitable methods. The majority of all direct measurements in magnetic lenses have been made with search coils. For the investigation of iron-free lenses, the e.m.f. at the search coil may be measured when the current of the lens coil is reversed. If the lenses contain iron, care must be taken to prevent errors from hysteresis effects; the search coil has to be brought into position, then snatched out of the field.

In the measurement of field distributions in microscope lenses, the main practical difficulty lies in the small physical dimensions of the field to be investigated. Thus, Dosse (1941) used a search coil with 100 turns of 0·02 mm wire whose outer diameter and length were only 0·4 mm. The position of the coil along the lens axis is measured with a microscope and a micrometer joined rigidly to the coil.

In the measurement of fields of larger lenses much time can be saved by rotating or oscillating the search coils. Cole (1938) used rotating coils. Langer and Scott (1950) refined this method to measure fields accurately down to $\frac{1}{100}$ gauss.

Klemperer and Miller (1939) have used the search coil oscillator, a small oscillating coil performing rapid reciprocating rotations through 180° about an axis at right angles to the axis of the lens, the induced current being picked up by a resonance amplifier. On the other hand, Gautier (1954) investigated the field and its derivatives along the lens axis only. He used various search coils performing small amplitude oscillations coaxially along the lens axis. The principle of this method is as follows: First, the current induced in a short search coil is proportional to B'. Secondly, the current from a very long search coil $\propto \int_{-\infty}^{z_1} B'\,dz$ is proportional to B at the point z_1 about which the end of the coil is oscillating. Thirdly, two identical short coils spaced in a distance Δz and with their output in opposition measure $B'(z+\Delta z) - B'(z) = B''\Delta z$, i.e. the second derivative.

The flux density measured by a search coil is strictly correct only when the field can be considered to be homogeneous over the area of the search coil. Correction formulae for inhomogeneous fields have been derived by Goddard and Klemperer (1944). Instead of applying such corrections, it is, however, more convenient to use search coils of smallest possible area and to increase the level of output current by sufficient amplification. Another method for overcoming the effects of local field inhomogeneity is due to Brown and Sweer (1945) and to Williamson (1947), who used the 'flux ball', which is a spherical search coil wound in co-axial layers of cylindrical windings, each layer having different length and diameter. Sweer showed by calculation that for this shape, the total flux is proportional to the field in the centre of the coil.

Herzog and Tischler (1953) and Pearson (1962), however, showed that also simple cylindrical search coils of certain length to diameter ratio (e.g. $L/D \approx 0.72$) measure the actual value of the field in the centre of the coil. Possible errors occur in the determination of the lens field which depend on the even order derivatives of the flux distribution.

In an alternative method, developed by Sandor (1941), the effects of field inhomogeneity are overcome, and still larger output currents are obtained by the use of a series of large co-axial search coils of very gradually varying diameter. The integral flux in axial direction is measured, and the flux density as a function of the radial distance from the axis is obtained from the difference in search-coil currents and search-coil areas respectively.

Particularly suitable for measuring the field in small microscope lenses is the magnetic field balance described by Ments and Le Poole (1947). It consists essentially of a very long coil evenly wound on a non-magnetic bar which is suspended horizontally so that its axis coincides with the lens axis. The coil winding has somewhere a point of reversal, so that it acts like two solenoids in opposition. The external poles of these two solenoids are far outside the lens field so that no force is exerted on them. The two internal poles are of equal sign, and they coincide with the point of reversal. Hence an axial force is exerted on the coil by the field of the lens, this force being proportional to the axial flux density of the lens at the point of reversal. The force is indicated by a small horizontal displacement of the bar which is read by a microscope.

We mention now some methods which are not based on the use of search coils. The method of 'peaking strips' can be employed for the measurement of field distributions in large lenses. There, a thin wire ($\sim \frac{1}{10}$ mm diameter ~ 5 cm long) of very high permeability is surrounded by a small coil in which a small a.c. current (say 60 Hz) is flowing permanently. The alteration of voltage output from this coil due to the lens field can measure flux densities of a few hundred gauss within a small fraction of 1 per cent.

The 'Hall probe' is based on the magnetic deflexion of conduction electrons of high mobility in a semiconductor, such as indium antimonide. The deflected electrons of a current, passing along a thin, narrow strip of the semiconductor, the so-called Hall probe, build up space charges at its edges. The potential difference across these edges, measured directly or registered by a d.c.

amplifier, is proportional to the magnetic flux at right angles to the strip and to the current through it, and it is inversely proportional to the thickness of the strip. Hall magnetometers employing probes of semiconductor or of metal film mounted on an insulator of say $1 \times 5 \times 25$ mm for measuring fields between 10^{-2} and 10^5 gauss are commercially available.

Of general interest is an experimental method for mapping distributions of the magnetic vector potential $A_\psi(r, z)$ produced by circular symmetrical current-carrying conductors and iron of arbitrary shape. According to Liebmann (1950b) these distributions can be determined with the help of a resistor network as described in §3.2. Any stud o of the network is joined to its four neighbour studs 1, 2, 3 and 4 by resistances R_1, R_2, R_3 and R_4, the values of which are functions of the three radial coordinates $r_0 (= r_1 = r_3)$, r_2 and r_4 of the respective studs. In distinction to the network described previously in §3.3, all studs are now connected to a common potential by a fifth resistance R_5, the magnitude of R_5 being proportional to the radial coordinate r_0. A current proportional to the current density flowing through the conductor (magnet coil) is fed into the studs at all positions occupied by this conductor. If iron is present, the network must be open-circuited along the boundaries of the iron in order to fulfil the condition of infinite permeability requiring that the normal derivative $\partial A_\psi/\partial n$ should vanish. Moreover, the values of the resistances R_5 have to be adjusted along this boundary. The network represents the operator

$$\left[\frac{\partial^2}{\partial z^2} + \frac{\partial^2}{\partial r^2} + \frac{1}{r} \frac{\partial}{\partial r} - \frac{1}{r^2} \right],$$

which, applied to the vector potential A_ψ, is proportional to the generating current density. It can be seen that the application of the resistor network method for magnetic field plotting is complicated and needs elaborate apparatus, but apparently, good accuracy has been obtained in its results. For details we have to refer to the original literature.

Relatively simple is the application of the electrolytic trough for magnetic field plotting in iron-shielded coils. Due to the large value of the permeability of the iron shell compared with that of free space, the pole-faces may be regarded as being at uniform magnetic potentials V_H and V'_H, and they may be represented by metal electrodes in the electrolytic trough (§3.3). Such electrodes could even be used to give approximate representation of the whole magnetic shell, if the two halves of the shell are assumed to be separated by a double-layer plane producing the difference of potential nI. The equipotentials measured in the trough correspond with surfaces of equal magnetic scalar potential.

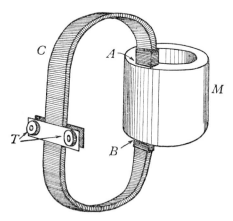

Fig. 5.18. Magnetic potentiometer and permanent magnetic lens.

If for a shielded coil the magnetic resistance of the iron part of the magnetic circuit is negligibly small in comparison with the resistance across the gap, the magnetic potential difference across the gap is near enough given by $V_H - V'_H = nI$, i.e. by the number of ampere-turns of the energizing coil. Generally, how-ever, the magnetic potential difference between the pole-pieces has to be obtained by measurement with a small magnetic potentio-meter, for instance. Such a potentiometer is shown in fig. 5.18. It consists of a long, narrow coil C made by winding wire in two layers in opposite directions upon a flexible non-magnetic strip of material. The terminals T of the coil have to be connected with a ballistic galvanometer. The ends A and B of the potentiometer are first set to the points between which the magnetic potential has to be measured, e.g. to the end-faces of the tubular permanent magnetic lens M shown in the figure, then withdrawn and set tightly against each other. The ballistic deflexion produced by this action is proportional to the magnetic potential difference; it may be calibrated in units of ampere-turns with an air coil in the usual manner.

§5.7. Ray tracing through magnetic lenses

According to the explanations of §§1.2, 5.1 and 5.2, axial rays may be traced through the refracting field in the rotating meridional plane, if the refractive index $N_m(r, z)$ is known as a function of the coordinates r and z. The vector potential $A(r, z)$ in (1.24) cannot be measured directly; but it has been pointed out in §5.2 how far A is related to the flux in the special case of circular symmetry. Apply Stokes's law (5.12) over a disc of radius r_0, set normal to the axial component of flux density B_z. The boundary of the disc is of the length $2\pi r_0$ and it encloses an area πr_0^2; hence

$$\oint A\,\mathrm{d}s = 2\pi r_0\,A = \int_0^{r_0} 2\pi r B_z\,\mathrm{d}r = \Phi, \qquad (5.34)$$

where Φ is the total flux through the considered disc. Using (1.24), the meridional refractive index in any field of circular symmetry is given by

$$N_m^2 = V_m(r, z) = V - \frac{e}{2m}\left(\frac{\Phi}{2\pi r}\right)^2 = V - 2\cdot25 \times 10^9\,\frac{\Phi^2}{r^2}, \quad (5.35)$$

where V is the initial electron energy in electron-volts which stays constant throughout the lens. Φ is the total flux through the circular area of radius r, which may be obtained either by direct measurements with a set of axial coils of various diameters, or according to (5.34) by integrating the flux-density distribution curves as given for instance in fig. 5.6. However, ray tracing through a meridional field plot cannot yield results of great accuracy for the reason that the rays cut the equipotentials at small angles, especially near the maximum elongation from the axis, where the ray meets a refracting equipotential at a grazing angle (cf. figs. 5.1 and 5.2).

Siday (1942) has used for tracing the integral (5.20) of the paraxial ray equation. If not only the integrals $\int_{z_1}^{z_2} B_z^2\,\mathrm{d}z$, taken between two points z_1 and z_2 on the axis, but also the initial values of r and $\mathrm{d}r/\mathrm{d}z$ are known, the rest of the ray can be traced. Other methods based on the paraxial equation of motion, but distinguished by the way in which this equation is integrated, have been described by Goddard and Klemperer (1944), Liebmann (1949b) and Durandeau et al. (1959).

Relatively simple is a method which uses the paraxial ray equation (5.19) in the form

$$\frac{\Delta^2 r}{\Delta z^2} = -2\cdot20 \times 10^{10} \left(\frac{r}{V}\right) B_z^2(r, z) \tag{5.36}$$

with substitution of the power series (5.14) for $B_z(r, z)$ disregarding higher terms. This yields

$$\frac{\Delta^2 r}{\Delta z^2} \approx -2\cdot20 \times 10^{10} \frac{r}{V} \left(B^2 - \frac{r^2}{2} B'' B + \frac{r^4}{32} B^{IV} B\right). \tag{5.37}$$

Now, since B as a function of z is known by measurement, B'' and B^{IV} as functions of z can be derived. Moreover, the initial ray aperture r_0 and the initial slope $\Delta r_0/\Delta z$ of the ray will be given. Hence, as we trace the ray, the first step supplies a new aperture r_1 and a new $\Delta r_1/\Delta z$. There, the new aperture is given by a McLaurin series* for r_0. Taking the length Δz of the step as a unit, this series can be written

$$r_1 = r_0 + \frac{\Delta r_0}{\Delta z} + \frac{1}{2}\frac{\Delta^2 r_0}{\Delta z^2} + \dots. \tag{5.38}$$

On the other hand, the new slope is given by

$$\frac{\Delta r_1}{\Delta z} = \frac{\Delta r_0}{\Delta z} + \frac{\Delta^2 r_0}{\Delta z^2}, \tag{5.39}$$

a relation which is simply derived from the definition of the second differential coefficient.†

After tracing the first step, r_1 and $\Delta r_1/\Delta z$ have become known quantities and in a second step, the new ray aperture

$$r_2 = r_1 + \frac{\Delta r_1}{\Delta z} + \frac{1}{2}\frac{\Delta^2 r_1}{\Delta z^2} \tag{5.40}$$

and the new slope

$$\frac{\Delta r_2}{\Delta z} = \frac{\Delta r_1}{\Delta z} + \frac{\Delta^2 r_1}{\Delta z^2} \tag{5.41}$$

---

* $f(x) = f(0) + xf'(0) + (x^2/2!)f''(0)$ where $x = \Delta z = 1$.

† $\dfrac{\Delta^2 r_0}{\Delta z^2} = \dfrac{(\Delta r/\Delta z)_1 - (\Delta r/\Delta z)_0}{\Delta z}$ with $\Delta z = 1$.

are easily obtained. The second differential coefficient $\Delta^2 r_1/\Delta z^2$ in (5.40) and (5.41) is calculated from (5.37) with $r = r_1$ where $B = B_z(0, z_1)$ is taken on the axis at $z_1 = z_0 + \Delta z$. By iterative application of the above procedure the ray can be traced step by step through the whole lens. Unavoidable inaccuracies of this method arise from the repeated differentiation of the field distribution curve, though the fourth derivative contributes but little to the result.

A method by Durandeau and Fert (1957) is very convenient for a quick estimate of the properties of the lens from the flux density distribution along the axis. There, the actual distribution is replaced by a rectangular distribution of constant flux density of a solenoid given by the well-known formula

$$B_0 = \frac{\mu n I}{L}, \qquad (5.42)$$

where L is the 'equivalent length' of the field. The latter is defined by

$$LB_0 = \int_{-\infty}^{+\infty} B_z \, dz \qquad (5.43)$$

so that the areas under the rectangular and the actual distribution curves are equal. The angle of rotation for the rays transmitted through the lens is according to (5.16)

$$\psi_L = \left(\frac{e}{8mV_r}\right)^{\frac{1}{2}} LB_0 \qquad (5.44)$$

and the paraxial ray equation (5.19) is integrated easily, since the coefficient of r has now become a constant $\left(= \dfrac{\psi_L}{B_0}\right)$. This leads to the effective focal length (cf. fig. 5.10)

$$f_1 = \frac{L}{\psi_L \sin \psi_L} \qquad (5.45)$$

and to the abscissa

$$OF_1 = z = \frac{L}{2} + \frac{L}{\psi_L \tan \psi_L}. \qquad (5.46)$$

Since ψ_L is proportional to the excitation, the minimum focal length of a microscope lens follows easily by using B_0 of (5.42) for the particular value of $(\mu n I)$.

A computational method by Goddard and Klemperer (1944) allows accurate tracing of paraxial and marginal rays through experimentally known magnetic fields. This method starts from the equations of motion:

$$\left.\begin{array}{l} \dfrac{\mathrm{d}^2 z}{\mathrm{d}t^2} = -\left(\dfrac{e}{m}\right)^2 A \dfrac{\partial A}{\partial z}, \\[2ex] \dfrac{\mathrm{d}^2 r}{\mathrm{d}t^2} = -\left(\dfrac{e}{m}\right)^2 A \dfrac{\partial A}{\partial r}, \\[2ex] \dfrac{\mathrm{d}\psi}{\mathrm{d}t} = \dfrac{e}{m} \dfrac{A}{r}. \end{array}\right\} \qquad (5.47)$$

The following steps are taken in the procedure of computation:

(i) The vector potential $A(r, z)$ has to be obtained which, according to (5.34), is given by

$$A(r, z) = \frac{1}{r_0} \int_0^{r_0} r B(r, z) \, \mathrm{d}r. \qquad (5.48)$$

The values of $B(r, z)$ in (5.48) may be taken from an experimental plot (such as shown, for instance, in fig. 5.6) and integrated numerically. A whole lattice of A values may be obtained.

(ii) The derivatives $\partial A/\partial r$ and $\partial A/\partial z$ are computed by numerical differentiation and two new lattices are constructed for these values.

(iii) With the aid of these three lattices the equations of motion (5.47) are integrated simultaneously using a step-by-step method.

Results of the various steps are conveniently put down in a table which may contain the following columns:

The time interval t, say in units of 10^{-9} s,

$$\frac{\mathrm{d}^2 z}{\mathrm{d}t^2}, \frac{\mathrm{d}z}{\mathrm{d}t}, z; \quad \frac{\mathrm{d}^2 r}{\mathrm{d}t^2}, \frac{\mathrm{d}r}{\mathrm{d}t}, r; \quad \frac{\mathrm{d}\psi}{\mathrm{d}t}$$

and

$$W = \left(\frac{\mathrm{d}r}{\mathrm{d}t}\right)^2 + \left(\frac{\mathrm{d}z}{\mathrm{d}t}\right)^2 + r^2 \left(\frac{\mathrm{d}\psi}{\mathrm{d}t}\right)^2.$$

The start of the table is given by the initial velocity components and coordinates of the given ray. The last column serves as a very convenient check; it contains the values for the energy W of the electron which should be constant in view of the energy equation. Thus a direct measure of the accuracy of the path is available for every step.

The method is of particular interest in so far as its procedure forms in every way a counterpart to the experimental electron-ray

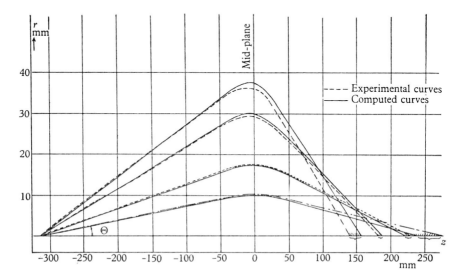

Fig. 5.19. Tracing of an initially homocentric bundle of 500 volt electrons through the meridional plane.

tracing method with the fluorescent target. Figs. 5.19 and 5.20 show, for example, the path of several rays through the lens field given by figs. 5.5 and 5.6. The good agreement of the computed trajectories with the experimental path may be recognized.

The experimental ray tracing with a sliding, fluorescent target differs for magnetic lenses in essential points from the corresponding procedure for electrostatic lenses which has been described in detail in chapter 3. In electrostatic lenses (restricting our attention to non-skew rays) the electron path is a plane curve; thus for tracing the path only two coordinates (y and z) need to be measured. In the case of the magnetic lens, however, owing to the

gyro-action of the field, the path is a twisted curve and it is necessary to measure three coordinates, viz. x and y, normal to the
z-axis, and the distance z along the axis itself. For this purpose a
cross-scale providing a metric in the xy-plane may be arranged in
the eyepiece of the measuring microscope with which the sliding
target is observed. The z-coordinate is measured by a pointer
attached to the moving system of the target. A second difference
arises from the fact that the path can only be measured outside
the electrostatic lens, since a fluorescent target placed inside the

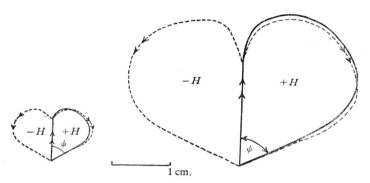

Fig. 5.20. Tracing of an initially homocentric bundle of 500 volt
electrons through the equatorial plane.

lens would disturb the electric field distribution. In the case of
the magnetic lens, however, there is no disturbance and the target
can be moved through the whole field of the lens, thus allowing the
full path to be traced.

Tracing of single rays with a sliding fluorescent target becomes
impracticable for the important case of electron-microscope
lenses, the dimensions of these lenses being too small for an application of the procedure. These lenses have been investigated
experimentally by measuring magnifications and object and image
distances. According to Ruska (1944) a thin metal foil with an
aperture of the order of 0·1 mm, illuminated from the back by the
electron beam, may serve as an object. This foil is mounted over
the end of a small tube, lowered by a pulley into the lens structure
in order to change the object position. A fine celluloid window
bearing a few particles of metal dust of colloidal dispersion covers

the aperture of the metal foil. This dust serves as an indicator for a sharp focus when the image is observed on a fluorescent screen. The magnification is obtained by a measurement of the diameter of the image of the foil aperture, the diameter of the foil aperture itself being known from measurements under a light microscope. The determination of the cardinal points of larger magnetic lenses does not need further explanation; the methods are quite analogous to those discussed in detail for electrostatic lenses (cf. chapters 2 and 3).

Finally, mention should be made of a practical method of ray tracing through magnetic fields. This has been developed for setting up systems for the channelling and transport of the relativistic heavy particle beams which are used in experimental high energy nuclear physics. This is the 'floating wire' or 'hodoscope' method (Loeb, 1947). The method is most frequently applied to systems consisting of large magnetic quadrupoles and deflecting magnets.

The method is based on the correspondence between the shape taken up by a freely suspended current carrying wire in a magnetic field and the trajectory of a charged particle through the same field. It can be shown that if the tension in the wire is T and the current passing is I, then the shape taken up by the wire is identical with the trajectory of a particle of electron-momentum $(Br)_e$ where

$$(Br)_e = T/I \text{ newton amp}^{-1}. \tag{5.49}$$

One end of the wire is attached to a fixed point outside the field of the element being investigated, and the other end, again outside the field, is a movable arrangement of frictionless pulley and weights to maintain the desired tension. For the practical difficulties, and precautions necessary to obtain a useful accuracy, the reader is referred to the literature (see e.g. Citron *et al.* 1959).

§5.8. Combined electrostatic–magnetic lenses

These lenses are constituted by any superimposition of electric and magnetic fields of circular symmetry. Only in very simple cases (e.g. superposition of homogeneous fields) can the effects of the two fields be separated completely. In general, the combined effect of the fields on the electron has to be considered at each point in space. In this way rays may be traced using the refractive index in the rotating meridional plane (cf. Sandor, 1941), given by

$$N_m^2 = V_0 - 2 \cdot 25 \times 10^9 \frac{\Phi^2}{r^2} + V_1. \qquad (5.50)$$

(5.50) follows from (5.35) by addition of the local electric potential V_1.

A method for numerical ray tracing starting from the paraxial ray equation

$$r''V_z + \frac{r'}{2} V_z' + \frac{r}{4} \left(V_z'' + \frac{eB_z^2}{2m} \right) = 0 \qquad (5.51)$$

can be found in a paper by Burfoot (1952).

If the computational method by Goddard and Klemperer (1944) is to be used for ray tracing, the equations of motion to be integrated are of the following form:

$$\begin{aligned}
\frac{d^2z}{dt^2} &= \frac{e}{m} \frac{\partial V}{\partial z} - \left(\frac{e}{m} \right)^2 A \frac{\partial A}{\partial z}, \\
\frac{d^2r}{dt^2} &= \frac{e}{m} \frac{\partial V}{\partial r} - \left(\frac{e}{m} \right)^2 A \frac{\partial A}{\partial r}, \\
\frac{d\psi}{dt} &= \frac{e}{m} \frac{A}{r},
\end{aligned} \qquad (5.52)$$

which follow from (5.47) by addition of the electrostatic acceleration. The method of ray tracing by digital computer of Carré and Wreathall (1964) (cf. §3.6) is directly applicable to the integration of equations (5.52), and provides a very accurate method of tracing both paraxial and marginal rays through combined fields. Clearly, equations (5.47) can also be integrated by the same method by treating them as a special case (V = constant) of (5.52).

The focal length of a thin combined lens is obtained by integration of the paraxial ray equation of a combined field. According to Zworykin et al. (1945) it is given by

$$\frac{1}{f'} = \left(\frac{V_0}{V_0'} \right)^{\frac{1}{4}} \int_z^{z'} \left[\frac{3}{16} \left(\frac{1}{V} \frac{dV}{dz} \right)^2 + 2 \cdot 20 \times 10^{10} \frac{B_z^2}{V} \right] dz, \qquad (5.53)$$

where V is the variable potential inside the lens, V_0 and V_0' are the potentials of object and image space respectively and z and z' are the axial coordinates of object and image. Thus the lens power $1/f'$

in the image space is simply the sum of the lens powers of the component fields. For $V_0 = V_0' = V$, (5.53) reduces to Busch's thin lens formula (5.21) while for $B = 0$ it reduces to an expression which is equivalent to the electrostatic thin lens formula (3.30). The rotation of the image by the combined lens is again given by (5.16), where $\sqrt{V}$ now, as a variable, is written under the integral sign.

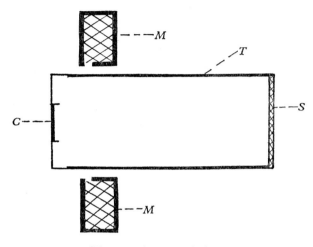

Fig. 5.21. Electrostatic–magnetic image converter.

Combined electric–magnetic lenses have found widespread application whenever slow electron emission of a thermionic or photoelectric cathode is accelerated and magnetically focused at the same time. For example, fig. 5.21 shows a combined lens used in the image converter of a television pick-up tube (McGee and Lubszynski, 1939). There, a flat cathode C is surrounded by a guard ring which curves the equipotentials in such a way that the electrons are accelerated towards the axis. The cylindrical tube T and the target S are at common potential. A magnetic field produced by the iron-clad coil M is superimposed on the whole arrangement.

§5.9. Magnetic electron reflector and magnetic bottle

Let an electron be injected into a magnetic field of slightly converging lines of force as indicated in fig. 5.22. We will show now that the electron must spiral along the surface of a tube of constant flux Φ, and that the magnetic moment of the circulating motion is

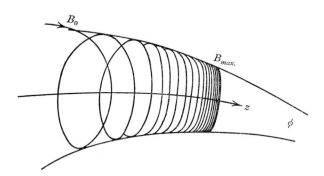

Fig. 5.22. Magnetic reflexion of an electron.

constant (cf. Post, 1956 or Spitzer, 1956). When the electron moves into regions of higher flux density, its velocity component $u_\perp$ at right angles to the z-axis of the flux tube increases. At the same time, the pitch of the orbital spiral is reduced until the velocity $u_\parallel$ parallel to the z-axis comes to a standstill at a certain flux density B_{max}, where it becomes reversed. After this reflexion $u_\parallel$ gradually increases again with the electron drifting now in the opposite direction. On the other hand $u_\perp$, and the pitch of the spiral increases again, but the sense of the circulating motion is the same throughout the whole process.

To show first the constancy of the moment M, consider the total energy T of the electron. T is unaltered in a purely magnetic field since the Lorentz force, which is always perpendicular to the velocity, cannot do any work. However, neither $T_\perp = \frac{1}{2}mu_\perp^2$ nor $T_\parallel = \frac{1}{2}mu_\parallel^2$ are constant, but owing to the conservation of energy

$$\frac{\mathrm{d}T_\parallel}{\mathrm{d}t} = \frac{\mathrm{d}}{\mathrm{d}t}(T - T_\perp) = -\frac{\mathrm{d}T_\perp}{\mathrm{d}t}. \qquad (5.54)$$

Now, the magnetic moment is given by the current I and the area a enclosed by it, namely

$$M = Ia = \left(-\frac{e\omega_c}{2\pi}\right)\pi r^2, \qquad (5.55)$$

where ω_c is the cyclotron frequency and r the radius of the spiral orbit. Introducing $\omega_c = B_z e/m$ and $\omega_c = u_\perp/r$ we obtain

$$M = -\frac{mu_\perp^2}{2B_z} = -T_\perp/B_z \qquad (5.56)$$

or

$$\frac{\mathrm{d}(MB_z)}{\mathrm{d}t} = -\frac{\mathrm{d}T_\perp}{\mathrm{d}t}. \qquad (5.57)$$

We consider now (in cylindrical coordinates) the identity

$$\mathrm{div}\, B = \frac{1}{r}\frac{\partial(rB_r)}{\partial r} + \frac{\partial B_z}{\partial z}. \qquad (5.58)$$

Owing to the circular symmetry, there is no angular component B_ψ, and $\mathrm{div}\, B = 0$ since there are no sources. Moreover, if the field lines converge slowly,

$$\frac{\mathrm{d}B_z}{\mathrm{d}z} \approx \frac{\mathrm{d}B}{\mathrm{d}z} \approx \mathrm{const.}$$

and integration of (5.58) yields

$$\int \partial(rB_r) = -\int r\,\partial r\left(\frac{\partial B}{\partial z}\right)$$

or

$$B_r = -\frac{r}{2}\frac{\partial B}{\partial z}. \qquad (5.59)$$

Now Newton's law $\qquad m\dfrac{\mathrm{d}u_\parallel}{\mathrm{d}t} = eu_\perp B_r$

with B from (5.59) yields

$$m\frac{\mathrm{d}u_\parallel}{\mathrm{d}t} = \frac{-er}{2}u_\perp\frac{\partial B}{\partial z}. \qquad (5.60)$$

Using (5.60) in the form

$$M = -\frac{e}{2}ru_\perp \qquad (5.61)$$

and (5.60) multiplied by $u_\parallel = dz/dt$ we obtain

$$m \left(\frac{du_\parallel}{dt}\right) u_\parallel = \frac{d}{dt} \left(\frac{mu_\parallel^2}{2}\right) = M \frac{\partial B}{\partial z} \frac{dz}{dt}$$

or
$$M \frac{\partial B}{\partial t} = \frac{dT_\parallel}{dt}, \qquad (5.62)$$

but comparison of (5.62) with (5.57) and (5.54) shows that $d(MB) = M\,dB$ or $M = const.$

Second, we show now that $M \propto \Phi$, where Φ is the flux enclosed by the orbit, and that the electron is reflected into the region of lesser field. Take $\Theta = \sphericalangle (u, z)$, then $u_\perp/u = \sin \Theta$. The constancy of M implies (cf. (5.56)) that $B \propto T_\perp \propto u_\perp^2$. Since for $u^2 = const.$ $u^2 \propto \sin^2 \Theta \propto B$, we obtain

$$\sin^2 \Theta = \frac{B}{B_0} \sin^2 \Theta_0, \qquad (5.63)$$

where Θ_0 is the angle between the total velocity u and the axis z of the flux tube at the point of injection.

Now, when B/B_0 rises to $1/\sin^2 \Theta_0$ then $\sin^2 \Theta = 1$ or $\Theta = 90°$ and $u_\perp$ reaches its maximum. Hence all the energy is transformed into transverse kinetic energy $T_\perp$ and $u_\parallel$ has fallen to zero. At this stage the electron has reached its highest flux density B_{max} and it will now be reflected into a region of lesser field.

The flux is given by $\Phi = \pi r^2 B$ or introducing the electron momentum $(Br)_e = mu_\perp/e$, we have

$$\Phi = \pi \left(\frac{mu_\perp}{e}\right)^2 \frac{1}{B}$$

and with (5.56)
$$\Phi = \frac{-2\pi m}{e^2} M. \qquad (5.64)$$

This shows that the flux is proportional to M, but as $M = const.$ it implies that the electron is moving along the surface of a flux tube.

The principle of reflexion of the electron has found an important application in thermo-nuclear research where the electrons of a plasma must be contained as long as possible in a finite volume. This is achieved by injecting the plasma into a 'magnetic bottle' which consists of two very strong co-axial magnetic fields enclosing a weaker field in their middle. Thus the electrons are reflected to and fro between the two strong fields, which are frequently called by a misnomer the 'electron mirrors'.

In geomagnetism huge magnetic bottles are formed by the lines of force extending into space from the north pole of the earth, the field being strongest near the poles. These bottles are filled with cosmic electrons from the solar wind, etc., forming the so-called 'Van Allen' belt. The electrons circulate in the belt, until they are deflected by some atomic collision into an angle Θ of such magnitude (cf. (5.63)) that the local flux density can no longer contain them.

§5.10. Magnetic quadrupoles (cf. Septier, 1961)

Fig. 5.23 shows a cross-section through a magnetic quadrupole lens which is produced by two pairs of magnetic pole pieces, N–N and S–S, the poles of each pair are opposite one another and of equal pole strength. Each pole piece is excited by nI ampere turns. The surfaces of the magnets are bounded, in the ideal case, by hyperbolae and the cross-section through the magnetic equipotentials, forming the lens field, consists of a series of hyperbolae, the rectangular asymptotes of which are chosen here* as the x- and y-axes of the coordinate system. Since the lines of force are hyperbolae which are orthogonal to the equipotentials, an electron beam moving the $-z$-direction ('into the paper') will be deflected at right angles to the lines of force, i.e. in a direction along the equipotentials as indicated by arrowheads in fig. 5.23. Thus, the beam will be converged in the y-direction and simultaneously diverged in the x-direction.

Since the equipotentials in fig. 5.23 are rectangular hyperbolae, referred to the axes as asymptotes, the equation for the magnetic potential V_H is given by

$$V_H = kxy. \qquad (5.65)$$

The surface of the pole pieces can be taken as an equipotential of

---

* In the electrostatic quadrupole lens (fig. 2.6), the coordinate axes were chosen to coincide with the asymptotes of the hyperbolic lines of force.

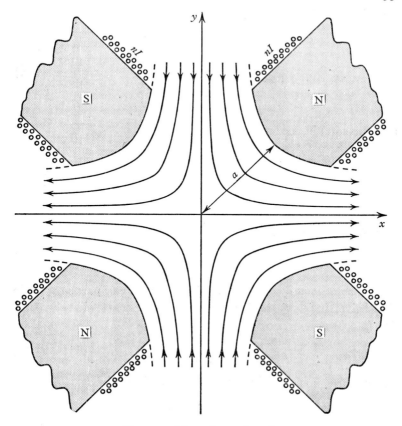

Fig. 5.23. Magnetic quadrupole.

the value V_a, at a distance a of its vertex from the origin. We then have

$$xy = a^2/2,$$

and with (5.64)

$$k = \frac{2V_a}{a^2},\tag{5.66}$$

a is the semi-aperture of the quadrupole as shown in fig. 5.23. Even at fairly strong excitation the pole piece surface can be considered as an equipotential, since saturation is known always to start first somewhere in the yoke. Expressed by the number of ampere turns for each pole piece, this magnetic potential is given by

$$V_a = nI.\tag{5.67}$$

The flux density has the components

$$
\left.\begin{aligned}
B_x &= -\mu_0 \frac{\partial V_H}{\partial y} = -kx, \\
B_y &= -\mu_0 \frac{\partial V_H}{\partial x} = -ky, \\
B_z &= 0,
\end{aligned}\right\}
\tag{5.68}
$$

and the equations of motion are

$$
\left.\begin{aligned}
-\frac{e}{m}(-u_z B_y) &= \ddot{x} = +\omega^2 x, \\
-\frac{e}{m} u_z B_x &= \ddot{y} = -\omega^2 y,
\end{aligned}\right\}
\tag{5.69}
$$

where, using (5.66)–(5.68) ω is given by

$$
\omega^2 = 2\frac{\mu_0 e}{m}\frac{nI}{a^2} u_z.
\tag{5.70}
$$

The solutions of the differential equations (5.69) are now again those given in (2.25) and (2.26) of §2.6 but the 'excitation constant' (cf. (2.27)) now becomes

$$
\beta = \frac{\omega}{u_z} = \frac{1}{a}\left(2\frac{\mu_0 e}{m}\frac{nI}{u_z}\right)^{\frac{1}{2}}.
\tag{5.71}
$$

We can use, analogously to the definition (5.43) in §5.7,

$$
L = \frac{\displaystyle\int_{-\infty}^{+\infty} B_r(z)\,dz}{B_r(0)}
\tag{5.72}
$$

as the 'effective length' of the field, where $B_r(z)$, is the transverse component of flux density at a distance r from oz, measured along lines which are parallel to oz. For a practical estimate, it is useful to calculate L from the mechanical length L_0 and from the aperture a by

$$
L = L_0 + 1 \cdot 1 a.
\tag{5.73}
$$

Quite analogously to the treatment in §2.6 of the electrostatic case we obtain, according to (2.34) and (2.37) for the two focal lengths in a first approximation,

$$
f_y = -f_x = \frac{1}{L\beta^2} = a^2/(L_0 + 1 \cdot 1 a)\left(\mu_0 nI \sqrt{\frac{2e}{m}} V\right), \tag{5.74}
$$

where eV is the electron energy. The trajectories are planar in the zox- and zoy-planes only, convergent towards z in the zox-plane and divergent in the zoy-plane. In all other cases, the magnetic force causes rotation of the trajectory which becomes a curved path. However, both the real focus F_y and the virtual focus F_y are straight lines parallel to the y-axis and to the x-axis respectively, i.e. these lines are inclined at $45°$ to the lines connecting equal poles. Once the excitation constant β of the lens is known, the cardinal elements of the magnetic quadrupoles can be taken as before from the graph in fig. 4.15 in §4.9, where the focal lengths f and the distances P of the two principal planes from the exit face are plotted for the two principal sections xz and yz.

Magnetic quadrupoles were originally employed by Hehlgaus (1935) in a line-focus tube producing sound tracks on sound films. There, the quadrupoles were produced by four mutually perpendicular straight iron cylinders, energized in alternating sense by solenoids and mounted like the spokes of a wheel inside an iron ring which served as a magnetic yoke. It is of some interest that the two focal lengths of this quadrupole can be calculated by a simple formula (Archard, 1954) namely:

$$f_q = \pm \frac{a^2 \sqrt{V_0}}{0.37 R \mu_r nI},$$ (5.75)

where V_0 is the accelerating voltage of the electron beam, a the distance from the lens axis of the cylinders of radius R, μ_r is the relative permeability of the iron cores and nI (ampere-turns) their excitation. If this system is compared with the electrostatic quadrupole consisting of four spheres (§4.9, (4.14)), of the same a and R, their focal lengths are equal when

$$\frac{nI}{V_a} = \frac{10^4}{\mu_r \sqrt{V_0}}$$ (5.76)

(typically mA-turns $\equiv \frac{1}{10} \times$ number of volts V_0).

Panofsky and Baker (1950) used a long powerful quadrupole magnet for focusing a 350 MeV proton beam which emerged from a cyclotron, since when magnetic quadrupoles have been used routinely for beam transport. The great practical interest in the optics of the quadrupoles, however, started only in 1952 when

Courant, Livingstone and Snyder proposed their use to guide fast ions in a new circular accelerator, the 'strong focusing' synchrotron. It has been explained already, in §2.3, that the combination of two quadrupoles is able to converge the beam in both principal sections and that a chain arranged with proper mutual distances of quadrupoles is able to canalize the beam so that even the outermost trajectory never exceeds a certain maximum aperture.

CHAPTER 6

LENS ERRORS:
GEOMETRICAL ABERRATIONS

§6.1. Classification of aberrations

The laws of Gaussian optics interpreting the behaviour of paraxial beams in electron lenses, which were developed in chapter 2, are valid only if certain restricting conditions are fulfilled. Any deviation of the real image from the ideal, perfect picture, due to lack of fulfilment of these restricting conditions, is termed an 'aberration' of the image. The radius vector, which can be drawn from any ideal picture point to the corresponding point of intersection of the actual ray with the Gaussian image plane, gives the value of the aberration which may be ascribed to every imaging ray. The total aberration of a ray may be regarded as the vector sum of a number of independent constituent aberrations. Each constituent aberration is due to the lack of fulfilment of particular conditions which are listed in table 6.1. The errors (1) and (2) of table 6.1 are purely geometrical, the errors (3)–(5) depend on properties of the electron beam, and will be discussed in the next chapter under the title 'Electronic errors'.

TABLE 6.1. *Classification of aberrations*

Condition	Error
1. No deviation from circular symmetry	1. Asymmetry error
2. Imaging aperture and image field must be relatively small	2. Geometric-optical error
3. Electron speed must be homogeneous	3. Chromatic error
4. Current density must be small	4. Space-charge error
5. Picture element and imaging aperture must be large in comparison with the wavelength of the electron	5. Diffraction error

The survey presented in table 6.1 may be further extended by giving the usual classification of the components of the geometric

[159]

optical error (2). It has been pointed out in chapters 1 and 2 that the application of Gaussian optics is restricted to paraxial rays. Gaussian optics is valid only if the rays are so close to the axis that terms of higher order than the first in the expansion of the sine of their slope angles (cf. (1.33)) can be neglected. However, in practically useful electron optical systems, the apertures are generally so large that paraxial rays constitute but a small fraction of the image-forming rays.

In 1856 L. Seidel extended the first-order theory by developing a trigonometrical analysis and by including the third-order term in the sine expansion (cf. Born and Wolf, 1964). In this third-order theory a series of five correction terms is applied to the first-order theory. These five terms, which characterize five independent lens errors, can also be developed in a completely different way starting from Hamilton's characteristic functions (cf. Synge, 1937). In this way it can be shown for the isotropic media of glass optics as well as for electrostatic electron optics that just these five and no more third-order aberrations exist.

For anisotropic refraction, however, which occurs in magnetic fields, Glaser (1933c) has shown that eight independent third-order errors have to be assumed. In table 6.2 we give a list of these errors with their causes and effects. These are the aberrations of circular-symmetrical lenses. For systems which are symmetrical about two mutually perpendicular planes like the quadrupole lenses, Burfoot (1954) has classified eleven new errors, three of which are third-order distortions, three give curved image surfaces, three are analogous to coma and two to spherical aberration (see §6.10).

According to the third-order theory, the isotropic geometrical errors of circular symmetrical lenses are represented by the following deviations Δx and Δy of the actual image point $(x'y')$ from the Gaussian image point with the coordinates x_0 and y_0, namely

$$\left.\begin{aligned}
\Delta x' &= -m_0 x_0 (S_5 r_0^2 + S_4 \rho^2 + S_2 r_A^2) + x_A (2S_3 r_0^2 + S_2 \rho^2 + 4S_1 r_A^2), \\
\Delta y' &= -m_0 y_0 (S_5 r_0^2 + S_4 \rho^2 + S_2 r_A^2) + y_A (2S_3 r_0^2 + S_2 \rho^2 + 4S_1 r_A^2),
\end{aligned}\right\}$$

$$(6.1)$$

TABLE 6.2

Error	Cause	Effect
2.1 Spherical aberration = axial or aperture error	Excessive or deficient refraction with increasing aperture	Paraxial and zonal rays have different intersection on the axis
2.2 Coma	Aberration for slanting incident parallel rays	Comet tails at image points
2.3 Astigmatism	Bundles of rays meet different curvatures of equipotentials	Off-axial object points are not imaged as points
2.4 Curvature of field	As 2.3	Image of plane object is curved
2.5 Distortion	Residual error of spherical aberration and position of aperture stop	Straight lines of object are curved at the margin of the image; squares are projected as pincushion or barrel-shaped images
2.6 Spiral distortion	Excessive rotation with increasing aperture	Tangential displacement of image points increasing with radial distance
2.7 Anisotropic coma	Rotation of orbits about individual axes	Asymmetrical deformation of individual image elements
2.8 Anisotropic astigmatism	As 2.7	Marginal object points projected into inclined line images

where $r_0^2 = x_0^2 + y_0^2$; $\rho^2 = 2(x_0 x_A + y_0 y_A)$; $r_A^2 = x_A^2 + y_A^2$, m_0 is the magnification, calculated from Lagrange's law (1.34), x_A, y_A are the coordinates of a small aperture stop, $S_1 \ldots S_5$ are the 'Seidel coefficients', for the particular errors listed in table 6.2 under (2.1) ... (2.5) respectively.

Although the mathematical theories supply a logical basis for classifying the lens errors, they have so far been of little practical assistance for the design of corrected electron-optical instruments. The only systematic way of improving the correction of electron lenses is to begin by measuring the lens errors separately, either by direct methods or by tracing the rays through plots of equipotentials. Then for electrostatic lenses, the arrangement and the shape of electrodes, for magnetic lenses the arrangement of the current conductors and the shape of the iron shielding has to be altered with a view to correction. In the following we shall not attempt to expound the general mathematical treatment of aberrations. Rather we will try to give some physical explanations, discussing the useful formulae for each error separately, and we will describe the important experimental methods for the detection and measurement of the aberrations.

§6.2. Deviations from circular symmetry

The effect of such deviations is frequently called 'Axial Astigmatism'. In common with the third-order error known as 'Astigmatism' (cf. §6.7) object points appear here in the image extended as lines. In distinction to the third-order error, however, the axial astigmatism also occurs in the image of axial object points. The consequences of any asymmetry of a lens is easily noticed, but the causes of the asymmetry are sometimes very difficult to detect. They may be given for instance by a misalignment, or by lack of circular symmetry of the lens electrodes. In electron-microscope objectives even traces of dust and dirt have frequently been observed to cause serious field asymmetries. In magnetic lenses, deviations from circular symmetry are often due to a lack of homogeneity of the magnetic material of the pole-pieces. Disturbances by spurious electric or magnetic fields are quite common causes of small deviations from circular symmetry, and the ways and means of avoiding them will be discussed below.

Small deviations from circular symmetry of the electrodes of electrostatic lenses are often very troublesome. If, for instance, a symmetrical two-tube lens (cf. §4.2) is used for focusing a spot in a cathode-ray tube, variations in tube diameter near the mid-plane may be less that 1 per cent; but they will still cause a distortion of the spot, which in this case is not round but of elliptical shape. The circular symmetry of relatively large electrostatic lens electrodes is tested in practice, for instance, with the help of a wedge gauge, i.e. a piece of metal sheet cut to triangular shape, the sides of the triangle having a slope of about 1:20. The depth to which the wedge gauge can be lowered into the lens electrode will change according to the asymmetry. Microscopic testing is necessary for small lens electrodes or small magnetic pole-pieces.

The smallest deviations from circular symmetry have been found to be very serious in the case of magnetic lenses of electron microscopes. Machining tolerances of the order of 200 Å have been quoted for high-quality magnetic electron-microscope objectives (Le Poole, 1948; Marton, 1950), but it has now been established that mechanical tolerances on symmetry which are required to reduce the astigmatism to negligible proportions for highest resolution are finer than can be met by the most exacting manufacturing methods (Archard, 1953). Under these circumstances the asymmetry error is not likely to be overcome by achieving ultimate accuracy of mechanical operation but rather by application of compensating elements which can be adjusted to reduce the field asymmetry to zero. A full discussion of such elements will be given at the end of this section.

A lens of perfect circular symmetry may show serious adjustment troubles if it is only slightly tilted with respect to the axis of the optical system. The image will be shifted in radial direction and also out of the Gaussian plane. In addition, even for a paraxial object, the image might be affected by field errors (cf. §§6.7–6.9). The tilted lens corresponds to an untilted lens with an object at a distance $s\epsilon$ off the axis, where s is the object distance and ϵ the angle of tilt.

Alignment errors will also occur when the lenses in an assembly are not strictly coaxial. Such errors can often be traced by simple tests. If, for instance, in a magnetic electron microscope the lens axes of objective and projector do not coincide, a change in objective current will cause the image to rotate about some point outside the image field. The relative lens positions may then be adjusted until rotation occurs about the centre of the image field.

Inhomogeneity of magnetic material, already mentioned as a serious cause of lens asymmetry, has to be avoided by the use of specially chosen mild steel for all pole-pieces or shields of electromagnetic electron lenses. In the manufacture of permanent magnetic lenses (§5.5), the inhomogeneities of an alloy of high remanence often cannot be avoided. The asymmetries, however, can be smoothed out sufficiently when the permanent magnets are used with suitable pole-pieces made from permeable material such as mild steel.

Spurious magnetic fields are sometimes troublesome in electro-static lenses if magnetic material is used in the electrode structure. For instance, nickel is very commonly used for electrodes in electronic engineering owing to its excellent properties in the evacuated tubes. If, however, a local discharge or spark has accidentally passed between two nickel electrodes, a spurious permanent magnetization is set up and distortions in the electron image are observed. The magnetization of the nickel electrodes is driven off by eddy-current heating or baking of the electrodes at sufficiently high temperature.

It need not be emphasized that all external disturbing fields have to be shielded off carefully from any electron lens. Spurious external electrostatic fields may easily penetrate through the gaps between electrodes unless these gaps are carefully protected. These fields may be set up, for instance, by charges on the wall of a glass tube enclosing the electron-optical arrangement or from the mechanical supports of the lens electrodes. Such supports consist of insulating material or of a conductor which may be at a potential different from that of a neighbouring lens electrode. The gaps between the electrodes have to be shielded either by an over-lapping lens electrode or by a specially designed 'penetration shield' connected to a lens electrode.

External spurious magnetic fields, as, for instance, the earth's magnetic field, can be shielded off sufficiently by a shield of material of high permeability. Theoretically (cf. Kaden, 1950) the 'shielding efficiency' of a long cylindrical shield is given by

$$\frac{B_0}{B_i} = 1 + \frac{\mu_r t}{r_o}\left(2 - \frac{t}{r_o}\right), \qquad (6.2)$$

where B_0 is a uniform flux density which drops inside the shield to B_i. Moreover, μ_r is the relative permeability of the shield, t is its wall thickness and r_o its outer radius. For a given material, the shielding efficiency thus depends only on the ratio t/r_o not on t or r_o alone. If $t \to r_o$ then $B_0/B_i \simeq \mu_r/4$ so that there is an ultimate limit, depending on permeability, to the degree of shielding obtainable. Equation (6.2) further shows that for single shields little improvement in shielding efficiency is obtained by increasing t/r_o beyond about 0·4. However, by using two or three shields separated within a distance of about a cm, ultimate efficiencies can be doubled or tripled respectively.

TABLE 6.3. *Shielding efficiencies of cylindrical shields*
(60 Hz)

Material	Permeability μ_r	Length (mm)	Thickness t (mm)	Outer radius r_o (mm)	t/b	Measured B_0/B_i	
						Axial	Transverse
Silicon iron	1 000	152	0·63	25·4	0·025	21	27
Silicon iron double shield spaced 6 mm	1 000	152	0·63	25·4	0·025	31	60
50 per cent nickel alloy	5 000	124	0·79	52	0·015	5·8	15·5
78 per cent nickel alloy	20 000	124	0·79	52	0·015	6·9	20
		152	0·63	25·4	0·025	94	355
		152	0·79	25·4	0·031	133	398
		152	1·57	25·4	0·062	316	447

Teasdale (1953) has measured the shielding efficiency for various values of permeability and thickness, and for shields of various shapes. We show in table 6.3 a selection of his results for some cylindrical shields. The shielding has reached these values within the shield at about a radius away from the open end of the shield. The values for a field B_0 at right angles to the cylinder axis agree well with the theoretical formula (6.2) but the shielding is poor for field components parallel to the axis. The permeability μ_r changes very much with increasing magnetic fields, reaching for the tabulated alloys a high maximum at very weak fields of a fraction of a Gauss. For example, 4 per cent Silicon iron $\mu_{max} \simeq 7000$, Mu-metal (75 per cent Ni) $\mu_{max} \simeq 10^5$, Supermalloy (78 per cent Ni) $\mu_{max} \simeq 10^6$ (cf. Hoselitz, 1952). In table 6.3 have been listed the initial values of μ_r as given by Teasdale (1953). The shielding off of strong fields (e.g. the deflecting fields of a beta spectrometer after the deflexion) is better achieved by the use of materials with relatively high μ at strong fields, e.g. by shields of mild steel.

'Helmholtz coils' may be used for compensation ('degaussing') of the spurious, external fields if these spurious fields are homogeneous enough. Helmholtz coils are two equal short coils placed co-axially and at a distance apart equal to one coil radius R. Near the axis, half-way between the two coils, is set up a very homogeneous field of flux density

$$B = (4/5)^{\frac{3}{2}} \mu n I/R = 0.9 \times 10^{-6}\, nI/R. \qquad (6.3)$$

For example, to compensate the total earth-magnetic field of 0.54×10^{-4} Wb m^{-2} in the centre of two Helmholtz coils of 1 m radius, 60 ampere-turns would be required for each coil. If field homogeneity is required in an annular zone rather than in a volume near the centre, the mutual distance of the coils has to be reduced. Craig (1947) has investigated the field homogeneity between the coils as a function of their mutual distance, and detailed information can be found in his paper. For instance, in an annular volume between $0.3R$ and $0.5R$, extending $\pm 0.2R$ from the mid-plane between the coils, homogeneity within ± 2 per cent is obtained with a mutual distance of the coils of $0.84R$.

Using one pair of degaussing coils, the axis of this pair must, of course, be oriented in the direction of the spurious field. In an alternative arrangement, three mutually perpendicular pairs of coils will, in any orientation degauss a particular volume if the energizing currents are adjusted in the appropriate manner.

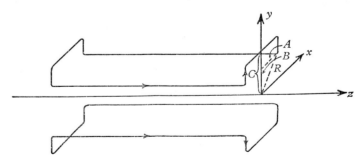

Fig. 6.1. Coils for producing a uniform magnetic field.

The use of rectangular degaussing coils is convenient if the earth's field has to be compensated over an elongated cylindrical volume. In particular, the 'bent-head' or 'tail-up' type of rectangular coils as drawn perspectively in fig. 6.1 produces a homogeneous field over a relatively large volume. The proportions of such coils have been investigated by Harris (1934) and by Lyddane and Ruark (1939). If the width (2A) of each coil equals 1·7 times the distance (2B) between them, and the distance (C) of the bent-up head from the z-axis equals 1·25 times the distance R of the rectangular part from the z-axis, then the region in which the field deviates less than 1 per cent from the value at the centre encloses a cylinder of about 0·05R radius and of 1·4L length where 2L is the coil length of the rectangular part.

An arrangement of three long, rectangular coils for the production of very homogeneous fields has been recommended by Haynes and Wedding (1951). The planes of these coils are parallel, and their sizes and spacings are defined by a circular cylinder, the axis of which lies in the plane of the mid-coil. The two outer coils are spaced by ±45° from the mid-coil on the surface of the cylinder.

Whenever a complete elimination of the causes of errors in circular symmetry of the lens proves to be difficult, the use of compensating devices is of great practical importance. Hillier and Ramberg (1947) corrected commercial magnetic microscope objectives in a practical way as shown by fig. 6.2. They arranged eight iron screws C in the non-magnetic metal spacer A of the pole-pieces B of the objective. The positions of these screws could be adjusted systematically until the microscope image was free from asymmetry errors.

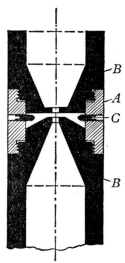

Fig. 6.2. Cross-section of microscope objective provided with compensating screws.

For an electrostatic correction of the asymmetry error, an arrangement of four electrodes is sufficient in principle. There, opposite electrodes are connected electrically and the two pairs thus formed are on adjustable potentials of opposite sign with respect to the anode potential. The whole arrangement is symmetrical about the lens axis, but it has to be turned bodily into the right azimuthal position in order to oppose the asymmetry of the bundle of rays due to the astigmatism in the objective lens.

In commercial microscopes a six-electrode 'stigmator' described by Rang (1949) and shown in fig. 6.3 is used in the following way: The two electrodes E and E' are permanently earthed, i.e. they are connected to anode potential. Referring to the remaining four electrodes, A is connected electrically to A' and B is connected electrically to B', each pair being charged by a separate potentiometer either above $(+)$ or below $(-)$ earth $(=$ anode$)$ potential.

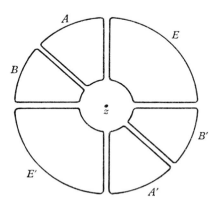

Fig. 6.3. Stigmator.

The two potentiometers are linked by mechanical gearing in such a way that complete azimuthal rotation of the correcting field can be performed by a single knob while the stigmator electrodes remain in their fixed position. The absolute potential values of the pairs AA' and BB' are controlled by another linked potentiometer pair. All stigmator electrodes are arranged symmetrically about the axis z which should strictly coincide with the lens axis. Strict alignment of the two axes during the operation of the stigmator is effected by means of a cross-slide on which the stigmator is mounted. The possibility of a quick and easy adjustment of the stigmator is a great asset, since the asymmetry of a microscope objective is often subject to daily uncontrollable changes which may require frequent readjustment.

§6.3. Spherical aberration

The axial error, called spherical aberration, occurs if rays emitted from a point object on the axis do not recombine to form a point image. The case of parallel incident rays (i.e. object at infinity)

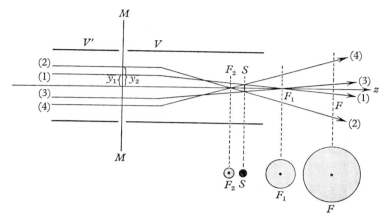

Fig. 6.4(a). Initially parallel rays, focused by lens with spherical aberration.

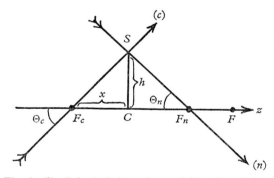

Fig. 6.4(b). Spherical aberration and disc of confusion.

which is of principal interest is illustrated in fig. 6.4. There is shown a symmetrical two-tube lens with the axis z and with the mid-plane M. Two pairs of parallel rays are drawn, the zonal rays (1) and (3) which are travelling at a 'medium' distance y_1 from the axis and the 'marginal' rays (2) and (4) at a relatively large distance y_2

from the axis. Paraxial rays (not drawn in the figure) would be focused in the paraxial focus F. For a lens with spherical aberration, the focus F_1 of the zonal rays and the focus F_2 of the marginal rays are found to be different from the paraxial focus F. The distances of FF_1 and FF_2 are called the longitudinal spherical aberrations of the rays (1), (3) and (2), (4) respectively. The focused spot as observed in different image planes at right angles to the z-axis is shown in the lower part of fig. 6.4(a). The spot S is known as the 'disc of least confusion' which will be fully discussed at the end of this section. If, as in the given example, the focal

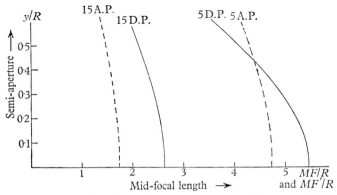

Fig. 6.5. Spherical aberration curves for symmetrical two-tube lens.

distance decreases with increasing aperture of the beam, the sense of the aberration is conventionally called positive. As a rule of experience, the spherical aberration of electron lenses is generally positive.

The spherical aberration for a given semi-aperture y changes with the voltage ratio of the lens. This is illustrated by fig. 6.5, where the mid-focal lengths of symmetrical two-tube lenses are plotted as a function of the semi-aperture of rays which are initially parallel to the axis. The two solid curves illustrate the spherical aberration of the symmetrical two-tube lens decelerating parallel rays; they refer to the voltage ratios 5 and 15. The corresponding curves for the same lens accelerating parallel rays are plotted as broken lines. It appears that the lens is of a much better quality if the parallel rays are at the low-voltage side.

The aberration for finite object or image distances is often, but not always, found to be intermediate between the two extreme cases of infinite distances. Quite generally it is of the same order of magnitude for all object positions. To obtain accurate results for any particular case, special measurements with the proper object and image distances have to be performed.

For a finite object distance s', the longitudinal aberration $\Delta J = JJ_n$ is defined by the distance of the paraxial image J from the zonal image J_n. In order to plot the aberration ΔJ of the divergent beam in accordance with the diagram shown in fig. 6.5, the location of the aperture, which is changing in size along the ray, has to be specified. This aperture (y) may conveniently be taken either at the principal plane (y_p) or at the thin equivalent lens (y_L). In these positions the apertures are connected, according to §2.2, with the slope angles Θ of the ray at the image side:

$$\tan \Theta_n \approx \Theta_n = \frac{(y_p)_n}{s_n} = \frac{(y_L)_n}{b_n}, \qquad (6.4)$$

where $s_n = PJ_n$ is the distance between principal plane P and image point J_n of the nth ray with aperture $(y_p)_n$, and where $b_n = LJ_n$ is the image distance from the thin equivalent lens L. In practical aberration diagrams, the longitudinal aberrations ΔF_n (change in focal distance) or ΔJ_n (change in image distance) are frequently plotted against Θ_n.

Aberration coefficients. The longitudinal aberrations ΔF or ΔJ respectively may be expressed in the form of a power series either of the linear aperture y or of the angular aperture Θ. Thus

$$\begin{aligned}\Delta J &= c_2 y^2 + c_4 y^4 + c_6 y^6 + \dots \\ &= a_2 \Theta^2 + a_4 \Theta^4 + a_6 \Theta^6 + \dots. \end{aligned} \qquad (6.5)$$

If ΔJ is plotted as a function of $\log \Theta$, straight lines with a slope $d(\Delta J)/d \log \Theta = 2$ will be obtained for small apertures; these lines will curve upwards at larger apertures indicating higher order aberrations, corresponding to the magnitudes of the coefficients $c_2, c_4, \dots$ or $a_2, a_4, \dots$ respectively. The odd power terms of the power series (6.5) vanish, since naturally ΔJ has the same value for $+\Theta$ and for $-\Theta$ corresponding to rays on either side of the axis. Thus ΔJ is a symmetrical function of Θ.

Neglecting the higher terms in (6.5) one obtains

$$\Delta J = cy^2 = C_s \Theta^2; \tag{6.6}$$

the factors c and C_s are called 'primary aberration coefficients'. Relations (6.6) hold for all real lenses to a certain approximation. They approximate the aberration curves, such as are given in fig. 6.5, by parabolae. The magnitude of c or C_s respectively is representative of all apertures of rays belonging to a given object position; it characterizes the quality of a given lens.

The representation of ΔJ by cy^2 is of course not quite equivalent to that by $C_s \Theta^2$. It can be shown, for instance, that, if all the higher terms c_4, c_6, ..., etc., in (6.5) strictly vanish, the first coefficient a_2 alone of the angular power series will describe the full aberration curve to some approximation only. For example, for the special case $c_4 = 0$, $c_6 = 0$, etc., the following relations are easily calculated:

$$\left.\begin{array}{l} a_2 = c_2 s_0^2, \\ a_4 = -c_2^2 s_0, \end{array}\right\} \tag{6.7}$$

where s_0 is the paraxial image distance. The difference between $c_2 y^2$ and $a_2 \Theta^2$, however, is generally of small order only.

From the general theory of third-order lens aberrations the 'Seidel term' S_1 is derived as a characteristic for the spherical aberration (cf. (6.1)). Since S_1 is frequently used to present experimental results, its connexion with the longitudinal spherical aberration ΔJ is of interest. This may be written as

$$S_1 = 2N\Theta^2 \Delta J, \tag{6.8}$$

where N is the refractive index. Comparison of (6.4) with Lagrange's law (1.34) suggests that the magnitude of S_1 remains unchanged when the longitudinal aberration ΔJ is projected by an aberration-free lens system into another image space. In this way S_1 is proportional to a quantity which may be termed an 'aberration invariant' (Conrady, 1929, p. 44).

Instead of observing the longitudinal aberrations ΔJ, it is often convenient to observe the distances Δr of the rays (h in fig. 6.4(b)) from the axis (z) in the Gaussian plane. Δr is called 'transverse aberration' and it is given by

$$\Delta r = \Delta J \tan \Theta. \tag{6.9}$$

Using (6.5) or (6.6) respectively, (6.9) can be written

$$\Delta r = a_2 \Theta^3 + a_4 \Theta^5 + \dots$$

$$= C_s \Theta^3. \qquad (6.10)$$

C_s of (6.10) and (6.6) which is the resultant spherical aberration, is obtained experimentally as an average. In the case of discrimination between the various components, the coefficient a_2 in (6.5) and (6.10) corresponds to the 'third-order' aberration, a_4 to the fifth order and so on (cf. also §1.5, (1.33)).

The optical quality of a lens is judged by the ratio of its aberration coefficient C_s to its paraxial focal length f. Thus the following dimensionless aberration coefficient is often used:

$$\frac{C_s}{f} = \frac{\Delta J}{f \Theta^2} = \frac{\Delta r}{f \Theta^3}. \qquad (6.11)$$

In the measurement of aberration constants for finite object and image distances, consideration must be given to the magnification y'/y. As Archard (1958) has pointed out, some confusion about this point exists in the current literature.

The circle of confusion in the Gaussian image plane is given by

$$(\Delta r)' = y'/y \, C_s \Theta^3, \qquad (6.12)$$

where Θ is the angular semi-aperture of the beam on the object side. If, however, the point object is set up in the previous image plane, then on the previous object side a disc of radius Δr will be obtained where

$$\Delta r = C_s \Theta^3 = \frac{y}{y'} C_s'(\Theta')^3, \qquad (6.13)$$

and using Lagrange's equation (1.34) with $N = N'$,

$$C_s' = \left(\frac{y'}{y}\right)^4 C_s.$$

The magnification must be included if aberration and beam aperture respectively are measured on different sides of the lens. It should be emphasized that C_s is not a lens constant but depends on the position of the conjugates, i.e. C_s is a function of object and image distance. Tables on the aberration constant C_s refer generally to parallel incident rays.

Disc of least confusion. For practical purposes the quantitative effect of the spherical aberration on the quality of any image point is most important. In the presence of spherical aberration, a perfectly sharp focus of a wide bundle of rays becomes impossible, since every zone of the bundle produces a different focal point.

The rays of a bundle of electrons focused by a lens with spherical aberration may be received on a sliding fluorescent target which is arranged so as to be always perpendicular to the optical axis. If the target is gradually moved along the axis, the radius of the fluorescent circle marking the cross-section of the bundle is found to pass through a minimum at a place which is some distance in front of the paraxial focus F. This minimum diameter corresponds to the best possible focus. The disc of electrons formed at this best focus is called the disc of least confusion.

Take, for instance, the rays (2) and (4) in fig. 6.4(*a*), which represent the wall of a cylinder, since this diagram shows an axial cross-section. This cylinder surrounds a bundle of parallel rays of electrons. As shown in fig. 6.4(*a*), and in an enlarged scale in fig. 6.4(*b*), the rays are focused and reach a minimum cross-section at the point C, CS being the radius of the disc of least confusion. The rays passing the periphery of the disc of least confusion are focused on the axis at F_c and at F_n respectively, and they intersect the axis at the angles Θ_c and Θ_n. The spherical aberration FF_n of a ray (*n*) may again be assumed to be proportional to the square of the angular semi-aperture Θ as given in (6.6). The neglect of higher order terms actually has little influence upon the result of the following calculation of the disc of least confusion.

The distance of the focus F_n (see fig. 6.4(*b*)) from the centre C of the disc of least confusion is given by

$$CF_n = F_cF - F_nF - F_cC$$
$$= C_s(\Theta_c^2 - \Theta_n^2) - x, \qquad (6.14)$$

if $F_cC = x$. The radius h of the disc of least confusion is

$$CS = h = x\Theta_c = CF_n\Theta_n$$
$$= C_s[(\Theta_c^2 - \Theta_n^2) - x]\Theta_n. \qquad (6.15)$$

From the third and the last term of this equation one obtains, by adding $x\Theta_n$, $\qquad x(\Theta_c + \Theta_n) = C_s(\Theta_c^2 - \Theta_n^2)\Theta_n$

or $$x = C_s\Theta_n(\Theta_c - \Theta_n),$$

and with (6.15) $$h = C_s(\Theta_n\Theta_c^2 - \Theta_c\Theta_n^2).$$ (6.16)

In order to find the minimum value of h if Θ_n is varied, (6.16) may be differentiated with respect to Θ_n and then put equal to zero:

$$\frac{dh}{d\Theta_n} = C_s(\Theta_c^2 - 2\Theta_c\Theta_n) = 0$$

or $$\Theta_c = 2\Theta_n.$$

Putting this value in (6.16) one finds

$$h_{min} = \frac{C_s}{4}\Theta_c^3.$$ (6.17)

Comparing (6.17) with (6.10) it appears that the radius of the disc of least confusion is only a quarter of the radius of the transverse aberration Δr_c of the ray (c) in the Gaussian image plane.*

Addition of spherical aberrations of two lenses. This is of considerable practical importance, since many electron optical instruments are made up of two-lens systems (e.g. emission system and focusing lens in the cathode-ray tube, or objective and projector lens in the electron microscope). First an intermediate image is formed between the two systems. Then a final image is produced which suffers from aberration contributions of both systems. The aberration coefficient C_s of the combination can be calculated from the coefficients C_{s1} and C_{s2} of the two systems (cf. Liebmann, 1949 a) to be approximately

$$\frac{C_s}{f} = \frac{C_{s1}f_2^2 + C_{s2}(D - f_1)^2}{(f_1 + f_2 - D)^2},$$ (6.18)

where f is the focal length of the combination and f_1 and f_2 are the focal lengths of the two lens systems which are a distance D apart.

§6.4. Spherical aberration: measuring methods

Spherical aberration is of fundamental importance in limiting the performance of every electron-optical device. It sets a limit to the resolution of electron microscopes, to the smallness of the focused

---

* Some authors define the aberration coefficient by h_{min}, i.e. by the radius of the disc of least confusion. This value would amount to one-quarter of the coefficient C_s in (6.6).

spot in the cathode-ray tube, to the line intensity in the electron spectrometer, etc. A great amount of work has been done on the exact investigation of spherical aberration. The tests which have been employed in these investigations may be divided into three groups: 'pepperpot' methods, 'shadow' methods and the remaining other methods.

The *pepperpot method* is well known in glass optics under the name of 'Hartmann test'. It has been introduced into electron optics by Epstein (1936). A modification of Epstein's method by Klemperer and Wright (1939), which uses a sliding fluorescent

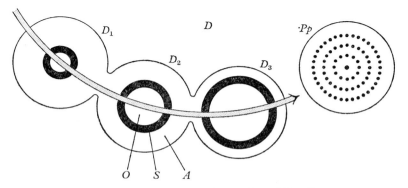

Fig. 6.6. Pepperpot diaphragm P_p and flap G.

target to trace the path of the electron pencils, has already been described in §2.1 (cf. fig. 2.1). In order to investigate the focal length for different apertures, a pepperpot diaphragm P_p containing a series of concentric circles of fine holes, as shown in fig. 6.6, intercepts the beam which has left the lens under investigation. The electron pencils transmitted through the fine holes converge into foci on the axial pencil which emerges through the fine central hole in P_p. Now, in order to be able to discriminate between the various foci belonging to the various circles of holes, a flap D consisting of several masking diaphragms $D_1, D_2, D_3, \ldots$ is introduced. Each of these can cover all but one circle of holes which is thereby left free to pass a batch of pencils through a circular slot, for example S in fig. 6.6, between the disc O and the annular disc

A. Tracing of the foci of the circles of pencils of various radii leads immediately to aberration curves as given in fig. 6.5.

For the accuracy of the pepperpot method it is essential that the pepperpot holes defining the thickness of the pencils Δy be small in comparison with the radii of the circles, i.e. the semi-apertures y of the pencils.

Diffraction methods. For relatively large electron lenses pepperpot diaphragms with mechanically produced apertures have proved satisfactory. However, for electron-microscopic objectives which are of the order of a millimetre in diameter, different means have to be used for producing sufficiently narrow zones ($\Delta y \ll 1$ mm) of pencils. Mahl and Recknagel (1944) have utilized for their aberration measurements the zones of electron rays which are produced by diffraction of a very fine pencil through a thin aluminium foil. The foil was arranged in the object plane of a

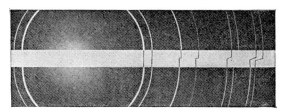

Fig. 6.7. Aberration measurement from the distortion of a diffraction pattern.

microscope objective. Debye-Scherrer diffraction rings were taken on a photographic plate in a large distance behind the objective. A comparison of the radii of these rings without lens and with lens gives quantitative information on the aberration of the objective. Diffraction patterns of a 50 kV beam are shown in fig. 6.7. The top half is taken without lens, and the bottom half with the saddle-field lens of an electrostatic electron microscope. Owing to the shorter focal length of the more marginal lens zones, the radii of the larger diffraction rings appear to be more contracted by the lens action than those of the smaller rings. In a similar way Hall (1949) determined the aberration of a magnetic microscope objective by measuring the displacement of a Bragg reflexion from the in-focus paraxial image of a diffracting ZnO crystal.

Shadow methods. These are related to the 'knife-edge test' of glass optics. In some early investigations on electron shadow-graphs, Boersch (1939) observed that spherical aberration does not lead to diffusion but to distortion. In fig. 6.8(*a*) K may represent a

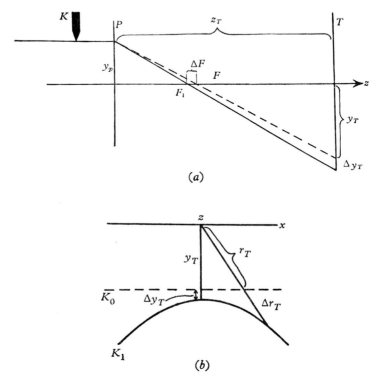

Fig. 6.8. Knife-edge test.

straight 'knife-edge' placed at right angles to the axis z of a lens with the principal plane P. A parallel bundle of rays limited by K crosses P with the semi-aperture y_p travelling at right angles to the straight edge. The limiting ray shown in the figure is refracted through the focus F_1 and meets the target T with the semi-aperture $y_T + \Delta y_T$. If the lens were free of aberration, all rays would be refracted through the paraxial focus F, and Δy_T would vanish.

Now, the radial semi-aperture of the limiting rays which is increasing along the straight edge may be given in the principal plane as $r_p = \sqrt{(y_p^2 + x_p^2)}$, where x is the coordinate parallel to the straight edge. Since the aberration is increasing with the aperture, the shadow of the edge on the target will be curved convex towards the z-axis as shown in fig. 6.8(b). If the longitudinal aberration is small and the target distance z_T is large in comparison with the focal length ($\Delta F \ll f \ll z_T$), the radial displacement of the curved shadow edge K_1 from the straight 'Gaussian' shadow edge K_0 can be calculated (cf. Liebmann, 1949a) to be

$$\Delta r_T = z_T \left(\frac{r_p}{f}\right)^3 \frac{C_s}{f}, \qquad (6.19)$$

where C_s is the aberration coefficient defined by (6.11). Assuming a square-law relationship (6.6) for the spherical aberration, it can be proved that the distorted shadow edge is a parabola, the origin of which is displaced by an amount given by (6.19), so that C_s can be determined from the shadow-graph if f is known from other measurements.

Spangenberg and Field (1942a, b) used for the measurement of spherical aberration a shadow method which has been described here (§2.4). As shown in fig. 2.4 two grids consisting of equidistant parallel wires are placed in front of and behind the lens. As above, from the known positions and dimensions of the grids and by measuring the magnifications of the shadows, the projection centres J and J' and hence the positions of the cardinal points are determined. For an investigation of the spherical aberration the position of J' has to be determined as a function of the angle of convergence Θ'. Since, owing to aberration, the centres of the paraxial projection (J') are nearer to the second grid than those of the marginal projection, the shadow of the grid would show a 'pincushion' distortion, i.e. the shadows of the wires shown in fig. 2.4(b) as straight lines would now appear to be convex towards the axis.

Heise (1949) used the shadow method for some exact aberration measurements of electrostatic microscope objectives of about 5 mm focal length. The shadow casting grid wires were of 0·03 mm diameter only and were adjusted within 5×10^{-3} mm.

The shadow method is not altogether satisfactory in electron optics, since the shape of the shadow edge is so easily changed by even the slightest spurious surface charges at the knife-edge. Such charges may be caused by insulating surface contaminations or by secondary emission. Moreover, if the aberration does not follow the pure square-law equation (6.6), i.e. if the higher terms in (6.5) are not negligible, the interpretation of the shadow-graph becomes rather complicated.

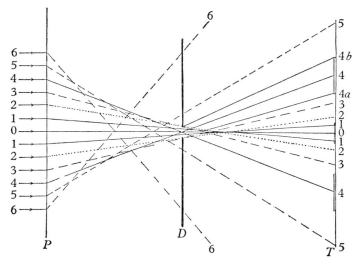

Fig. 6.9. Halo formation in presence of aberration.

Halo test. A useful qualitative aberration test has been introduced by Klemperer (1951). In presence of aberration, a focus of rays appears as a spot surrounded by a discrete halo ring, if a diaphragm with a fine circular aperture is placed across the electron beam in front of the focus. The geometry of rays forming the halo is shown in fig. 6.9. o, ..., o represents the axial beam. (1), (2), ..., (6) are rays which initially are equidistant and parallel to the axis, *P* is the principal plane of the lens. The marginal rays (6) are focused first, the paraxial rays (1) are focused last. A diaphragm *D* with a narrow circular aperture is fixed at right angles to the axis in a plane containing the focus of the zonal rays (4). At some distance from the diaphragm D on the fluorescent target T, two discrete groups of rays are intercepted which have passed the aperture in D:

(i) A paraxial group forming a spot.

(ii) A zonal group surrounding ray (4) and producing the picture of an annular ring on the target.

The two regions on the target illuminated by electron rays are marked in the cross-sectional fig. 6.9 by double lines parallel to the target. The rays (2) and (3) are intercepted by the diaphragm so that a dark zone is left between ring and spot. Also rays (5) and (6) are intercepted so that the space outside the bright ring is dark again.

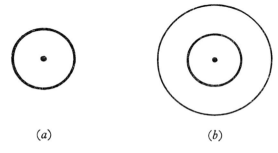

(a)   (b)

Fig. 6.10. Single halo (a) and double halo (b).

Fig. 6.10(a) shows the picture at the target near the minimum diameter of the electron beam, a spot being surrounded by a halo. This pattern belongs to an electron lens with simple positive spherical aberration, e.g. to a symmetrical two-tube lens. The other pattern, fig. 6.10(b), has been taken with a saddle-field lens containing a space charge near the axis as will be described in §8.5. This lens shows positive aberration for marginal rays but negative aberration for zonal rays. Consequently two separate halo rings are discovered by a halo test as shown in fig. 6.10(b). The smaller ring is produced by zonal rays with negative aberration, the larger one by marginal rays with positive aberration.

§6.5. Spherical aberration: discussion of results and possibility of correction

Spherical aberration of electrostatic lenses. A wide range of results is available on the aberration of symmetrical two-tube lenses. Values of the coefficient C_s of (6.6) measured in units of a tube radius R are plotted in fig. 6.11 against the voltage ratio V'/V. The curves are derived from measurements by Klemperer (1953) such as

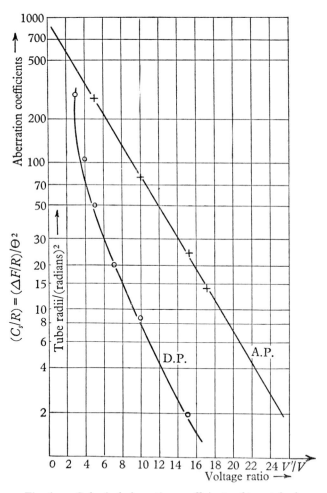

Fig. 6.11. Spherical aberration coefficients of two-tube lens.

recorded by fig. 6.5 in §6.3. A.P. and D.P. refer to parallel rays accelerated and decelerated respectively. These curves refer to the longitudinal spherical aberration ΔF for relatively small angles Θ. For large Θ the higher coefficient a_4 in (6.5) has to be taken into account. Measurements on the symmetrical two-tube lens lead to $a_4 = -a_2/f_0$, where f_0 is the paraxial focal length, since (cf. (6.7)) the longitudinal aberration of this lens is found to increase fairly accurately with the square of the aperture y.

The spherical aberration of asymmetrical two-tube lenses has been investigated by Gundert (1939). He measured the coefficient C_s in units of the radius R' of the large tube as a function of the ratio of the tube radii R'/R. As a result C_s/R' is found to pass through a distinct minimum at $R' = R$. A reduction of the radius of the high-voltage electrode has a smaller adverse effect than the reduction of the radius of the low-voltage electrode. Results of digital computation by Firestein and Vine (1963) show that the spherical aberration of all two-tube lenses decreases monotonically as the voltage ratio is increased. Unlike focal length versus voltage ratio curves, which become almost flat at about $V'/V > 20$, the aberration curves continue to drop quite steeply at high voltage ratio. Tube lenses with diameter ratios $\frac{1}{5}$, $\frac{1}{2}$, 1 and 5 have been investigated by computation. Typical values for the symmetrical two-tubes lens are given as $C_s/f_1 = 30$, 3 and 0·5 for $V'/V = 4$, 10 and 50 respectively.

The spherical aberration of a tube, followed by a diaphragm (cf. fig. 4.3(g) of §4.3) has been investigated by Gobrecht (1961). Taking the aperture in the diaphragm equal to the tube diameter, the aberration appeared to be about 50 per cent smaller than that of a symmetrical two-tube lens for accelerated beams and about 10 per cent smaller for decelerated beams. A conclusion that the tube-diaphragm lens should be generally superior to a two-tube lens is, however, not justified, since, in order to obtain adequate electrostatic shielding, the diaphragm containing the aperture would need to have a diameter substantially larger than that of the tube. The advantage of the tube-diaphragm lens would thus to some extent be lost if in the available space a larger two-tube lens could be fitted.

Spherical aberration coefficients of some symmetrical two-diaphragm lenses of the type shown in fig. 4.3(e) have been computed by Read (1969a, b). For given voltage ratio, C_s is found to decrease with increasing distance between the diaphragms and with decreasing object distance.

Spherical aberration of large, weak electrical saddle-field lenses has been investigated by Gobrecht (1941) and by Klemperer (1942). Initially accelerating saddle fields have very much less spherical aberration than the initially decelerating ones. However, the latter type is used nearly exclusively for practical reasons. The aberration largely depends upon size and shape of the electrodes. The following results of Gobrecht referring to three-tube einzel lenses are of practical interest:

(1) When the ratio of radii (R_i/R_o) of inner and outer tubes is varied, the aberration measured in units of the larger radius is found to reach a minimum for equal tube radii ($R_i = R_o$).

(2) For given object and image position, and for a given total length of the lens, the aberration coefficient decreases with increasing tube radius. The rate of decrease in aberration with increasing tube radius is rapid for short lenses, but it is rather small for total lengths of the lens tubes greater than one tube radius.

(3) Aberration decreases with increasing total length of the lens. It depends, however, not only upon the total length but also upon the ratio of the lengths of inner and outer tubes separately. Minimum aberration is obtained when each of the outer electrodes takes about 30 per cent of the total length.

(4) The aberration increases with increasing focal length of the lens, but it decreases with increasing image distance.

The dimensionless spherical aberration coefficient of electrostatic univoltage lenses is represented approximately, according to Liebmann (1949a), over a wide range of modifications of lens design or lens potentials by the following empirical formula:

$$C_s/f = K(f/R_i)^2, \qquad (6.20)$$

where f/R_i is the focal length expressed in terms of the radius of the intermediate tube. K is a constant which depends upon the shape of the electrodes; K is of the order of 2·5 for plain-tube lenses, but it is increased, to about 5, if diaphragms are introduced in the tube electrodes.

It will be pointed out in §8.4 that many initially decelerating three-tube saddle-field lenses suffer from disturbances by space charge and should therefore be used for focusing rays of relatively small intensity only. Three-diaphragm saddle-field lenses are comparatively free from space-charge errors.

TABLE 6.4. *Spherical aberration of einzel lenses with equal aperture radii R*

	$S = R$			$S = 2R$		
$\dfrac{V_i - V_o}{V_o}$	$\dfrac{C_s(f)}{R}$	$\dfrac{C_s(\infty)}{R}$	$\dfrac{f}{R}$	$\dfrac{C_s(f)}{R}$	$\dfrac{C_s(\infty)}{R}$	$\dfrac{f}{R}$
29	3·0	0·24	3·2	106	0·68	6·8
9	9·0	0·44	3·4	8·2	0·28	3·8
3	280	0·96	8·4	78	0·40	7·2
−0·6	118×10^3	2·86	44	96×10^2	1·42	24
−1·0	27×10^2	3·56	11·4	147	2·16	5·1
−2	9·6	32	0·84	—	—	—

Kanaya *et al.* (1966) calculated the aberration of three diaphragm einzel lenses represented by Glaser's bell-shaped field distribution (4.11) in §4.6. The results of these calculations are plotted in fig. 6.12 where the spherical aberration constants C_s divided by the semi-half width w of the potential distribution curve, or respectively divided by the focal length f are plotted against $\overline{V_i}/V_o$. The ratio of the potential $\overline{V_i}$ of the saddle point to the accelerating voltage V_o is obtained from fig. 4.7 in §4.6 as a function of S/R, i.e. the ratio diaphragm spacing S to aperture radius R. (The curves f/w and C_c/w in fig. 6.12 are discussed in §4.6 and §7.1 respectively.) Spherical aberration coefficients for einzel lenses with equal apertures in plane, thin diaphragms ($T \ll R$ in fig. 4.6) have been computed by Read (1969*b*) using a computed potential distribution. Table 6.4 shows a selection of $C_s(f)$ values for an object point near the focus and $C_s(\infty)$ values for parallel incident rays. These values are shown as a function of the voltages V_i of the intermediate diaphragm, with respect to V_o the voltage of the outer diaphragms which corresponds to the accelerating voltage. C_s is measured in units of the aperture radius R. Corresponding values for the focal length f are shown for comparison.

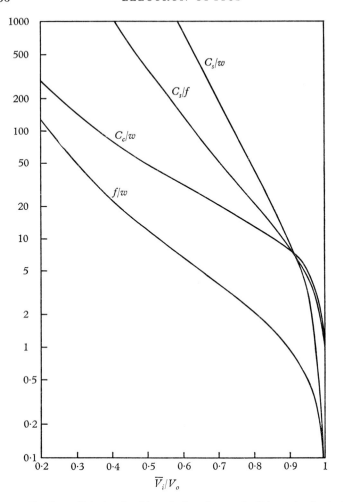

Fig. 6.12. Relative focal length f/w, chromatic C_c/w and spherical
aberration C_s/w, C_s/f as a function of $\overline{V}_i/V_o$: see fig. 4.6.

The small, strong three-diaphragm einzel lenses are of great
importance for electrostatic electron microscopes and have thus
been investigated in detail. Chancon (1947) showed from field-
plotting and ray-tracing studies that at high voltage ratios the
dimensionless aberration coefficient C_s/f, (6.11), hardly depends
upon diaphragm spacings, aperture diameters and electrode shape

or thickness. A reduction of the focal length, however, nearly always leads to a reduction of C_s/f. This focal length may, according to §4.6, be decreased by scaling down the geometrical dimensions or by increasing the applied voltages. The shortest obtainable focal length is eventually reached when sparking limits the field that can be maintained between inner and outer diaphragms. Up to about 200 kV cm^{-1} can practically be applied to the lens electrodes. Surfaces which stand highest fields lead to einzel lenses of shortest focal length, and hence to a reduction of the aberration of the electrostatic-microscope objective. Practical lenses have therefore electrodes with highly polished and well-rounded surfaces. Their general design for avoiding disruptive discharges has been explained in §4.6.

Einzel lenses with focal lengths of a few millimetres have been found to have aberration coefficients of the order of $C_s/f = 10$. Heise (1949) found an optimum voltage ratio yielding a minimum aberration coefficient C_s/f for which simultaneously a minimum focal length was experienced. This minimum was found to be a function of the geometrical lens parameters; it occurs for all given spacings S (see fig. 4.6 of §4.6) when T_i, the thickness of the central electrode, is of the order 0·8 to 1·6 aperture radii R. A useful approximate formula for thin einzel lenses, namely

$$C_s/f = 3 \cdot 5 \left(\frac{f}{S + T/2} \right)^2, \tag{6.21}$$

has been given by Haine and Cosslett (1961).

Spherical aberration of magnetic lenses. The iron shielded lens, specified by fig. 5.5 which is used for a television pick-up tube has an aberration coefficient $C_s/f = 2 \cdot 4$ for object and image distance shown in the ray tracing diagram of fig. 5.19. The reader can verify this from the trajectories shown in the figure. Large weak lenses formed by iron-free coils of a few layers of wire have been investigated by Becker and Wallraff (1938, 1940b). They found the aberration coefficient to decrease with increasing length ($2L$) of the coil according to the empirical relation

$$C_s/f \propto L^{-0 \cdot 4}. \tag{6.22}$$

The length-to-diameter ratio (L/R) of the coils was varied in Becker and Wallfraff's experiments between 0·1 and 5. The focal length f was kept between R and $2R$, R being the mean coil radius. For instance, $C_s/f = 3·0$ is found for an air coil with $L/R = 0·13$ and $f/R = 1·6$. This applies to initially parallel beams.

Relatively much larger C_s values are obtained for diverging beams. Graham and Klemperer (1952) measured aberration coefficients of monolayered coils for symmetrical object and image positions and presented the C_s-values in terms of the object to image distance z_f as a function of the coil radius R and of the coil length $2L$. They found for instance $C_s = 9z_f$ for $R = 0·1z_f$ and $2L = 0·1z_f$. The C_s/z_f values decrease rapidly with increasing L/z_f and with decreasing z_f/R. For $L \gg R$, the field approaches homogeneity. In the homogeneous field the longitudinal aberration depends on the cosine of the initial semi-aperture Θ of the ray, since according to §5.1, the distance between object and image point is given by
$$z_\Theta = z_f \cos \Theta.$$

Hence
$$\Delta J = z_f - z_\Theta = z_f(1 - \cos \Theta) = z_f \left[1 - \left(1 - \frac{\Theta^2}{2!} + \frac{\Theta^4}{4!} \cdots \right) \right]$$
$$\approx z_f \frac{\Theta^2}{2}. \tag{6.23}$$

With (6.11), the aberration coefficient is found to be
$$C_s = z_f/2. \tag{6.24}$$

The spherical aberration of an iron-shielded coil is appreciably larger than that of the corresponding air coil. Becker and Wallraff (1938) investigated this aberration as a function of the width of the gap in the shield. Their curves for a coil of inner radius $R = 4$ cm and length $2L = 6·5$ cm are reproduced in fig. 6.13. The dimensions of coil and iron shield are large enough to avoid the saturation region of the iron at the used number of ampere-turns. The aberration is least for the smallest gap width; however, the current to obtain the given focal length is excessively large at the smallest gap widths (cf. §5.4). The aberration reaches a maximum at about 1 cm gap width and then continuously decreases with larger gap widths, while the current to produce the given focal length reaches a minimum at about 4 cm gap width.

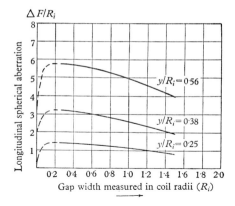

Fig. 6.13. Variation of spherical aberration with width of gap of iron shielded coil for three different semi-apertures y of rays.

Iron-shielded lenses of shortest possible focal length are of great interest for use in electron microscopes. Their aberration has therefore been studied in great detail. The coefficient C_s/f has been found to decrease with increasing lens power. The lens power is increased by increasing the field strength and reducing the lens dimensions. The limits set to both these procedures by the saturation properties of the iron have been discussed in §5.4. Aberration coefficients of magnetic-microscope objectives have been measured by Ruska (1934), Ardenne (1939), Dosse (1941), Liebmann (1946, 1955) and by Durandeau and Fert (1957).

The decrease in aberration with focal length is borne out by the fact that C_s/f values of magnetic-microscope objectives are several times smaller than the corresponding coefficients of large, weak magnetic lenses. Haine and Cosslett (1961) proposed two simple formulae for an approximate calculation of the aberration coefficient as a function of the focal length f, of the pole piece diameter D and the gap width S, namely

$$\frac{C_s}{f} = 5 \cdot 0 \frac{f^2}{(S+D)^2}, \qquad (6.25)$$

or as a function of the excitation coefficient $nI/\sqrt{V_r}$ (cf. §5.4) namely

$$\frac{C_s}{f} = 3130 \frac{V_r^2}{(nI)^4}. \qquad (6.26)$$

The smallest obtainable aberration coefficients are found to be about

$$C_s/f \approx 0.6.$$

These will occur at the minimum value of $f_1 \approx 0.7$ mm (see fig. 5.11) for $S = D = 0.3$ cm and $nI/\sqrt{V_r} \approx 14$. For 100 kV electrons an excitation of 4 500 ampere-turns is then required.

According to Ruska (1965) an optimized conventional magnetic objective should achieve a spherical aberration coefficient $C_s = 0.55$ mm at 100 keV. This value of C_s seems to be a practical limit which arises from the magnetic properties of pole pieces. Mulvey and Wallington (1969) obtained by computation values of minimum possible aberration. They found that $C_s B_p/V_r^{\frac{1}{2}}$ ($B_p =$ maximum flux density in the gap) decreases from 16 to 4·8 when S/D decreases from 10 to 0·2. Only the initial decrease of C_s, down to $S/D \approx 1$ is steep, but the decrease of C_s is small for a further reduction of S/D. At $S/D \approx 1$ the minimum aberration occurs at an excitation coefficient $nI/V_r^{\frac{1}{2}} \approx 25$. The small reduction of $C_s B_p/V_r^{\frac{1}{2}}$ is achieved at the cost of a substantial increase in lens excitation. It is, however, not always advisable to use lenses with the absolute minimum of spherical aberration; Mulvey and Wallington (1969) have shown how to obtain optimum performance by compromising with the effects of other aberrations which do not reach a minimum at the same excitation. We shall come back to this point in §7.7.

Special requirements for small C_s-values are indicated for lenses used to form electron probes for scanning microscopes, for X-ray micro-analysers, etc. Accurate data for these lenses cannot be taken from the literature on short focus microscope lenses because of the rapid increase of C_s with decreasing excitation parameter. This follows from the empirical relation (6.26). Moreover, the angle of illumination in electron microscopes is 10^{-3} to 10^{-4} radians while 10^{-2} is needed for electron probe analysers. Barnes and Openshaw (1968) found by experiment and by computation, that at a given S/D ratio, C_s is greater for large bore D, large gap S and for high nI rather than for small bore and gap and for lower values of nI. C_s depends on the working distance L between specimen and top of probe forming lens. A few significant values are shown in table 6.5.

TABLE 6.5. *Spherical aberration of probe forming lenses with gap S and bores D_1 and D_2 of the two pole pieces*

C_s (cm)	5	10	50	S/D_2	D_1/D_2
L (cm)	$\begin{cases} 1 \\ 1\cdot5 \end{cases}$	$\begin{matrix} 1\cdot5 \\ 2 \end{matrix}$	$\begin{matrix} 4 \\ 3 \end{matrix}$	$\begin{matrix} 1\cdot0 \\ 0\cdot25 \end{matrix}$	$\begin{matrix} 0\cdot2 \\ 0\cdot2 \end{matrix}$

Discussion. The large positive spherical aberration of nearly all electron lenses is not unexpected. In electrostatic lenses with electrodes at the periphery, the marginal field is generally so much stronger than the field near the axis that the marginal rays are over-focused as compared with the paraxial rays. Magnetic lenses usually suffer from the cosine effect (6.23) which has been shown to shorten the focal length of marginal rays in homogeneous fields. Quite generally, the longitudinal component of velocity of homo-centric rays will always be less for the marginal rays than for the paraxial rays.

Searching for the type of lenses with smallest aberration Ramberg (1942) investigated numerically a few electrostatic and magnetic lenses for which approximate analytical presentations were available. Of these, accelerating and decelerating symmetrical two-tube lenses showed for a given focal length, the smallest and the largest aberration respectively. Symmetrical three-diaphragm lenses and a magnetic lens with cylindrical pole pieces ranged between these extreme cases. The aberration of all these lenses appeared to decrease with increasing lens power as given approximately by

$$C_s/f \propto f^{1\cdot7}. \tag{6.27}$$

The conclusion that least spherical aberration will occur with each lens structure at about its shortest focal length is quite generally justified by experience. Moreover, improvements can be obtained by changing the geometry of the fields.

Electrostatic lenses. Best results are obtained with three-diaphragm lenses when the apertures are smaller than their mutual distances and when the radial extension of the diaphragms is relatively large. The aberration of three-diaphragm einzel lenses with equal apertures depends, according to digital computations by Vine (1960), upon the thickness of the central diaphragm. It is relatively large for very thin diaphragms, e.g. for a spacing of one aperture radius ($S \approx R$), $C_s/f \approx 300$ but it comes down at this spacing to $C_s/f \approx 6$ when the thickness of the central diaphragm is increased to $T \approx 1 \cdot 5R$. For still greater thickness the C_s/f seems to increase again gradually.

Attempts to overcome aberration by application of axial electrodes, in order to increase the paraxial field, are not new. Such 'co-axial' electrostatic lenses have been described first by McGregor-Morris and Mines (1925) and by Nicoll (1938). An electrostatic co-axial lens may consist, for instance, of a straight cylindrical wire in the axis surrounded by some annular electrodes, say, for instance, by two tubes. Potential is applied to the wire in order to provide increased acceleration of the paraxial electrons towards the axis and thus to decrease the paraxial focal length until it equals the zonal or even marginal focal length. In the co-axial magnetic lens, a straight conductor is mounted in the axis, and current is passed through this conductor by flat strips connected radially to both ends of the conductor. When the electron flow through the conductor is in the same direction as that in the electron beam, a converging co-axial force is produced upon the electrons which is inversely proportional to their distance from the axis of the conductor. Evidently it is not possible to utilize a complete annular aperture in these lenses, since the wire must be supported, and the support will interfere with the beam. Also the supports must be arranged in field-free zones so that the lens field is not disturbed in its circular symmetry. Gabor (1946) first suggested a practicable three-electrode arrangement for such co-axial lenses. Dungey and Hull (1947) have shown by extensive numerical calculations that in such an arrangement, having a focal length of 3 mm at 60 kV, zones of an angular range of $0 \cdot 09 \pm 0 \cdot 006$ radian, can be corrected to have an error of less than 10^{-7} radian. Though these lenses are in principle of much interest, they have found no practical application.

The possibility of correcting spherical aberration by shaping the equipotentials in electrostatic lenses by the introduction of wire meshes (cf. §4.8) or thin conducting films (cf. Gianola, 1950) have obviously an application only in such cases where the scattering of the electron rays by these correcting elements does not matter. The possibility of inserting a correcting electron mirror into the electron microscope has been studied by Ramberg (1949). He calculated the proper concave shape for the mirror electrodes, but found that for an effective correction of the usual microscope objectives the dimensions of the mirror would have to be unreasonably small (of the order of 0·1 mm). On the other hand, the correction of electron lenses of larger focal length by means of electron mirrors still appears a hopeful task, especially since some of these mirrors have been found experimentally to show large negative aberration.

Magnetic lenses. Glaser calculated the aberration of magnetic lenses as represented by the bell-shaped field distribution (5.26) of §5.3 as a function of the half-maximum width $2w$ of the distribution curve. Minimum aberration was obtained when the dimensionless lens parameter of (5.28) in §5.3 reached the value

$$K_H^2 = 2·8.$$

Liebmann (1952 *a*, *b*) arrived at practically the same result with the help of his resistance network (§3.3).

Now, the deflecting force for the electron acts at right angles to the axial component of the flux density B_z. We have represented B_z in §5.2 by a power series which may be written in detail as follows:

$$B_z(r, z) = B_z(0, z) - \frac{r^2}{4} \frac{d^2 B_z(0, z)}{dz^2} . \tag{5.14}$$

There, the second derivative of B_z is usually negative (cf. the bell-shaped distribution fig. 5.3) so that the second term of the power series (5.14) becomes positive. Hence, the marginal rays which pass through the region of larger r, in which B_z is increased, will suffer a stronger deflexion, so that positive aberration will result. Consequently—and this was first pointed out by Glaser (1941 *a*)—if we desire to reduce positive aberration, the axial distribution of flux density should not be, as in figs. 5.3 and 5.6, bell-shaped or concave towards the axis, but rather convex towards the axis as

shown in fig. 6.14. Siegbahn (1946)
succeeded in producing th ʒse particular
fields in a large iron-shielded lens
which is shown here diagrammatically
in fig. 6.15. There the co l C is placed
in an iron cylinder F of 6o cm length,
and has internal and external radii of
10 and 16 cm respectively. The iron
end-pieces E_1 and E_2 are 5 cm thick
having pole-pieces with apertures A_1
and A_2 of 1·4 cm radius. Object and
image are placed at these apertures

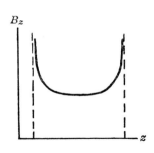

Fig. 6.14. Dip-field with no
spherical aberration.

where the flux density B_z is largest. Towards the middle, B_z
decreases to a few per cent of its value near the apertures.

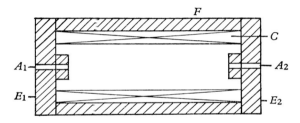

Fig. 6.15. Corrected lens with closed iron ends.

Reasoning along sirr ilar lines, Quade and Halliday (1948) have
used a segmented air coil. This consists essentially of two short
individual coils which are arranged sufficiently far away from the
mid-plane to produce a dip in flux density near the mid-plane.
However, for reducing spherical aberration the latter arrangement
is less effective than S egbahn's iron-shielded lens (fig. 6.15). The
field of the segmented air coils in this particular arrangement
decreased substantially before reaching object and image planes.
For best correction, object and image have to be placed in the
respective mid-planes of each coil. There, however, the coils have
to be spaced so far apart that a relatively very high power consumption is required for energizing the field. A powerful dip-field
arrangement of this kind has been described by Agnew and
Anderson (1949); it was made up of two coils of 74 cm diameter,

their centres spaced within 96 cm. These coils consumed 10 kW for focusing 2 MeV electrons. An application of the dip-field principle for reducing the aberration of microscope lenses has been discussed by Marton and Bol (1947), but the result promises little practical advantage.

Annular focusing correction of spherical aberration in magnetic lenses can be achieved to some extent by placing a small magnetic coil on the axis of symmetry at the centre of the field of a large lens (cf. Hubert (1951); Dolmatova and Kelman (1959)). Such a 'paraxial baffle coil' intercepts the paraxial rays, but it acts as a lens of negative focal length for the zonal and marginal rays. By suitably diverging these rays it makes them substantially homocentric.

Finally, we mention an interesting proposal by Hoppe (1963) for correcting the spherical aberration of a magnetic microscope objective by the introduction of a zone plate at its exit plane. Such a plate would, for instance, consist of annular stops which cut out certain zones at the image side of the wave surface from which elementary wavelets of opposite phase are emitted towards the image point. Thus the annular apertures in the phase plates are arranged in such a way that equal phases (multiples of 2π) only can reach every image point. This creates a situation equivalent to the spherical aberration free case in which all optical path lengths between axial conjugate points are equal and where, in consequence, there is zero phase difference between elementary wavelets arriving at the image point. Lenz (1964) has given formulae for the calculation of the zone radii. Langer and Hoppe (1967) have shown by numerical calculation for the focusing by objectives with zone plates that aberration could be corrected completely. We shall discuss in §7.6 the importance of zone plates for influencing contrast in electron microscope pictures of highest possible resolution.

A very hopeful way to correct spherical aberration was indicated by Scherzer (1947, 1949); it relies on the introduction of line focus elements such as quadrupole lenses and stigmators. We shall deal with it in §6.11.

§6.6. Field aberrations: coma

If points lying off the axis are reproduced by rays of large semi-aperture, a lens error called coma causes an asymmetrical deformation of the individual image points to a comet-like appearance. The coma is usually determined as the distance of the focus of the marginal pair of rays from the principal ray. To illustrate this, in fig. 6.16, parallel rays are shown crossing the axis at an angle α.

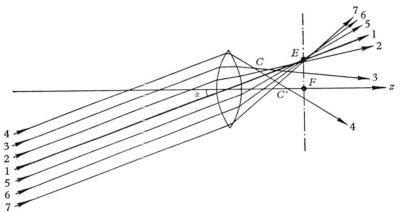

Fig. 6.16. Formation of comatic image.

If the lens were perfect, all the beams would be focused at the point E which lies in the plane intersecting the axis z at the focus F. If the lens suffers from coma only, the rays (5), (6), (7) with small angle of incidence to the equipotential (lens surface) will be focused close to point E on the principal ray (1) but rays (2), (3), (4) with larger angle of incidence will be focused at different points, e.g. C, C', the zonal variation CE being a function of the semi-aperture of the beam. In distinction to the coma, spherical aberration would shift the marginal focus of rays of the same aperture, say, rays (7) and (4) to a common intersection situated on the central beam (1) at some distance from E. If both coma and spherical aberration are present, the intersection of the marginal beams can happen anywhere off the central beam and off the focal plane. Thus in a picture a general lack of definition would result, and the blurring would increase with increasing distance from the axis. There are several possible ways of specifying the coma of a lens, and one which we shall discuss here has been used for electron lenses.

The simplest condition for the absence of coma is due to Abbe. For large apertures, Lagrange's equation (1.34) has to be replaced by Abbe's sine law (1.32), according to which the lateral magnification y'/y should be a constant for all zones if $\sin \Theta'/\sin \Theta$ is a constant. For an infinitely distant object $\sin \Theta'$ is proportional to the height y of the incident rays, so that $y/\sin \Theta$ should be constant for incident parallel rays and should be identical with the focal length PF of the system

$$\frac{y}{\sin \Theta} = PF. \tag{6.28}$$

In a Gaussian system, each of a bundle of parallel incident rays (if continued) intersects the corresponding emergent ray in a plane. This is commonly known as the principal plane. According to

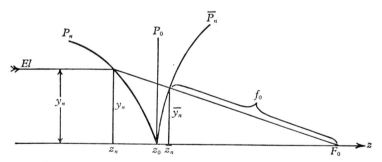

Fig. 6.17. Coma and the curvature of the principal surface.

(6.28), however, this is no longer true for large apertures. For a lens which is free from coma and spherical aberration, these intersections should take place on a spherical surface, the radius of curvature of this 'principal surface' being the focal length. If no other aberrations were present, the deviation of the principal surface from the spherical shape would reveal coma. In the case of a lens suffering also from spherical aberration the coma could be found by subtracting the spherical aberration from the paraxial focal length along the various rays. This is shown in fig. 6.17.

There, $\bar{P}_n$, a spherical surface of a radius f_0 about the centre F_0, represents the principal surface of a coma-free lens with the focal length f_0 and with the focus F_0. On the other hand, P_0 would represent the principal plane according to the first-order theory, while P_n may be the actual principal surface of a lens which suffers from coma. All three surfaces (P_n, P_0 and $\bar{P}_n$) cut the lens axis (z) at z_0.

The electron ray El which is initially parallel to the z-axis is, in the absence of spherical aberration, focused through the paraxial focus F_0. The intersection of the direction of the ray at the focus F_0 and of its initial direction leads to a point with the coordinates (y_n, z_n) on the principal surface P_n. On the other hand, the direction of the ray at the focus F_0 cuts the ideal principal surface at a distance f_0 from F_0 at a point with the coordinates ($\bar{y}_n$, $\bar{z}_n$).

Now, the coma may be expressed either by

$$\Delta y_n = y_n - \bar{y}_n, \tag{6.29}$$

or by

$$\Delta z_n = z_n - \bar{z}_n. \tag{6.30}$$

For the ideal principal surface $\bar{P}_n$ one would obtain

$$z_0 - \bar{z}_n \approx \bar{y}_n^2 / 2f_0. \tag{6.31}$$

(6.31) follows from geometry, since $z_0 - \bar{z}_n$ represents the deviation of a circle of radius f_0 from its tangent. For the representation of the actual principal curve there is frequently used an expression similar to (6.31), namely,

$$z_n - z_0 = \frac{y_n^2}{f_0} \left(\frac{C_T}{f_0} - \frac{1}{2} \right), \tag{6.32}$$

where C_T is called the coma coefficient which is proportional to the coma and vanishes for coma-free lenses.

If a lens suffers from spherical aberration, the focus of a ray of semi-aperture y_n is shifted by ΔF_n (cf. §6.3). For a lens with spherical aberration and coma, the change in focal length Δf_n is composed of the shift in focus (ΔF_n) and of the deviation of the principal surface from its ideal shape ($-\Delta z_n$), hence

$$\Delta f_n = f_n - f_0 \approx \Delta F_n - \Delta z_n - \Delta \bar{z}_n, \tag{6.33}$$

but since, according to (6.6),

$$\Delta F_n = C_s(y_n/f_0)^2,$$

(6.32) with (6.33) yields

$$\Delta f_n = \frac{y_n^2}{f_0^2}(C_s - C_T). \qquad (6.34)$$

This zonal difference in focal lengths causes a zonal difference in magnification which owing to (6.34) is proportional to $(C_s - C_T)$.

There are few experimental results on foci and principal surfaces which are sufficiently accurate for a determination of Δf_n. Heise (1949) published curves for principal surfaces and focal surfaces (location of F_n against y_n) of some microscope-objective einzel lenses as a function of the voltage ratio. The results represented by these curves appear to be suitable for precise evaluation which, however, has not been given by this author. Apparently, C_T is always smaller than C_s, but both coefficients are of the same order of magnitude. Liebmann (1949a) gave some experimental results on the coma of weak electrostatic univoltage lenses, the coefficients are obtained with ± 20 per cent accuracy, and for the various lenses they range between $2 < C_T/f_0 < 14$. The coma of magnetic microscope lenses has been determined from field plotting and ray tracing investigations by Liebmann and Grad (1951). It was found to increase with increasing ratio S/R of pole-piece spacing over bore radius (cf. §5.4). Compared with the spherical aberration, its influence becomes more pronounced with increasing lens excitation.

§6.7. Field curvature and astigmatism

A most serious lens error in electron optics is the curvature of field. It causes the image of a plane object to be formed upon a curved surface instead of the usually desired flat image plane. When astigmatism is also present, there will exist two curved image surfaces, corresponding to the sagittal and the tangential or meridional foci. This may be illustrated in the following introductory manner. Take as object a wheel with its axle lying on the axis of the lens system. The rim of the wheel will be projected sharply on to the tangential image surface, whilst the spokes of the wheel will

appear to be focused sharply in the sagittal image surface. More
exact explanations may be given by means of fig. 6.18, where z is
the lens axis and (1) is the central ray of a bundle of rays crossing
the axis at O. The two tangential beams (2) and (4) and the two
sagittal beams (3) and (5) are also shown. These pairs of beams
indicate that the whole bundle of rays has two longitudinally

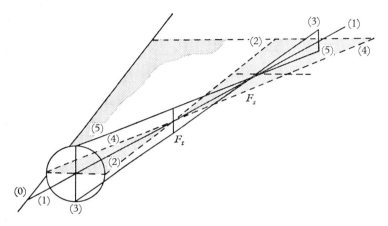

Fig. 6.18. Astigmatism of a bundle of rays.

separated constrictions, the tangential or primary focus F_t and the
sagittal or secondary focus F_s. In the absence of coma the two
constrictions become sharp focal lines. Nowhere along the central
ray exists a point focus; the nearest approach to it is about midway
between the two astigmatic foci, in a disc of least confusion.

The four image surfaces crossing the axis at the Gaussian image
point are shown in fig. 6.19. G is a section through the Gaussian
image plane in which the paraxial rays are focuses. T and S
represent the tangential and the sagittal image surface respectively
and D is the surface of least confusion. There is also shown one
central ray OO (which intersects the z-axis) and a bundle of rays
1, 2, 3 and 4 passing through apertures of a pepperpot diaphragm
on a circle about the axis. In all actual electron lenses the image
surfaces are strongly curved; they are all concave towards the lens.
The astigmatism is caused by bundles of rays in the tangential and
sagittal planes meeting equipotentials of different curvature.

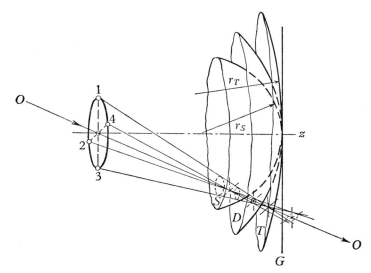

Fig. 6.19. Field curvature and astigmatism:
tangential and sagittal image surfaces.

Some measure of the astigmatism is given by the radii of curvature r_T and r_S of the tangential and of the sagittal surface near the axis. In experimental investigations, for instance, let the image of a flat cathode serving as an object be projected by a lens on to a movable target. If the cathode contains a pattern of concentric circles and of radial lines, the target positions in which sharp images of circular zones of the cathode appear can be measured as a function of the mean zone radii. In this way, the tangential image surface is located by observation of the circular pattern, and the sagittal surface is located by observation of the radial pattern.

More accurate measurements are obtained if a point object is moved across a flat object plane and its images are located point by point on a movable target. The images of the point object will appear as two short lines projected into mutually perpendicular (tangential and radial) directions with an intermediate disc of least confusion. If the astigmatism of a lens is measured in this way by observing the images of a point object off the axis, the divergent rays emerging from the point object are usually confined within a given aperture. For example, as seen in fig. 6.19, the rays may emerge through pepperpot holes arranged in concentric circles so that co-axial cones of pencils are formed just as in the measurement of spherical aberration. Now, however, the cones are not co-axial with the lens. The axis of the cones, i.e. the principal ray, intersects the lens axis in the lens centre (thick lenses in a nodal point). Usually, the deviations Δz_S and Δz_T of the tangential and of the sagittal surfaces respectively from the Gaussian plane are plotted against the slope angle of the principal ray. Such measurements can be taken for cones of different apertures of rays from the object point.

Electrostatic lenses. Morton and Ramberg (1936) applied the method of location of the image of a large, flat cathode for the investigation of a symmetrical two-tube lens. They used a photo-cathode on to which a light-optical image was projected. The electron-optical pictures, however, were so strongly curved that only a minute axial part of them could be obtained on a flat screen. The axial radii of curvature for the tangential and for the sagittal image surfaces were found to be $r_T = R/23$ and $r_S = R/13$ respectively, where R is the radius of the tubular lens electrodes.

The image curvature of a symmetrical two-tube lens has also been investigated by Gobrecht (1958a). He traced the image position, size and shape with a sliding fluorescent target, the 'point object' being a small movable aperture in a diaphragm, the rear of which was exposed to an electron beam. It was found that the tangential and the sagittal images respectively were small, mutually perpendicular lines, the length of which was equal in both image surfaces and increased linearly with increasing lens aperture; it also increased with increasing focal length (decreasing voltage ratio) and with increasing object distance. The astigmatism was found to be greater for the decelerating than for the accelerating lenses of the same focal length.

The lengths of the image lines for given positions of a point object is of particular interest. In the case of accelerating lenses numerical values could be derived from these for the 'Seidel coefficient S_4' (cf. §6.1) which is a measure for the magnitude of the field curvature. S_4 was found to decrease from 4×10^{-3} to 10^{-3} mm^{-2} when the focal length increased from 13 to 17 mm.

Some measurements by Johannson (1933, 1934) on the large field curvature of two-diaphragm objectives will be discussed in connexion with the field curvature of emission systems (§9.4).

Exact measurements on three-tube einzel lenses have been taken by Gobrecht (1942) with a point source off the axis. The tangential image surface was found to have stronger curvature than the sagittal. The magnitude of the field curvature can be judged from the fact that the image deviation Δz from the Gaussian plane for a given slope angle of the principal ray is about 5 to 7 times greater for electrostatic einzel lenses than for uncorrected spherical glass lenses of similar focal length in light optics. The image deviation decreases with the focal length of the lens; moreover, it becomes smaller the longer the inner lens tube, and the larger the tube radius R. For instance, for an inclination of $7°$ of the principal ray, for a focal length of $f = 2 \cdot 6R$ and with both outer lens tubes having a length $L_o = 0 \cdot 9R$, deviations of the tangential image surfaces $\Delta z_T = 1R$ and $2 \cdot 2R$ have been observed for inner lens tubes of lengths $L_i = 2 \cdot 1R$ and $0 \cdot 9R$ respectively. The image distance in the quoted example was of the order of $4R$. Generally, the image deviations slightly decrease with increasing image distance. The image curvature does not depend on the aperture of the cone of rays from the object point.

Magnetic lenses. Becker and Wallraff (1939, 1940a, c) investigated the astigmatism of various magnetic lenses with a point source off the axis. Again, the tangential image surface had the greater curvature in all cases. And quite generally, the longer the lens, the smaller the curvature of the image field. For air coils this curvature was found to increase greatly with the focal length ($\propto f^2$ or f^3). Compared with the electrostatic tube lenses the field curvature of the air coils is small. Becker and Wallraff quote, for instance, for the radius of curvature $r_D = R/3 \cdot 6$ for the mean image surface of an air coil of inner radius $R = 5 \cdot 6$ cm of semi-length $L = 0 \cdot 18R$ and focal length $f = 1 \cdot 5R$.

Iron-shielded coils show, according to Becker and Wallraff (1940c), much greater field curvature than air coils. The field curvature depends upon the width of the gap in the iron shield. It is smallest for very narrow and very wide gaps and reaches a maximum for approximately the same gap widths as those for which maximum spherical aberration is observed (cf. fig. 6.13).

The problem of correcting field curvature can be attacked with the help of the so-called Petzval theorem which was introduced in glass optics about a hundred years ago. If the system consists of a number of media, of refractive indices N_0, N_1, N_2, ..., having spherical faces whose radii of curvature are r_0, r_1, r_2, ..., the Petzval curvature of the image is

$$\frac{1}{r_P} = -\sum_j \frac{1}{r_j} \left(\frac{1}{N_{j-1}} - \frac{1}{N_j} \right). \qquad (6.35)$$

The Petzval surface is in general slightly less curved than either the sagittal or the tangential surface. If these two surfaces coincide, they also coincide with the Petzval surface, and the field will then be flat, if the Petzval curvature vanishes. For electron-optical purposes the Petzval theorem has been formulated in a convenient form by Klemperer and Wright (1939). The field of an electrostatic lens may be subdivided as indicated in §3.4 in such a way that the ratio of the square roots of subsequent equipotentials $\sqrt{(V_j/V_j')}$ is a constant (see (3.20)). If V_n' is the voltage of the last equipotential and r_j is the radius of curvature of the jth equipotential, the radius of curvature r_P of the image surface is given by

$$\frac{1}{r_P} = \sqrt{V_n'} \left(1 - \sqrt{\frac{V_j}{V_j'}} \right) \sum \frac{1}{(V_j)^{\frac{1}{2}} r_j}. \qquad (6.36)$$

Take both an image field and an equipotential concave towards the cathode. Then, the value of $1/r_P$ and that of $1/(V_j)^{\frac{1}{2}} r_j$ are assumed to have the same sign if an electron passing the equipotential surface is decelerated.*

The extreme difficulty of trying to flatten the image field may be realized from a consideration of (6.36). If, for instance, a two-tube lens is used to project the electrons from a flat cathode on to a flat screen, the electrons are accelerated in passing the equipotentials of the lens, and these are convex towards the cathode at the low-potential side of the mid-plane (see fig. 3.5), i.e. so long as V_j is

---

* In glass optics, a field curvature concave towards the object is conventionally defined to be positive. Assuming the object in air, a glass surface, however, concave towards this object, has, according to the usual conventions, a negative sign of curvature.

small. It therefore follows that the first semi-lens will produce a large part of the sum $\Sigma(V_j)^{-\frac{1}{2}}r_j^{-1}$. However, when the curvature, of the equipotential has changed sign (in the second semi-lens), the electrons have already gained so much velocity that there are only a few equipotentials with opposite sign contributing to the sum in (6.36), and V_j is so large that the terms $1/(V_j)^{\frac{1}{2}}r_j$ are small. Hence $1/r_P$ will be a large sum giving r_P as a small radius corresponding to a steeply rounded field concave towards the cathode.

Some theoretical discussion of the Petzval curvature of electron lenses can be found in papers by Glaser (1935) and by Goddard (1946b). A useful expression for the Petzval curvature of purely magnetic lenses is derived in the latter paper:

$$\frac{1}{r_P} = -\frac{e}{4m}V^{\frac{1}{2}}\int_{-\infty}^{+\infty}B_z^2\,dz, \qquad (6.37)$$

where e/m is the charge to mass ratio, V the constant electric potential (electron volts) in the lens and B_z the axial magnetic flux density along the axis. Since the integrand is always positive, it follows that $1/r_P > 0$. Thus the Petzval field curvature of a purely magnetic lens cannot be eliminated. Combining (6.37) with Busch's lens formula (5.21), the field curvature of a thin lens is found to be

$$\frac{1}{r_P} = \frac{2}{f\sqrt{V}}, \qquad (6.38)$$

i.e. the radius of curvature of the image surface is proportional to the focal length of the lens and to the electron velocity.

It is of much practical interest that image curvature can be corrected in some planar systems (cf. §2.5, §10.2) with the help of fine meshes. Lubszynski et al. (1969) have successfully corrected the image curvature in a 'vidicon' television pick-up tube by introducing a plane mesh immediately after the scanning deflexion plates. If these plates are at potentials V_0+V_p and V_0-V_p respectively, the mesh at a potential V_0 causes a strong bending of the diverging equipotentials of the fringe fields into a direction parallel to the mesh. By variation of the distance between mesh and deflector plates, the image curvature can be controlled and even over-compensated. It is remarkable that the quality of the image hardly suffers by the introduction of the mesh.

§6.8. Distortion

This error is again of major importance in electron optics. It is somewhat different from the four isotropic errors in so far as it has no effect upon the sharpness of the image; it only detracts from its faithfulness. Distortion merely displaces the image points towards or away from the optical axis. According to Seidel's third-order approximation (cf. §6.1) the amount of displacement $\Delta r'$ is calculated to be proportional to the cube of the distance r of the object point from the axis. In the paraxial part of the image, the magnification y_0'/y is fixed by Lagrange's law (1.34). Thus, the distortion is described by

$$\Delta r' = C_d \frac{y_0'}{y} r^3 = -S_5 m_0 r_0^3. \tag{6.39}$$

where the distortion constant C_d corresponds to the Seidel coefficient S_5 of (6.1) in §6.1 and $\Delta r'$ is the distance of the actual marginal image point from its ideal position. It can be seen from (6.1) that the term S_5 is independent of x_A and y_A the coordinates of the aperture of the pencil. Hence, the amount of distortion must be independent of the diameter of the aperture stop. In the experimental investigation of distortion, it should thus be useful to keep the beam aperture small in order to minimize disturbances by other aberrations present.

In practical distortion measurements it has been found conconvenient to compare the marginal magnification r'/r directly with the paraxial magnification y_0'/y, and the following dimensionless distortion factor is found in the experimental literature:

$$D_d = \frac{r'}{r} \frac{y}{y_0'} - 1. \tag{6.40}$$

This factor is generally measured as function of the slope angle of the principal rays. But since $r = y_0$ and $\Delta r' = r' - y'$, it would be expected that

$$D_d = C_d r^2. \tag{6.41}$$

If the distances of the actual image points from the axis are greater than the distances of the ideal image points, the outer parts of the image are on too large a scale, the factor D_d in (6.40) is positive and the distortion is called 'pin-cushion distortion'. This is explained by fig. 6.20(*b*). There, a broken line indicates the Gaussian image of a square-pattern, the solid line shows the stretched, actual image. If, on the other hand, the outer parts of the image are on too small a circle, the distortion factor D_d is negative and so-called 'barrel distortion' will occur. This is illustrated by fig. 6.20(*a*), the Gaussian image again being indicated by a broken line.

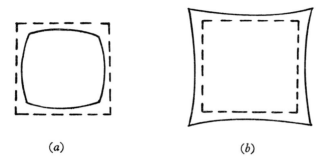

(*a*)   (*b*)

Fig. 6.20. Barrel- and pin-cushion-shaped distortion.

All field aberrations affecting an image point off the axis depend upon the part of the lens field that is utilized for image projection. For example, field aberrations can be changed by shifting an aperture stop along the axis. This effect is very pronounced in the case of distortion. In glass optics, distortion is known to be a function of the effective position of the aperture plane. This is the plane in which the principal rays of the imaging pencils intersect the axis. As shown in fig. 6.21(*a*), barrel-shaped distortion results if the aperture plane is in front of a thin lens. With this very same lens, pin-cushion distortion can be produced as shown in fig. 6.21(*b*) if the aperture plane is shifted behind the lens.

Electrostatic lenses. The distortion error of some two-tube lenses has been measured by Gobrecht (1958*b*). He used a fine aperture in the anode of an electron gun as an object point. Tracing the image with a sliding fluorescent target, he found that the distortion factor D_d increases with decreasing focal length of the lens. Table 6.5 which contains the result for two tubes, each 45 mm long and of 30 mm tube radius, shows the great change of the distortion

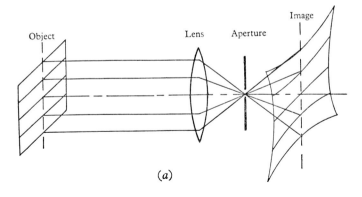

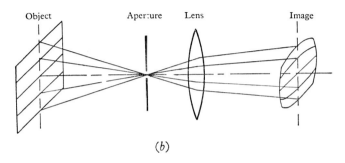

Fig. 6.21. Effect of aperture position on character of distortion.

factor (D_d in per cent) resulting from an alteration of focal length (which is controlled by the voltage ratio). By measuring the deviation $\Delta r'$ of the actual image points and their distance from the axis, the Seidel coefficient can be calculated. From (6.39) one obtains $S_5 = 7\cdot9 \times 10^{-4}$ mm^{-2}.

TABLE 6.6. *Distortion factor (D_d) and focal length (f) of a two-tube lens for pencils of different inclinations (α) towards the lens axis*

f (mm)	160	154	150	140	135
$100D_d \begin{cases} \alpha = 11\cdot3^\circ \\ \alpha = 9^\circ \end{cases}$	5 —	— 4	16 8	29 19	36 24$\cdot$5

Rang (1948) and Heise (1949) measured the distortion error of various einzel lenses as a function of the voltage ratio V_i/V_o in the neighbourhood of the focal-length minimum which has been discussed in §4.6. As long as f decreases with decreasing V_i/V_o, the distortion is found to be positive (pin-cushion), but for f increasing again with further decreasing V_i/V_o, the distortion is found to be negative. At the minimum focal length a distortion-free lens is obtained. It is of particular interest that the region of vanishing distortion coincides with the region of minimum spherical aberration which is also reached at minimum focal length. Of practical importance is the distortion-free univoltage lens (i.e. the einzel lens with $V_i = 0$). After the above experience this lens can be found by varying the radius R_i of the intermediate electrode until the focal-length minimum is reached just at the voltage ratio $V_i/V_o = 0$.

Distortion can be rather troublesome in microscope projector lenses, since these lenses have to focus rays of relatively large distances from the axis. Hence, much attention has been drawn to the distortion error of strong electrostatic einzel lenses. Jacob and Mulvey (1949) measured the distortion of strong symmetrical einzel lenses with three apertures of equal radius R as a function of the spacing S and of the thickness T_i of the intermediate aperture (cf. fig. 4.6). They found the parabolic law (6.41) to hold up to nearly $r = 0.5R$, and they found pin-cushion distortion mainly for small values of T_i/R but barrel distortion for large values of T_i/R. For $T_i/R = 1.5$, where with decreasing spacing S the distortion is found to change over from barrel to pin-cushion, a corrected lens is obtained at $S \approx 2R$.

An example for correcting the distortion of an electron lens has been described by Morton and Ramberg (1936). A two-tube lens was used in an image converter for projecting a picture of a flat photoelectric cathode on a flat fluorescent screen. The resulting strong pin-cushion distortion could be cured by curving the cathode surface concave towards the lens. In this way the principal rays from the cathode were slightly converged so that their intersection with the axis was shifted, in the direction of the cathode, closer to the effective centre of the lens until an undistorted image resulted.

Magnetic lenses. Becker and Wallraff (1940*a*, *c*) compared the distortion of various weak magnetic lenses for a given object distance and for a fixed focal length. For instance, with an air coil of the mean radius $R = 75$ mm, of the axial width $2L = 20$ mm and of focal length $f = 100$ mm, they found a distortion coefficient

$$C_d = 3 \cdot 6 \times 10^{-4} \, (\text{mm}^{-2}).$$

The cube law equation (6.39) for the displacement was found to hold well for all air coils. However, for iron-shielded coils the displacement $\Delta r'$ was found to increase with a higher power of the axial distance, namely, with $r^{3 \cdot 5}$ to $r^{3 \cdot 7}$, which apparently was caused by the presence of 5th order aberrations. C_d was found to have a slightly higher value for iron-shielded coils than for air coils. Varying the gap of the iron shield, a maximum of C_d was reached at those gap widths at which, according to §§6.3 and 6.7, maximum spherical aberration and maximum field curvature have been observed. For a lens of 100 mm focal length, shielded by 4mm thick iron and having a gap width near that value which results in a maximum distortion, Becker and Wallraff measured $C_d = 5 \times 10^{-4}$ mm^{-2}. There, the factor D_d was found to increase rapidly with the slope angle of the principal rays. For example, D_d values of $-0 \cdot 4$ and $-0 \cdot 8$ were found for slope angles of $13°$ and $17°$ respectively.

Accurate measurements of the distortion factor D_d of weak magnetic lenses have been carried out by Gobrecht (1958*b*). Again increasing distortion was found with decreasing focal length. E.g. for an air coil of mean radius $R = 75$ mm and of $L = 20$ mm length, D_d increases by 20 per cent for a pencil of $10°$ inclination when the focal length f is decreased from 8 to 2 mm, D_d being a linear function of f. Moreover, at a given f, the distortion could be reduced if either at a constant coil radius R, the length L was increased, or if at constant L, R was increased. Thus when the coil current was adjusted to produce a given focal length, the values of the distortion factor given in table 6.6 were measured. The Seidel coefficient for the above magnetic lens with $L = 60$ and $R = 75$ mm was determined as $S_5 = 3 \cdot 8 \times 10^{-4}$ mm^{-2}. For large object distances a distinct deviation of about 6 per cent from the third-order law (6.41) could be observed which

TABLE 6.7. *Distortion factors D_d for magnetic coils of length L (mm), radius R (mm) and focal length f = 80 mm for an inclination $\alpha = 15°$ of the electron pencil*

$R = 75 \begin{cases} L \\ 100 D_d \end{cases}$		20	100	$L = 60 \begin{cases} R \\ 100 D_d \end{cases}$		75	100
		31	20·5			24·8	19·8

shows that fifth-order distortion aberration of this lens is not negligible.

From all these measurements it follows that the distortion error of electrostatic as well as of magnetic lenses is always incomparably greater than that of a reasonably good photographic glass objective.

Independent of the proper distortion error just discussed is a distortion-like displacement of image points from their Gaussian position which is observed when the aperture of rays is sufficiently small but spherical aberration is sufficiently large. A variation of focal length with aperture will appear as a differential shift of each image point as the outer pencils are more strongly refracted than those nearer to the axis. Hence the magnification of the outer zones is increased. However, no confusion of the individual image points is noticed owing to the great depth of focus of the narrow pencils. This aberration distortion has been discussed in §6.4 in connexion with the shadow method for the measurement of spherical aberration. Though the causes of true distortion and of spherical aberration distortion are quite different, the effects are similar in both cases. A comparison of (6.19) with (6.39) shows that in both cases the radial displacement increases with the cube of the distance from the axis. For relatively thin lenses spherical aberration alone is responsible for nearly all the distortion of the image.

Spherical aberration distortion plays some part in projector lenses of electron microscopes. There the image produced by an objective constitutes the 'object'. The area of this object is large compared with the area of the projector lens, but the individual pencils utilized in the projector lens have an extremely small angular aperture of the order of 10^{-5} radian. None of the third-order aberrations appears as an observable diffusion of the image.

According to Liebmann (1952 a) the pin-cushion aberration distortion of a projector lens drops steeply with increasing lens excitation, it vanishes in the region of minimum focal length (§5.4), and for strongest excitations it changes into barrel distortion. Hence, in principle, all projector lenses can be operated free from aberration distortion.

§6.9. Anisotropic aberrations

Magnetic electron lenses show three anisotropic errors in addition to the five isotropic errors, which we have just discussed. The general theory of these anisotropic errors has been worked out by Glaser (1933 b). For the case of a magnetic microscope objective Glaser and Lammel (1943) have given some quantitative expressions for all eight errors which they derived from the bell-shaped analytical distribution curve of the lens field which we discussed in §5.3.

The anisotropic distortion is often called rotational distortion or spiral distortion. It results in tangential displacement of the real image point from the ideal image point which generally increases with the cube of the axial distance of the point. In §5.2 we have discussed the rotation of the image by a magnetic lens. In the case of spiral distortion, the angle of rotation (ψ in (5.16)) is no longer independent of the radial distance of the image point from the axis. The effect of spiral distortion is seen in fig. 6.22, which shows the images of a cross of two mutually perpendicular lines. The broken lines represent Gaussian images while the solid lines show the actual distorted images; these are shaped like spirals from which the name of the aberration is derived. When the current through the lens is reversed, the direction of the field B and the sense of the circulation of the electron is changed (cf. $\breve{\omega}_c \parallel \breve{B}$). As a consequence the image of fig. 6.22(a) changes into the image of fig. 6.22(b), the latter being the mirror image of (a) by reflexion in the axes.

A coefficient C_x of spiral distortion is usually defined by the increment $\Delta\psi$ in rotation of the image point, namely

$$\Delta\psi = C_x r^2, \tag{6.42}$$

where r is the distance of the object point from the axis. (6.42) leads to the following expression for the lateral displacement of the image point

$$r'\Delta\psi = C_x(y_0'/y)r^3, \qquad (6.43)$$

where r' is the distance of the image point from the axis and y_0'/y is the magnification. (6.43) is quite analogous to the radial displacement (6.39). In magnetic lenses, spiral distortion always occurs

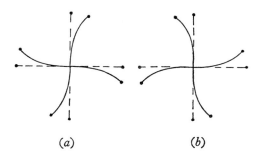

(a)   (b)

Fig. 6.22. Spiral distortion.

together with the isotropic distortion. According to Gobrecht (1958b), however, C_x does not show the strong dependence on the length L of the lens which was experienced for the isotropic distortion (cf. §6.8).

Liebmann (1952a) found that the coefficient C_x of a microscope projector lens increases with increasing lens excitation but decreases with increasing ratio S/R of pole-piece spacing over bore radius.

According to Becker and Wallraff (1940b), spiral distortion is not greater for large, weak iron-shielded lenses than for air coils. Spiral distortion can be avoided, in principle, by using two equal magnetic lenses in opposition forming the rotation-free lens type described in §5.3. However, this type of lens suffers to such an extent from other lens errors that it is useless for most practical purposes. On the other hand, it appears feasible to improve spiral distortion by decreasing the strength of the marginal part of a lens field in comparison to its axial part. For instance, the use of co-axial coils of different radius with their fields in opposition has been recommended.

Anisotropic coma occurs when not only the whole image is rotated around the axis of the lens but every single element of the picture is also rotated round its proper centre. There results an asymmetrical deformation of the individual image elements to a comet-like appearance. The tail of the comet is not, however, radially directed as with isotropic coma, but lies in a direction perpendicular to the radial coordinate.

Anisotropic astigmatism causes a marginal object point to be projected in a line focus which is inclined at an angle of 45° to the radial coordinate. The image surface where these line foci are sharp is generally curved. Therefore, a general confusion will result in the image plane, especially in the marginal zones.

The last two anisotropic errors have so far been unimportant, and no quantitative experimental investigations have been made about them. However, Diels and Wendt (1937) demonstrated the existence of all eight third-order errors in a magnetic lens by investigating the image of an electron-point source.

In spite of the occurrence of the anisotropic errors, magnetic lenses are often found to be superior to the best electrostatic lenses. Magnetic lenses can be made with shorter focal length than electrostatic lenses. But even if lenses of equal focal length are compared, magnetic lenses can be superior with respect to every one of the five isotropic errors. The disadvantage of anisotropic errors in magnetic lenses is not serious enough to reverse the general judgement.

§6.10. Aberrations of line focus lenses

The important errors in focusing systems with symmetry in two mutually perpendicular planes are of the third order, i.e. the transverse errors increase with the third power of the beam aperture. We shall start with their classification as introduced by Burfoot (1954) (cf. Septier, 1961; Hawkes, 1966).

In analogy to the ordinary spherical aberration of round lenses (§6.3) the aperture error of line focus systems is represented by the two transverse deviations Δx and Δy from the Gaussian image where x and y are the coordinates parallel and at right angles to the line focus (cf. figs. 2.7 and 5.23). These deviations are given by

$$\Delta y = C_1 r_0^3 \sin^3 \alpha + C_2 r_0^3 \sin \alpha \cos^2 \alpha,$$
$$\Delta x = C_3 r_0^3 \cos^3 \alpha + C_4 r_0^3 \cos \alpha \sin^2 \alpha,$$

$$(6.44)$$

r_0 and α are polar coordinates of the trajectory in the aperture plane, α being the angle with the y-axis.

Fig. 6.23. Aperture aberrations of line focus lenses.

In systems of circular symmetry there exists only one spherical aberration, but here we have three different versions of the aperture error. Burfoot (1954) found that rays emitted from a point object into a cone are projected into the Gaussian plane, not as a point image, but, as shown in fig. 6.23 as lying either inside a circle, when in (6.44)

$$C_1 = C_2 = C_3 = C_4$$

or in a star, when

$$C_3 = -C_1, \quad C_2 = C_4 = 0$$

or in a rosette, when

$$C_1 = -C_3, \quad C_2 = -C_4.$$

The various distortions occurring in line focus lenses are represented by

$$\Delta x = D_1 x_0^3 + D_2 xy^2,$$
$$\Delta y = D_3 y_0^3 + D_4 x^2 y.$$

$$(6.45)$$

The six types of possible distortions are shown in figs. 6.20 and 6.24 respectively. For $D_1 = D_2 = D_3 = D_4 < 0$ and > 0 we have as in fig. 6.20 the ordinary barrel and pin-cushion distortions respectively. Moreover, for $D_1 = -D_2 = D_3 = -D_4 > 0$ and respectively < 0, the corresponding 'inverted' distortions occur. These are shown in fig. 6.24 (a) and (b). In distinction to fig. 6.20, the object (drawn by broken lines) is now (in fig. 6.24) smaller than

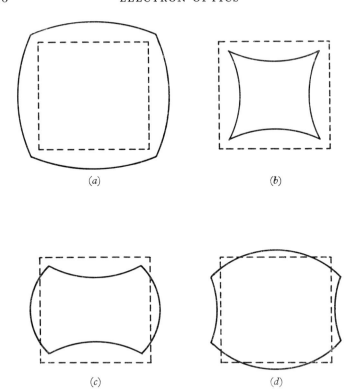

Fig. 6.24. Distortions: (a) inverted barrel distortion; (b) inverted pin-cushion distortion; (c) and (d) hammock distortions.

the image for the barrel distortion, but larger than the image for the pin-cushion distortion. In the third case, namely where

$$D_1 = -D_2 = -D_3 = D_4 > 0,$$

and $$D_1 = D_2 = -D_3 = -D_4 > 0$$

the two versions of the so-called 'hammock distortions' occur, these are shown in fig. 6.24(c) and (d).

With respect to astigmatism, two versions may be present in addition to the 'ordinary' type, these are called 'conjugate' and 'right' astigmatisms. Also, two field curvatures can be distinguished, the 'ordinary' and the 'cosine' curvature. Finally four versions of coma have been derived. The theory of aberrations of doubly symmetric systems is highly complex and we have to refer to the relevant literature (cf. Hawkes, 1966). However, many of these aberrations produce figures which are strongly characteristic when a point object and a circular aperture are used, and if one aberration predominates it should be possible to recognize it from the image.

Few exact measurements exist on the magnitude of aberrations of single quadrupoles. Deltrap and Cosslett (1962) adapted the shadow method (cf. §6.4) to the measurement of spherical aberration of some magnetic quadrupoles. They measured the longitudinal spherical aberration, which is connected with the transverse aberration of (6.44) and is given by

$$\Delta z = C_1 \alpha^2 + C_2 (\tfrac{1}{2}\pi - \alpha)^2. \tag{6.46}$$

As an object, a series of equidistant straight wires was set up parallel to the x-direction (i.e. the direction of the line focus). The images of these wires generally appeared as a series of curved lines as shown in fig. 6.25. From the varying magnification in the y-direction along a line $x = 0$ (i.e. for $\alpha = \tfrac{1}{2}\pi$) C_1 could be found by using (6.46). On the other hand, by measuring the variation in magnification in the x-direction along a line $y = 0$ the constant C_2 was determined.

C_1 and C_2 were plotted against the focal lengths f of the quadrupoles which were determined in separate measurements at various energizing currents. Within the range of the measurements, C_1 appeared approximately proportional to f^2, C_2 proportional to $f^{2.5}$. It could also be verified that the aberration decreases with increasing semi-aperture of the quadrupole. By inserting one octopole in the path of the rays, either C_1 or C_2, but not both could be corrected. Simultaneous correction of both coefficients, together with the correction in the x-direction can be achieved by applying octopoles in regions of different astigmatism of the beam (Scherzer, 1947; Seeliger, 1951; Hawkes, 1966; Archard, 1955: cf. §6.11).

Fig. 6.26 shows some results obtained by Deltrap (1964) with the above shadow method. He investigated single magnetic quadrupoles of effective lengths 10, 20 and 30 cm. For a semi-aperture of 12 mm and an object distance of 21 cm, the coefficients C_1 and C_2 are plotted in the graph against mid-image distances

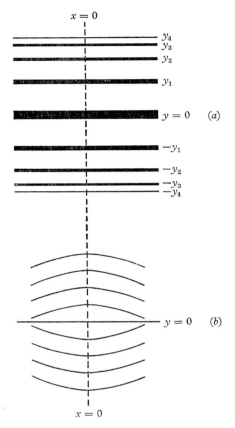

Fig. 6.25. Image of a series of parallel equidistant wires projected by a single quadrupole with spherical aberration; (a) if $C_1 \neq 0$, $C_2 = 0$; (b) if $C_2 \neq 0$, $C_1 = 0$.

(measured from the centre of the quadrupole). The parameters of the curves are the effective lengths (cf. §4.9) of the quadrupoles. The longest quadrupoles are seen to have the least aberration.

A short magnetic quadrupole with a toroidal iron core has been investigated by Markovitsh and Zukerman (1960, 1961). As an object they used two pairs of parallel slits, illuminated by an electron beam. The image on the fluorescent screen consisted of four curved lines, the length and width of which gave an indication of the aperture aberration. The conclusion was reached that the

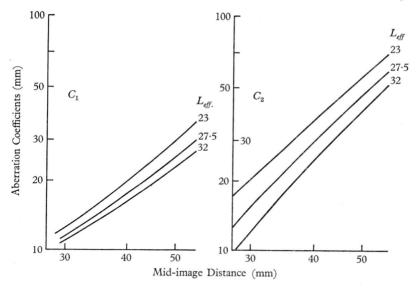

Fig. 6.26. Aperture aberration coefficients of quadrupoles of different effective length L_{eff} for a mid-object distance 21 cm.

aberration was positive and could not be corrected over the whole length of the line focus. It could, however, be predicted that a change of sign of aberration should be possible in the case of certain asymmetries of the field.

Grivet and Septier (1960) investigated the overall width of some line-foci of magnetic quadrupoles, the result giving an indication of the magnitude of the total aberration. They used hollow incident beams parallel to the axis, of circular cross-section and of diameter $2r_0$ which was variable. The beam left the lens with a hollow elliptical cross-section, perturbed by the aberrations. On the screen, certain elongated focal patterns were observed, the width (2δ) of which was measured as a function of the beam aperture $2r_0$ and of the focal length f, adjusted by the energizing current. For a quadrupole of $L = 19\cdot5$ cm length, δ/r_0 was found to change in inverse proportion to the focal length f, with, for instance, $\delta/r_0 = 1 \times 10^{-2}$ for $f = 17$ cm and $r_0 = 1$ cm. The semi-aperture a of the quadrupole was 6 cm. When a was reduced to 4 cm, δ/r_0 of the larger beam apertures had increased by a factor of about 2. For this lens with $a = 4$ cm, the results were well represented by

$$\frac{\delta}{r_0} = C \left(\frac{r_0}{f}\right)^2 + C' \left(\frac{r_0}{f}\right)^4 \qquad (6.47)$$

with $C = 3$ and 25 and $C' = 170$ and 10^5 for $f = 17$ and 130 cm respectively. For $f < L$ these values are comparable to those of good magnetic round lenses.

It was found that the aberration of magnetic quadrupole lenses could be greatly reduced by slipping rods of mild steel tangentially along the pole pieces, i.e. parallel to the z-axis at a certain distance r_1 from it. E.g. for the above quadrupole at $f = 130$ cm, aberration was minimized for $r_1 = 5 \cdot 9$ cm. The correction could be effected in another way by shaping the profile of the pole pieces.

The aberrations of some electrostatic quadrupoles have been studied by Septier and Van Acker (1961) measuring, as above, the width 2δ of the focal line. The electrostatic types seem to have slightly less aberrations than the magnetic ones, in particular since it is possible to correct the electrostatic quadrupoles to some extent, not only by shaping the electrodes but also by a somewhat unequal distribution of the voltages, i.e. by applying $-V_1$ on two opposite electrodes and $+V_2$ on the other pair.

Orr (1963) succeeded in largely correcting the aperture error of an electrostatic quadrupole by laminating each electrode in 8 sections with boundaries parallel to the xy-plane and applying voltages to the laminae which were increasing along the axis towards the xy-plane. It was, however, not possible to reverse the sign of aberration by this method.

We have shown in §4.9 that line focus lenses can be constructed from 'lipped' tubes (fig. 4.17). According to measurements by Klemperer (1953) these lenses suffer very little from aberrations and produce excellent, sharp line images. On the other hand, line focus lenses made up by slots of finite length show a large aperture error at the two ends of the line image. According to Klemperer (1953) this can be corrected if the width of the slot is gradually reduced towards its two ends.

§6.11. Design of stigmatic lens systems with the help of line focus lenses

Various authors have shown experimentally that reasonably good stigmatic (= point focus) lens systems can be obtained by a combination of single quadrupoles. Experiments by Siegel and Reisman (1957) with magnetic quadrupoles and by Septier (1961) with electrostatic quadrupoles demonstrated that a doublet of two crossed quadrupoles is quite satisfactory for focusing homocentric rays into a nearly round spot, though this doublet does not make a lens system of high performance. For short focal lengths, magnetic doublets are superior to symmetrically loaded electrostatic doublets, i.e. with electrode voltages, say $(+\Delta V_1)$ and $(-\Delta V_1)$ with respect to the beam voltage. For long focal lengths, the aberration of both these doublets is of the same magnitude.

In an unsymmetrically loaded electrostatic quadrupole doublet, large positive spherical aberration is produced when the beam is decelerated, say, with electrode voltages o and $(-2\Delta V_1)$. When the beam is accelerated, however, say, with electrode voltages o and $+2\Delta V_1$ small negative aberration results. Hence by sufficient asymmetry some correction of the spherical aberration is possible.

A disadvantage of the doublet arises from the fact that the position of principal planes and the degree of astigmatism vary greatly with excitation. In the symmetric triplet the principal planes are symmetric with respect to the mid-plane thus this triplet can be treated as a thin equivalent lens in a fixed position. Bauer (1966) has shown that the symmetric triplet can be truly stigmatic, for in the two principal planes the beam aperture ratio is equal to unity.

Quadrupole quadruplets have already shown much promise as a microscope projector. The spherical aberration of such combinations has been calculated by Dymikov *et al.* (1966) from the aberrations of the component quadrupoles. Kawakatsu, Vosburgh and Siegel (1968) constructed a magnetic quadruplet projector for a high energy electron microscope. These authors found that their combination of quadrupoles was sensitive to rotational misalignment with a required tolerance of ± 2 m radians, to a displacement of the members of 0·5 mm and to a tilt of 10 m radians. However, the third-order aberrations of their quadruplet were lower than those of the best set of rotationally symmetrical lenses designed to cover the required range of focal lengths ($0·7 < f < 3·6$ mm) for electron energies up to 100 keV, where the magnetization of the quadrupoles was still well below saturation.

Details of a microscope objective consisting of five electrostatic quadrupoles have been given by Bauer (1966). There, an intermediate image of unit magnification was formed between the first triplet and the following doublet. In this way the high magnification of the doublet ($\sim 700 \times$) could be used to project an image sufficiently far outside the system. The whole objective was 8 cm long and the voltages at the five single quadrupoles were 4·8, 5·9, 4·8; 9·0 and 9·0 per cent respectively of the beam voltage. Excellent images of 70 Å resolution obtained with 50 keV electrons are shown in Bauer's publication. Spherical aberration constants in the two directions were determined to be $C_s = 0·16 \times 10^{-3}$ and 2×10^{-2} mm respectively. It should be remarked, though, that a combination of five quadrupoles cannot be easy to operate owing to the sensitivity of quadrupole systems to misalignments.

An early proposal by Scherzer (1947, 1949) for the correction of the third-order errors of a round (circular symmetric) microscope objective by the introduction of two quadrupole and three octopole elements has led to some complicated experimental arrangements in which the image was first split up into line foci and then combined again at infinity.

The correcting octopoles in this arrangement were stigmators similar in effect to those discussed in §6.2. The correcting action of the stigmator, as required for the compensation of spherical aberration, is explained by fig. 6.27. There are eight alternately charged electrodes arranged symmetrically about the z-axis which here extends at right angles to the plane of the drawing.

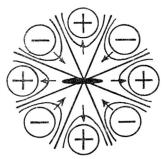

Fig. 6.27. Stigmator action for correction of spherical aberration.

The potential V and the field E_r in the stigmator are given by

$$\left.\begin{aligned} V &= kr^4 \cos 4\psi, \\ E_r &= -4kr^3 \cos 4\psi, \end{aligned}\right\} \tag{6.48}$$

where k is a constant, r is the radial distance from the z-axis and ψ is the azimuth angle. The force on the electron in the radial direction changes sign four times when ψ is continuously increased from o to 2π. For a suitable potential difference of the $+$ and $-$ electrodes, a given third-order aberration is cancelled in the xz-plane, i.e. for $\psi = 0$ and in the yz-plane, i.e. for $\psi = 90°$, while it is doubled in the two orthogonal sections $\psi = 45°$ and $135°$. In order to cancel the aberration for all azimuth angles simultaneously, stigmators have to be arranged in the planes of the astigmatic intermediate images.

In fig. 6.27, there is shown inside the stigmator a line-focus which is parallel to the x-axis. The forces on the electron beam which are proportional to the third power of the distance of the electron from the axis, are indicated by arrows. However, the fields which are directed inwards are not effective, because there are no rays in the regions where the inward fields are acting.

The above scheme of Scherzer (1947, 1949) has been tried out by Seeliger (1949, 1951) and by Möllenstedt (1956). It was found to be possible to correct to a great extent by such an arrangement most of the third-order errors. However, the adjustment of the system is so complicated that no practical use of it can be expected. Moreover, Meyer (1961) has shown that if the system is arranged at some distance from the objective to be corrected, substantial fifth-order errors are introduced.

To avoid these difficulties, Scherzer and Typke (1967) proposed placing the quadrupole field inside the round lens. This is shown in fig. 6.28(a) for the correction of an einzel lens. There, the middle electrode carries four, symmetrically arranged round studs which need not be insulated from it and may be at the same potential. For the corresponding correction of a magnetic objective shown in fig. 6.28(b) the pole piece nearest to the image carries studs of paraboloidal shape which can be made, for instance, from mild steel. The forces due to the additional field introduced by the studs increase with r^3, the third power of the distance from the axis and are

therefore able to correct third-order aberrations. An octopole stigmator behind the corrected lens will serve for the fine adjustment
of astigmatic errors, while the coarse adjustment is given by the size
and shape of the studs. Scherzer and Typke (1967) have calculated

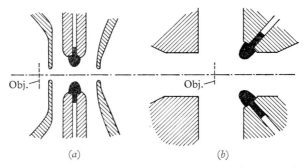

(a)   (b)

Fig. 6.28. Electron lenses corrected by 4 symmetrically arranged studs:
(a) electrostatic; (b) magnetic.

the intensity distribution in the image projected by the corrected
objective with the result that a gain in resolving power by a factor
about 3 could be effected by the correction; however no experimental results have been obtained so far.

CHAPTER 7

ELECTRONIC ABERRATIONS

In addition to the geometrical lens errors studied in the previous chapter, some other kinds of errors occur in electron-optical image formation. These errors are not primarily caused by properties of the lens field, but they are rather produced by some properties of the particular electron beams. Unavoidable inhomogeneities in electron velocity will lead to chromatic aberration. The mutual repulsion of the electrons in sufficiently intense beams will lead to a space-charge error. The de Broglie wavelength of the electron will lead to a diffraction error.

§7.1 Chromatic aberration

So far, the electron beams passing through electron lenses have been assumed to consist of electrons of strictly homogeneous velocity. This simplifying assumption, however, is never fulfilled exactly. Electron rays start from a cathode. There, the original range of emission velocities always covers a certain band. The relative band width as compared with the actual electron velocity is, of course, narrowed down as the electron rays are accelerated. But there are more sources of inhomogeneities of electron velocities. For instance, there may be small fluctuations in the accelerating voltage. Or, for example in electron microscopy, inelastic scattering of the electron rays at the atoms of the object leads to a 'straggling' of velocities.

Chromatic aberration in light optics arises from the fact that light of different colour (wavelength) is refracted to a different degree, leading to a glass lens of shorter focal length for blue light than for red. For electrons, the wavelength changes inversely with the velocity. An increase in velocity (shorter wavelength) results in electron optics in a reduced refractive power of the lens. In addition to this, the magnification and, for magnetic lenses also the rotation, will be different for electrons of different velocities.

The small change in focal length caused by a small change ΔV

in electron energy (electron volts) and by small fluctuations ΔI of the current energizing a magnetic lens is called the longitudinal chromatic aberration Δf_{cr}. According to (5.22) of §5.3, the focal length of an iron free lens is given by $f \propto V/I^2$. By differentiation we obtain

$$\Delta f = f \left(\frac{\Delta V}{V} - \frac{2\Delta I}{I} \right). \tag{7.1}$$

In order to include lenses which contain iron in which f is no longer strictly proportional to V/I^2, a chromatic aberration constant C_{cr} was introduced by Glaser (1940) replacing f in (7.1) so that the chromatic aberration is expressed by

$$\Delta f_{cr} = C_{cr} \left(\frac{\Delta V}{V} - \frac{2\Delta I}{I} \right). \tag{7.2}$$

The lateral chromatic aberration is then, analogously to the lateral spherical aberration (cf. §6.3) given by

$$\Delta r_{cr} = \Delta f_{cr} \Theta. \tag{7.3}$$

In electrostatic lenses, V in (7.1) is the arithmetical mean of electron energies in the object space and in the image space. At high energies V is superseded by the relativistically corrected V_r of (1.9) in §1.2. C_{cr} is frequently measured in terms of the focal length f of the lens. $C_{cr}/f = 1$ applies for weak magnetic lenses. For the strong magnetic objectives of electron microscopes, C_{cr}/f has been found to come down to a value of about 0·7.

C_s/f appears to depend slightly upon the shape of the field. Glaser (1940) calculated that a minimum of C_{cr} should occur at a value $K_H^2 = 1\cdot2$ of the lens strength parameter K_H of (5.28) in §5.3. There, the minimum of the spherical aberration coefficient C_s was found at $K_H^2 = 2\cdot8$ (cf. §6.5). This was confirmed experimentally by Dosse (1941) who determined the minimum of C_{cr} and the minimum of C_s for various magnetic microscope lenses and found that $(C_{cr})_{min}$ and $(C_s)_{min}$ occur at field distributions of different half width $2w$. It is, however, fortunate that the conditions under which both C_{cr} and C_s are minimized are close to those for a minimum focal length f. According to computations by Mulvey and Wallington (1969) the value of $C_{cr}B_p/\sqrt{V_r}$ falls steadily towards a limiting value of $5\cdot3 \times 10^4$, when S/D, the gap to bore ratio of the magnetic microscope lens is gradually increased to a large value ($\gtrsim 2$). B_p is the maximum flux density (Wb m^{-2}) in the gap. On the other hand, when at $S/D \simeq 2$, the excitation of the lens is adjusted to produce the minimum spherical aberration (cf. §6.5), the chromatic aberration will exceed its minimum value given above by about 30 per cent.

Chromatic aberration affects the image projected by magnetic lenses in two ways, namely by the chromatic differences in magnification and rotation, so that the coefficient C_{cr} is composed of C_{crM} and C_{crR}. Both cause a deviation of the image point from the Gaussian image point. This is hardly noticeable in the axial part of an electron microscope picture where they produce a small lack of sharpness only. At the marginal part produced by inclined rays, C_{crR}, however, draws the picture points out into small lines by a spiralling rotation. The latter effect is particularly noticeable in dark field illumination (Watanabe and Morito, 1955). On the other hand it is helpful in the adjustment of the lenses of an electron microscope since misalignment is detectable as a spiralling appearance of the picture (Leisegang, 1954).

C_{cr}/f is larger for electrostatic than for magnetic lenses. In electrostatic lenses, $\Delta V/V$ changes along the electron path, where V is the local value of the electron energy. Indeed, $\Delta V/V$ is much larger in regions of low voltages V than in other parts of the lens with high voltage. Changes in $\Delta V/V$ are to some extent responsible for the relatively large values of C_{cr}/f. Ramberg (1942) found $C_{cr}/f \approx 2$ for all weak electrostatic lenses. With increasing lens power C_{cr}/f is found to decrease in initially accelerating saddle-field lenses only. In all other types of electrostatic lenses, however, C_{cr}/f increases with increasing lens power.

Decelerating two-electrode lenses have the smallest chromatic aberration coefficients, accelerating two-electrode lenses the largest ones. Initially decelerating saddle-field lenses are intermediate between the latter and the former types. For saddle-field lenses of electron microscopes in which the centre electrode is connected to the cathode, C_{cr}/f is found to be of the order of 4. The chromatic aberration of einzel lenses can be minimized by increasing the thickness T_i of the middle diaphragm and the diameter $2R_i$ of its aperture (cf. fig. 4.6). This is shown in the fig. 7.1 by Lippert and

Pohlit (1952) where R_i and T_i for a minimum of C_{cr} are plotted as a function of the focal length f. The magnitudes of R_i, T_i and f are measured here in units of the spacing $(2S + T_i)$ between the two outer electrodes (cf. fig. 4.6).

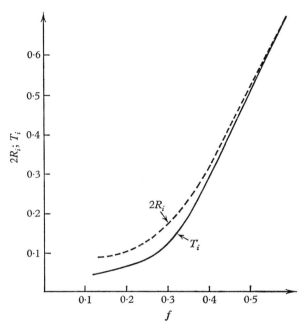

Fig. 7.1. Geometry and focal lengths of einzel lenses
with minimum chromatic aberration.

The correction of chromatic aberration is not easy, since all ordinary electron lenses have chromatic errors of the same sign. Though electron mirrors have chromatic errors of the opposite sign, their application for chromatic correction meets, according to Ramberg (1949), with serious difficulties (cf. also §6.5).

The chromatic aberration of particular combinations of quadrupoles can vanish completely, and lenses of this type have in consequence been much studied. Dymikov and Yavor (1964) showed theoretically that a quadrupole doublet can have zero and even negative chromatic aberration in the xz-plane (cf. figs. 2.7 and 5.23), if the first quadrupole is magnetic and the second quadrupole is electrostatic. If the sequence of the two quadrupoles is interchanged, the system becomes achromatic in the yz-plane. In this set up, the two quadrupoles are sequentially separated. If, however, the two quadrupoles are superposed, a completely achromatic multipole can result. This has been demonstrated theoretically and experimentally by Yavor, Dymikov and Ovsyanova (1964 a, b). In their experiment an electrostatic quadrupole was set up inside a vacuum tube; it was surrounded by a magnetic quadrupole outside the tube so that the mid-planes of the two quadrupoles coincided, but the electrostatic and magnetic lines of force were crossed, acting in the opposite sense. By control of the relative strength of the two fields, the system could be rendered completely achromatic.

Some practical causes of chromatic aberration. In electron microscopy, effects from chromatic aberration can be caused by fluctuations in the driving voltage or by fluctuations in the energizing currents of magnetic lenses. This kind of chromatic aberration is eliminated in practice by the use of carefully designed voltage and current stabilizers respectively. In high-resolution microscopes stabilization tolerances of 0·01–0·001 per cent are required both for voltages and for currents.

Another kind of chromatic error appearing in electron microscopes arises from energy losses of the electrons in the object. Normally, this does not seem to affect the resolution; it even appears to increase the contrast in the picture. Different intensities in the picture are produced by elastic and inelastic scattering of the electrons by the different thicknesses of the object. There, a certain fraction of the electrons is scattered outside the physical aperture of the objective. These scattered electrons no longer contribute to the density of the picture elements. A relatively large amount of the electrons, however, is scattered through small angles and still remains inside the physical aperture. Many of these electrons have suffered a noticeable loss of velocity in the scattering process. These can be eliminated by filter lenses which will be discussed in §7.3.

An important source of velocity variations is given by the distribution of emission velocities of the electrons leaving the cathode. In the case of a thermionic cathode, the velocities are distributed according to the Maxwellian law. If $n(V)\mathrm{d}V$ is the fraction of electrons emitted with an initial kinetic energy between eV and $e(V + \mathrm{d}V)$, we have

$$n(V)\mathrm{d}V = \frac{2}{\sqrt{\pi}} \left(\frac{e}{kT}\right)^{\frac{3}{2}} V^{\frac{1}{2}} \exp\left(-\frac{eV}{kT}\right)\mathrm{d}V, \qquad (7.4)$$

where k is Boltzmann's constant and T is the absolute temperature of the cathode. Of more interest in electron optics is the corresponding current density distribution, i.e. the fraction of current carried by electrons in a given initial energy range (cf. Möllenstedt and Lenz, 1963), namely

$$\frac{\mathrm{d}i_z}{i_z} = \frac{eV}{kT} \exp\left(-\frac{eV}{kT}\right) \mathrm{d}\left(\frac{eV}{kT}\right). \qquad (7.5)$$

This distribution has a maximum at

$$V_m = \frac{kT}{e} = \frac{T}{11\,600} \text{ [eV]}. \qquad (7.6)$$

For example, for a tungsten cathode with $T \approx 2900°$, $V_m = 0.25$ V. Or for an oxide cathode with $T \approx 1000°$, $V_m = 0.1$ V. Taking V_m for the voltage variation ΔV in (7.1) and (7.2) and assuming an electron-microscope objective with $f \approx 3$ mm, with an angular aperture $\Theta \approx 10^{-3}$ and with 50 kV electron rays, the chromatic disc of confusion is calculated to be below 1 Å, i.e. negligibly small. The thermionic energy distribution can be reduced substantially by application of a filter lens as will be shown in §7.3. For instance, for a 50 keV beam a band width of 12 meV has been obtained. By this energy selection, the current density was, however, reduced to the order of 0.1 A m^{-2} only.

In the case of the chromatic error of an electron gun, the upper limit of current density that can be focused into a spot is set by the velocity distribution of thermionic emission. The spot is an electron-optical image of a cross-over of rays or of the cathode itself. Thus the radius y' of the spot may be calculated from the radius y of the object (cross-over or cathode surface) by application of Abbe's sine law (1.31) which by squaring leads to

$$\left(\frac{y}{y'}\right)^2 = \frac{(N')^2}{N^2} \frac{\sin^2 \Theta'}{\sin^2 \Theta}, \qquad (7.7)$$

N and N' being the refractive indices and Θ and Θ' the corresponding angular semi-apertures of the beam at the cathode and at the spot respectively. We now multiply y^2 and $(y')^2$ by πI, where I is the emission current which is conserved from the cathode to the target. Then, the left-hand side of (7.7) gives i'/i, the ratio of current densities at cathode and target respectively. On the right-hand side of this equation the squared refractive indices may be replaced by the voltages of the beam according to (1.8). Thus

$$\frac{i'}{i} = \frac{V_A + V_{em}}{V_{em}} \frac{\sin^2 \Theta'}{\sin^2 \Theta}.$$

Now the maximum angular aperture of the emission is given by electrons which leave the cathode surface tangentially, so that $\Theta = \frac{1}{2}\pi$, hence

$$i' = i(V_A/V_{em} + 1) \sin^2 \Theta'. \tag{7.8}$$

This important equation was derived first by Langmuir (1937 b). According to it, the current density i' in the spot depends on the ratio of anode voltage V_A to the energy V_{em} which corresponds to the effective band-width of thermionic emission velocities. Moreover, i' is proportional to the emission density i and to the squared sine of the angular semi-aperture Θ' of the beam at the spot. It may be concluded that the chromatic aberration is one of the most important limiting factors for the performance of an electron gun. In table 7.1 maximum possible current densities have been calculated with (7.8) using V_m of (7.6) for V_{em}. The half angle Θ' is limited by the maximum useable aperture of the focusing lens which should be reasonably free of spherical aberration.

TABLE 7.1. *Theoretical maximum current density in the spot of focused cathode rays. Final voltage* $= V'$. *Half angle subtended by beam* $= \Theta'$. *Cathode temperature* $= 1160° K$. *Cathode current density* $= 10^3 A\ m^{-2}$

	sin Θ'		
V' volts	0·01 A m^{-2}	0·032 A m^{-2}	0·10 A m^{-2}
100	100	10^3	10^4
1 000	1 000	10^4	10^5
10 000	10 000	10^5	10^6

The chromatic aberration due to the range of emission velocities is even larger for electron-optical arrangements employing photocathodes. While the band-width of the thermionic cathode is practically of the order $\Delta V \approx 0\cdot 1$ eV, that of a photocathode is at least $0\cdot 5$ eV. Photocathodes have to be used in the technically important image converters. In these instruments a light-optical image is converted by a photocathode into an electron emission image which is projected by an electron lens system on to a fluorescent target. In order to keep the chromatic aberration of an image converter within practicable limits the voltage V_A at the target has to be sufficiently high. For example, for $V_A = 5$ kV, $\Delta V/V_A \approx 10^{-4}$. The maximum angular aperture at the cathode is again $\Theta = \frac{1}{2}\pi$, due to the occurrence of tangential emission.

It should be noted, however, that the chromatic error depends very much upon the projecting electron optics. This has been calculated by Henneberg and Recknagel (1935), who have compared the performances of three technically important cases of image converters. Assuming unit magnification they found the following expressions for the discs of least confusion:

(i) $\Delta r = 4l \left(\dfrac{\Delta V}{V_A} \right)^{\frac{1}{2}}$ for a homogeneous electric field projecting the photoelectrons on to the target.

(ii) $\Delta r = 2l \dfrac{\Delta V}{V_A}$ for superimposed homogeneous electrostatic and magnetic fields.

(iii) $\Delta r = 2/3l \dfrac{\Delta V}{V_A}$ for a short magnetic field, the electrons being accelerated by a potential rising linearly up to the lens.

In all three cases, l represents the distance between cathode and target. In practice l can be kept very much shorter in case (i) than in cases (ii) or (iii). However, assuming a given l the least disturbance by chromatic aberration occurs in the converter employing the short magnetic lens.

§7.2 **The use of chromatic aberration in electron-lens spectrometers** (Siegbahn, 1966; Gerholm, 1956; Klemperer, 1965)

Chromatic aberration as a desirable property of electron-optical systems has found an important use for the purpose of sorting electron velocities in 'lens spectrometers'. A first model of this device has been described by Klemperer (1935), and a detailed discussion of it has been published by Deutsch, Elliot and Evans (1944). Spectrometers which utilize the chromatic aberration of solenoidal focusing have been described by Tricker (1924) and by Witcher (1941).

In order to utilize chromatic aberration for separating different electron velocities, an annular slot is set up near the mid-plane of an electron lens, and a small electron source is arranged on the lens axis. Rays of a given velocity will be focused into a small spot which forms the electron-optical image of the source at a given distance from it. Hence rays of a given velocity can be collected by a small aperture subtending the area of the spot. Rays of greater or of smaller velocity are focused on the axis at greater or smaller distances from the source respectively. In the plane of the collecting aperture, however, the rays of these different velocities are spread over annular discs of confusion and are not able to enter the selecting aperture.

In other words, a lens of given focal length with chromatic aberration, collecting hollow, co-axial cones of rays into a paraxial aperture, will select a small interval of momenta between p and $(p + dp)$. The value of the collected momentum p depends on the power of the lens. The selected p can be varied, say in the case of a magnetic lens, by varying the coil current. Hence the number of selected electrons as a function of the coil current provides the momentum spectrum or velocity spectrum of the investigated electron source, and in this way a simple electron lens can be used as an electron spectrometer.

A scheme for such a simple lens spectrometer is shown in fig. 7.2. A cone of electrons emitted from a source S enters through an aperture A. A ring baffle B arranged in the mid-plane of a magnetic lens coil L selects the external part of this cone of rays which is focused by the lens on to the collector aperture C. A Geiger counter T registers the number of electrons passing through the aperture C as a function of the coil current of the magnetic lens. If the spherical aberration of the lens is negligible, relatively wide cones of rays from the source can be collected.

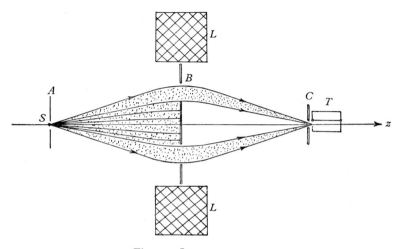

Fig. 7.2. Lens spectrometer.

The fraction of the total emission from an isotropic source that reaches the collector aperture defines the 'collecting efficiency' Ω of the spectrometer.* In the special case of a point source being used in combination with a lens of negligible spherical aberration, the collecting efficiency is equal to $\Omega_0/4\pi$, where Ω_0 is the solid angle subtended by the rays at the source, i.e.

$$\Omega_0 = 2\pi \int_{\Theta_1}^{\Theta_2} \sin \Theta d\Theta = 2\pi(\cos \Theta_1 - \cos \Theta_2) \approx \pi(\Theta_2^2 - \Theta_1^2) \quad (7.9)$$

if the emission selected by the baffles is enclosed between the cones of semi-vertical angles Θ_1 and Θ_2.

If the collecting power Ω for an extended source is expressed as the percentage of the total emission of rays that reaches the

* In the literature about electron spectrometers, the terms 'transmission' or 'gathering power' or 'light intensity' are sometimes used instead of the term 'collecting efficiency'.

collector, it can be used for defining an 'effective solid angle' or 'solid angle of acceptance' $\Omega_0 = 4\pi\Omega$, which may be measured in steradians. For instance, a collecting efficiency of 1 per cent corresponds to an effective solid angle $\Omega_0 = 4\pi \times \frac{1}{100} = 0\cdot126$ steradian.

A great collecting efficiency is always desirable, especially if spectra from weak sources have to be investigated. If a point source is used with an aberration-free lens, the maximum semi-vertical angle Θ is limited by the aperture and by the power of the available electron lens only. Now, with an extended source of given emission density, the collected current can be increased by increasing the source area. On the other hand, the selected angular interval $d\Theta$ and hence the collecting efficiency increases with increasing window area. The momentum spread of the collected rays, however, increases both with source and with window area. This may be shown now for the example of a symmetrical spectrometer with a very narrow ring aperture in the mid-plane. The simple geometrical discussion is due to Cosslett (1940).

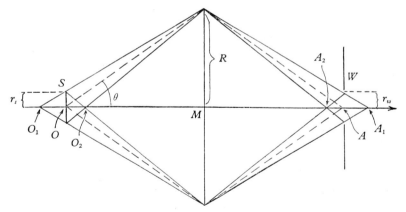

Fig. 7.3. Geometric limit to the resolving power of a lens spectrometer.

In fig. 7.3 rays from an extended source S at O are focused by an aberration-free lens and collected by an extended window in the diaphragm W. The slowest electrons to pass through the window, cross over at O_2 and A_2, the fastest electrons cross over at O_1 and A_1. Let the radii r_s and r_w of the circular source and window respectively be small compared with the radius R of a ring aperture in the mid-plane M of the thin lens by which the mean semi-vertical angle Θ of the transmitted rays is defined. From similar triangles

$$\frac{O_1 O}{MO} \approx \frac{O_2 O}{MO} = \frac{r_s}{R}, \qquad (7.10)$$

and

$$\frac{A_1 A}{MA} \approx \frac{A_2 A}{MA} = \frac{r_w}{R}. \qquad (7.11)$$

Now, the focal length f of the thin lens, and hence the object-to-image distance OA are, according to §5.2, proportional to the square of the momentum p of the rays, i.e.

$$OA = 4f \propto p^2. \qquad (7.12)$$

Hence, on differentiating,

$$\frac{\Delta OA}{OA} = \frac{2\Delta p}{p}. \qquad (7.13)$$

Adding (7.10) and (7.11) and comparing with (7.13) gives

$$\pm \frac{\Delta p}{p} = \pm \frac{r_s + r_w}{4R}. \qquad (7.14)$$

It can be shown that if a solenoid is used instead of the short lens, $\Delta p/p$ will be twice as large as the value given by (7.14), since for the solenoid a proportional relationship, (5.5), holds between OA and p instead of (7.12). For thick lenses $\Delta p/p$ will be of an intermediate value between the limits assigned to the thin lens and to the solenoid.

The total momentum spread of electrons $\pm \Delta p$, admitted to the collector at a given, constant coil current, could also be obtained as the total base width of a 'transmission curve'. Such a curve gives the intensity distribution obtained with a monochromatic emission by varying the coil current, i.e. this curve shows the collector currents against the coil currents measuring apparent momenta. A relative momentum range $\Delta p/p$ which indicates the least difference of momenta which can still be resolved is called the 'resolution', and its reciprocal $p/\Delta p$ is called the 'resolving power'.*

---

* In practice, Δp is obtained as the half-maximum width of the experimental transmission curve. However, a change in momentum of less than $0.1\Delta p$, can frequently still be detected.

In practical spectrometry, the available intensity generally sets a limit to that increase in resolving power which can be obtained by reduction of the aperture of a given apparatus. For instance, for a lens spectrometer with a source of negligible size, the resolving power is, according to (7.14), inversely proportional to the radius of the collecting aperture, namely, $(p/\Delta p) \propto 1/r_w$, but the collecting efficiency is $\Omega \propto r_w^2$, hence

$$\left(\frac{p}{\Delta p}\right)^2 \times \Omega = C, \tag{7.15}$$

where C is a constant for a given spectrometer, characterized by a fixed object-and-image distance and by a fixed baffle arrangement. C can be increased, for example, by using a lens which utilizes a greater solid angle of rays. The greatest obtainable resolving power will always be required simultaneously with greatest possible collecting efficiency. Hence, the value of C can be considered to be a figure of merit for a particular type of lens spectrometer.

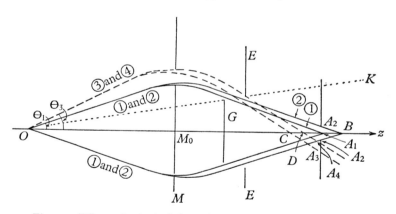

Fig. 7.4. Effects of spherical aberration on a lens spectrometer. Hole collector and ring collector.

A most important limitation to this figure of merit C of (7.15) is given by the spherical aberration of the focusing lens. This may be explained with the help of fig. 7.4, showing a meridional cross-section through some selected bundles of rays emitted from a point source O. Let the rays (1) contain electrons of momentum p_1 travelling along the surface of a cone of semi-vertical angle Θ_1.

These rays pass the mid-plane M of the lens and are focused on the lens axis (z) at a point A_1 which is the centre of an aperture in a diaphragm A. Let another bundle of rays (2) be emitted at the same angle Θ_1 as the rays (1) but with slightly greater momentum $p_2 = p_1 + \Delta p$. Hence they will cross the z-axis at a point B slightly farther removed than A_1. These rays (2) will reach the diaphragm A at a circle through A_2 by which the radius ($A_1 A_2$) of the aperture in A is defined. Again, electrons of emission angle Θ_1 but with momentum ($p_1 - \Delta p$) would cross over before reaching A_1, but they too would pass the plane of the diaphragm A at the above circle through A_2.

Now consider the bundles of rays indicated in fig. 7.4 by (3) and (4). Both bundles start along the same conical surface of semi-vertical angle Θ_3, and they may have again the above momenta p_1 and $p_1 + \Delta p$ respectively. If spherical aberration is non-existent, the positions of the foci are independent of the emission angle Θ, and all rays within the interval of momenta $p_1 \pm \Delta p$ would again pass through the aperture in the diaphragm A.

In presence of spherical aberration, however, the focal length of the system depends upon the slope angle Θ of the rays. Hence the rays (3) and (4) are now shown to cross the z-axis at the points C and D respectively, and to reach the diaphragm A at the respective distances $A_1 A_3$ and $A_1 A_4$ off the axis.

The bundles represented by (3) and (4) are intercepted by the diaphragm and do not pass the collector aperture of radius $A_1 A_2$; hence it can be seen that the collecting power Ω for a fixed interval ($p_1 + \Delta p$) of momenta must be very much smaller than the solid angle Ω_0 defined according to (7.9) by the range of emission angles between Θ_1 and Θ_3. On the other hand, rays emitted at the angle Θ_3 with momenta somewhat greater than ($p_1 + \Delta p$) would cross the axis at somewhat increased focal distance and thus would be able to pass the collecting aperture in the diaphragm A. According to these considerations a natural limitation appears to be set to the design of lens spectrometers by an unavoidable amount of spherical aberration inherent in every electron lens.

A quantitative relationship between the resolving power, the chromatic aberration and the spherical aberration of an electron lens can be obtained by the following argument which is due to

Grivet (1950): Let a source be assumed to emit two mono-chromatic electron energies V and $V+\Delta V$ of equal intensities. Each of these energies will be focused in a separate image point surrounded by a disc of confusion of radius Δr_s due to spherical aberration. We can use (6.10) for Δr_s, but (7.3) will give too low a value for Δr_{cr}. In order to take into account the spherical aberration in the simplest possible way, we should use, instead of Δr_{cr}, the modified value

$$\Delta r'_{cr} = \Delta r_{cr} + \Delta r_s.$$

This assumes that the spherical aberration varies negligibly between the positions A_1 and A_2, and it leads to

$$\frac{\Delta r'_{cr}}{\Delta r_s} = \frac{\Delta r_{cr} + \Delta r_s}{\Delta r_s} = 1 + \frac{C_{cr}}{C_s} \frac{1}{\Theta^2} \frac{\Delta V}{V} \qquad (7.16),$$

where Θ is the semi-angle of convergence at the image side. If the collector aperture subtends only a small area of the disc of confusion, the current received will be inversely proportional to $(\Delta r)^2$. Now, two monochromatic velocities of the same intensity will be separated if $\Delta r'_{cr}/\Delta r_s = 2$ since the transmitted intensity, $I \propto (\Delta r_s/\Delta r'_{cr})^2$ will fall in relative value from unity, when the spectrometer is adjusted exactly on one or other of these lines, to $1/\sqrt{2} = 0.71$ when it is adjusted to the average momentum p of the two monochromatic beams. Thus we can estimate a value for the resolving power of the spectrometer to be

$$\frac{p}{\Delta_p} = \frac{2V}{\Delta V} = \frac{2C_{cr}}{C_s \Theta^2}. \qquad (7.17)$$

Great improvements in the design of lens spectrometers have been made after Frankel (1948) showed the advantages of using a 'ring collector' in front of a relatively wide aperture in the diaphragm A shown in fig. 7.4. Already Witcher (1941) recognized that the bundles of rays (1), (2), (3) and (4) form a ring-shaped cross-over before they reach the axis. All these rays which are selected by an annular slot in the baffle at the mid-plane M can be collected by an annular slot enclosed by the two diaphragms E and G as shown in fig. 7.4. By the use of this ring collector, the collecting efficiency appears to be much increased without a corresponding loss in resolving power. This can be seen clearly in fig. 7.5 where the resolution $\Delta p/p$ against the collecting efficiency Ω is plotted for a thin lens and for a solenoid with and without ring collector (the curve of the two-directional focusing deflexion also plotted in this figure will be discussed later in §10.9). It is seen that

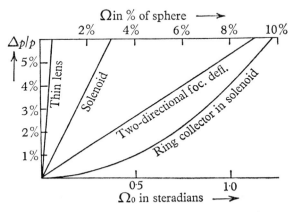

Fig. 7.5. Resolution and collecting efficiency in various electron ray spectrometers.

a proportional relationship holds between $\Delta p/p$ and Ω for the thin magnetic lens and for the solenoid, and that by the introduction of a ring collector slit the resolution is greatly improved, particularly for small Ω.

For the practical design of the ring collector, Frankel recommended that its defining edges formed by E and G should lie on a cone (indicated in the figure by the dotted line K) co-axial with the lens and with its vertex at the source, so that no straight rays (e.g. γ-radiation, etc.) can pass the slot. For this purpose, the inner diaphragm G has been placed closer to the source than the outer diaphragm E.

Let the emission from a point source be selected by an aperture subtending a relatively small angular interval which is enclosed between the two cones of semi-vertical angle Θ and $\Theta + \Delta\Theta$. The corresponding solid angle is then given by

$$\Omega_0 = 2\pi \sin \Theta \Delta\Theta. \qquad (7.18)$$

Hence for fixed $\Delta\Theta$, the collecting efficiency of the lens spectrometer increases with increasing Θ. On the other hand, position and sharpness of the ring cross-over depend upon Θ. For fixed Ω, an optimum Θ can be determined for which the sharpness of the ring cross-over is best so that the collecting ring aperture can be made of minimum width. In every case, the optimum angle of emission from a point source is found to be $\Theta_{opt} = 45°$. This optimum, however, is not very pronounced, and in practical spectrometers much smaller angles (10–30°) have been used on account of finite-source dimensions, for reasons of economy in weight of copper wire and to save energizing power.

Detailed theoretical discussions on the ring cross-over in arbitrary magnetic lens fields have been given by Verster (1950) and by Grivet (1950, 1951). There, formulae can be found for the position and for the thickness of the ring cross-over and for resolving and collecting efficiencies as functions of the coefficients of spherical and chromatic aberrations and of the angular apertures of rays.

If extended sources have to be used, e.g. in investigations of preparations of low-emission density, an optimum source diameter can be determined, for which the product of source area and collecting efficiency, i.e. the so-called 'luminosity', reaches a maximum. This optimum source diameter is found to be of the order of four times the width of the ring selector (cf. Grivet, 1950). With this optimum source diameter the resolving power is about five times smaller than with a point source. Even with large sources, however, the use of the ring selector always definitely improves the resolving power at a given collecting power.

For instance, Keller, Koenigsberg and Paskin (1950) studied the ring cross-over of a thin lens spectrometer with extended sources. They determined a set of trajectories by numerical integration of the equations of motion. At a given resolution, a comparison between collecting efficiencies of ring selector and of ordinary circular hole collector showed a superiority of the ring-selector instrument by a factor two for the extended source. Again, Graham and Klemperer (1952) investigated experimentally the use of the ring-selector aperture for a lens spectrometer with relatively small source and window and with initial beam apertures

between 8° and 12°. Under otherwise similar conditions the best
gain in resolving power effected by the ring selector was 75 per cent.

A further improvement in resolution without loss of collecting
efficiency has been obtained by Hubert (1952) with a modification
of Frankel's ring collector. He used the caustic baffle which is
shown here in fig. 7.6. There, z is the lens axis, and 1, 2 and 3 are

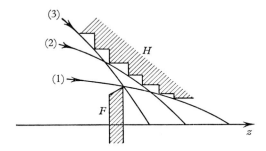

Fig. 7.6. Hubert's caustic baffle.

electron trajectories. F is a circular internal baffle positioned as in
Frankel's ring-selector arrangement (see fig. 7.4). H is a specially
shaped external baffle which follows the outline of the caustic
envelope of the converging rays (cf. §7.7). This ensures a very sharp
cut-off on the low energy side of spectral lines. (The line profile
on the high energy side is identical with that from a conventional
ring collector, since this profile is governed by the location of the
internal baffle F.) The Hubert caustic baffle has been used by
Jungermann et al. (1962) in a solenoidal spectrometer in which high
resolution could be obtained using small sources (see table 7.2).

The resolving power of every lens spectrometer (with or without
ring selector) is increased at a given collecting efficiency whenever
the spherical aberration of the lens is reduced. For this reason,
Siegbahn (1946) developed his spectrometer with the iron-shielded
dip-field lens which we described in §6.5. Other authors have
shown the merits of spectrometers with dip-fields produced by
two separate lenses spaced by about 3 to 6 times the half-maximum
width of their field distributions.

In a spectrometer by Slätis and Siegbahn (1949) a coil with
closed iron shield produces a dip-field with two maxima which act

like two separate lenses. Spaced iron shielded lenses have been used in a spectrometer by Agnew and Anderson (1949). The theory of the dip-field spectrometer consisting of two thin lenses with a ring cross-over between the lenses has been worked out by Bothe (1950 a, b).

The spectrometers described so far, employ magnetic lenses. They are generally used for beta-ray spectrometry. The separation of the different velocities is effected in a direction along the axis.

A highly sensitive velocity analyser which works in quite a different way, makes use of the chromatic aberration of the initially decelerating saddle-field lens (cf. §4.6). Introduced by Möllenstedt (1949, 1952), it has found very wide application for electron beam spectrometry in the region of 20–80 keV and its optical properties have been studied by various authors, e.g. by Lippert (1955), by Dietrich (1958) and also by Metherell and Whelan (1965) who investigated its performance by numerical trajectory tracing. In practical instruments, einzel lenses with line focus, i.e. lenses with slots rather than with circular apertures are used.

In fig. 7.7(a), there is shown an einzel lens with two outer electrodes L_o connected to an anode potential (say $V_o = +30$ kV) and an inner electrode L_i at cathode potential ($V_i = 0$). Let an incident beam, parallel to and at distance y_0 from the axis, be limited by a fine slit (say of width $\Delta y = 5\mu$) but extending in the x-direction at right angles to the plane of the drawing. When a negative bias is applied to the central electrode L_i the convergence of the lens first increases and the beam (of the assumed 30 keV) moves in fig. 7.7(a) from position 1 to position 2. With further increase in bias the convergence passes through a maximum, returns to the axis, intersects it, moves away from it, then returns and oscillates as indicated by the positions 3, 4 and 5. The corresponding y-ordinates of the beam on a photographic plate (P in fig. 7.7(a)) are plotted in fig. 7.7(b) against the voltage V_i of the central electrode. Conversely, V_i can be set to make the emergent beam parallel to the axis and the spectrum of an inhomogeneous beam will be obtained on the photographic plate. Velocity analysers with einzel lenses can have very high resolving power, e.g. $p/\Delta p \lesssim 10^5$ ($\sim$ 0·1 eV in 30 keV) have been achieved.

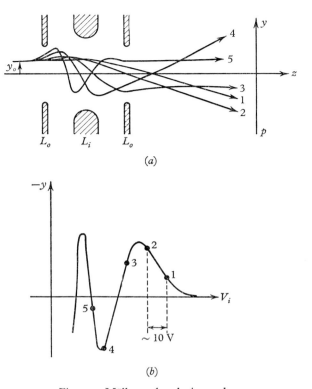

Fig. 7.7. Möllenstedt velocity analyser.

Some significant data on the performance of lens spectrometers are given in table 7.2. These will be compared with other types of electron beam spectrometers in §10.9.

§7.3. Filter lenses

Electrostatic saddle fields (cf. §4.6) are called 'filter lenses' when they are employed for sorting velocities. For instance, an einzel lens will act as a filter when a relatively high negative bias is applied to the central diaphragm. In this case, the negative potential across the aperture in this diaphragm gradually decreases towards its centre (cf. fig. 4.8). Paraxial electrons of a beam of finite diameter will therefore still be able to pass through the lens, when

TABLE 7.2. *Comparison of electron lens spectrometers*

Spectrometer	Max. semi-aperture (cm)	Source-collector distance (cm)	Source (mm)	$\Omega \times 10^{-3}$	$p/\Delta p$	Reference
Thin lens (air coil or iron shielded)	3·5	120	3	0·8	14	Klemperer (1935)
Thin lens (iron)	4	60	4 × 0·2	0·8	100	Siegbahn (1942)
Thick lens (iron)	4·5	60	4 × 0·2	0·3	150	Siegbahn (1944)
Thick lens, spherically corrected	7	50	4 × 0·2	20 6	20 35	Siegbahn (1946)
Thick lens, spherically corrected with ring diaphragm	7	50	4 × 0·2	80	25	Slätis and Siegbahn (1949)
Thin lens (air coil)	7·5 7·5	100 100	4 5	1 8	59 17	Deutsch, Elliott and Evans (1944)
Solenoid (air coil)	12·5	90	20	10	18	Witcher (1941)
Solenoid with ring collector	40	100	0	10	≈ 80	Frankel (1948)
Spaced coils (iron)	50	96	4 25	14 110	91 11	Agnew and Anderson (1949)
Thick air coil for 16 MeV rays with ring collector	7	100	1	6	125	Graham and Klemperer (1952)
Air coil, uniform field Hubert caustic baffle	28	230	<0·5	40 360	55 20	Jungermann *et al.* (1962)
Einzel lens	10^{-3}	70	$(5 \times 10^{-3}) \times 4$	10^{-2}	7×10^4	Möllenstedt (1949)

peripheral electrons are already reflected (cf. electron mirror in fig. 4.9). Trajectories illustrating this property of the einzel lens filter have been calculated by Regenstreif (1951); they are plotted in fig. 7.8 for a symmetric einzel lens with thin electrodes. There, a uniform parallel beam enters the lens from the left. Nine trajectories are shown and the initial semi-apertures (r) in terms of

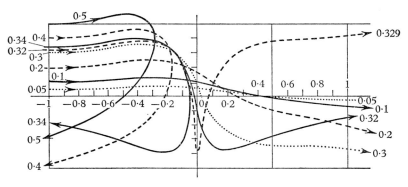

Fig. 7.8. Electron trajectories through an einzel lens/mirror.

the spacing (s) between adjacent diaphragms are marked at each trajectory. Up to $r/s = 0.33$ all electrons are transmitted but rays with larger semi-aperture are reflected.

The voltage distribution in the lens is characterized by the ratio between the minimum height V_s of the potential saddle and the energy V_{El} of the incident electrons which, for fig. 7.8, is

$$V_s/V_{El} = 0.1.$$

If the negative bias of the middle electrode is gradually increased, the transmitting area of the filter contracts like an iris and it vanishes for $V_s = V_{El}$ when all incident electrons are reflected.

The essential properties of einzel lens filters as energy analysers are well illustrated by some results obtained by Frost (1958). He used an arrangement which is shown schematically in fig. 7.9(a). Three diaphragms D_A, D_M and D_B are placed in front of the Faraday collector F. The aperture D_A is small and it defines the diameter of the electron beam. The middle diaphragm D_M, to which the negative control potential is applied is relatively thick

$(2z_M = 4$ mm) and has a wider aperture $(2R_M = 2$ mm) than D_A. The aperture in the third diaphragm is again wider than that of D_M. It will be seen from fig. 7.8 that the trajectories of electrons leaving the aperture D_M will always diverge. Both the outer diaphragms D_A and D_B are at earth (anode) potential. Fig. 7.9 (b)

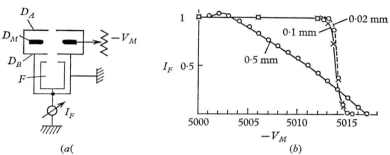

(a((b)

Fig. 7.9. (a) Einzel-lens filter and Faraday collector. (b) Characteristics of an einzel-lens filter for electron beams of different diameters.

shows three curves representing the energy analysis of an approximately parallel beam from a gun with thermionic cathode. The current I_F collected in the Faraday cage F is plotted against the voltage V_M of the middle electrode D_M. The three curves refer to three different beam diameters which were collimated by aperture stops of 0·5, 0·1 and 0·02 mm respectively in the diaphragm D_A.

The resolution of the filter lens is seen to increase with decreasing beam diameter, in the best case (for the 0·02 mm beam) it was found to be 1 in 10^4 volt. The electron beam was accelerated by 5000 V, the curves in fig. 7.9 (b), however, reach the abscissa at $V_M = 5017$ V. The voltage V_M applied to the middle electrode, for just stopping the passage of the beam, always exceeds the energy of the beam electrons by the difference $V_M - V_S$ where V_S is the voltage of the lowest point of the potential saddle which for the symmetrical einzel lens occurs in the centre of the middle aperture. The negative potential across this aperture gradually decreases towards its centre, hence the paraxial electrons of the beam of finite diameter will still be able to pass through the lens when the peripheral electrons are already reflected.

The potential V_S of the saddle point with respect to a given V_M, and also the radial variation of the potential across the surface of maximum retardation in the middle diaphragm, depend very much on the geometry of the filter. The saddle field becomes flatter with increasing thickness of the middle electrode D_M and with decreasing diameter of the aperture in D_M. The flattest saddle field, stopping electrons of a given homogeneous energy with the smallest range of V_M, would lead to the best energy resolution. Moreover, under optimum conditions, all electrons of the beam should meet the stopping equipotential of the field saddle as perpendicular as possible.

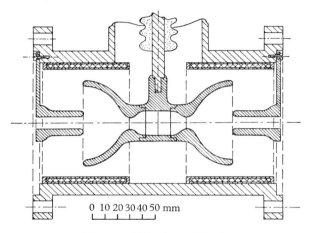

0 10 20 30 40 50 mm

Fig. 7.10. Filter lens of Brack.

A special design of electrodes in order to achieve these conditions has been developed by Brack (1962). A longitudinal cross-section of his einzel lens is shown in fig. 7.10. The saddle point was found to be only 1 volt lower than the potential at the middle electrode, when 40 kV were applied to it, with the outer electrodes being earthed. The expected energy resolution was

$$\Delta V = 6 \times 10^{-3} \text{ eV}.$$

The lens was used as an energy filter for a diffraction pattern.

The design of a spectrometer in which the essential conditions can be clearly recognized, has been developed by Boersch and Miessner (1962). As shown in fig. 7.11, an electron gun projects a narrow pencil of 20–50 keV and of very small intensity (10^{-9}–

10^{-11} A) on to a thin film, the energy loss spectrum of which is being investigated. By means of a double condenser, the scattering angle is selected and the pencil is realigned with the electron optical axis, while the measuring aperture selects an angular range of 0·15 mrad only. A pencil of 0·05 mm diameter enters the filter lens. This consists of the saddle field of an einzel lens with a

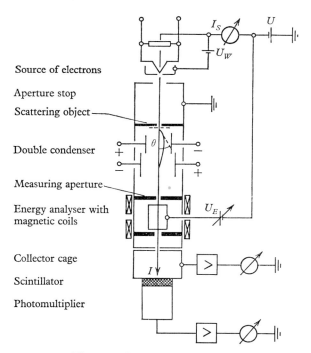

Source of electrons

Aperture stop

Scattering object

Double condenser

Measuring aperture

Energy analyser with magnetic coils

Collector cage

Scintillator

Photomultiplier

Fig. 7.11. Spectrometer of Boersch and Miessner.

relatively long tubular middle electrode, combined with a strong axial magnetic field (the cross-section of the coils is seen in the figure) which prevents spreading of the beam due to small un-avoidable transverse velocity components. Only electrons with energies greater than the electrostatic potential of the field saddle pass into the post-accelerating potential of a collector cage. There, they are registered by a scintillator and a photomultiplier. The spectrometer resolved 0·4 eV, this being the limit given by the energy distribution in the primary beam. The einzel lens itself should have a substantially higher resolution.

Filter lenses have proved to be very suitable for selecting a narrow interval from the Maxwellian energy distribution of a beam from an electron gun. In this case, Hartl (1966) has studied the use of the filter as a monochromator. He used the einzel lens of Brack (fig. 7.10) but he superimposed an axial magnetic field over it. The power $1/f$ of the einzel lens is known to oscillate between positive and negative values when the potential of the middle electrode V_i is varied (cf. §4.6, also fig. 7.7). These oscillations accumulate when V_i approaches the cut-off point. In the present einzel lens, the voltages V_i for the various states of zero lens power ($1/f = 0$) are very closely spaced. It was found, however, that with the superimposed magnetic field, the distances between these zero power states could be increased. Moreover, when the incident beam is slightly tilted (≈ 1 mrad) towards the lens axis, the various foci of the lens will be displaced sideways, and by means of a fine aperture, a pencil of suitable energy could be chosen.

In this way it was possible to monochromatize thermionically emitted electrons of 30–50 keV so as to obtain a pencil of 12 meV half width. The resulting energy distribution could be checked by measurements with a spherical deflexion analyser which will be described later in §10.4. It must be remarked, that the current density in the pencil was of the order of 10^{-7} A m^{-2} only, and it is of fundamental importance for any monochromator that the minimum obtainable band width ΔV of electron energies depends upon the current density i. (Owing to a space charge effect, discovered by Boersch (1954) and confirmed by various later investigators (cf. Hartwig and Ulmer, 1963 a, b), ΔV increases proportional to $i^{\frac{1}{3}}$.)

A filter with a fine wire mesh was used by Boersch (1949, 1953) in order to clear inelastically scattered electrons from electron microscope pictures or from diffraction patterns. His filter, shown in fig. 7.12, is again an einzel lens, the negatively charged electrode being a fine wire gauze G with 40 meshes/mm and of 70 per cent transmissivity. The vacuum tube A which is on anode (earth) potential acts as the two outer electrodes of the einzel lens. The electrons enter the filter as a slightly divergent beam, but the tubular extension T, connected to a diaphragm on which the gauze G is mounted, curves the equipotentials in front of the gauze so that the trajectories are approximately perpendicular to the gauze. In this way the total energy of the electrons can be measured by the potential in the middle of the gauze meshes which is just

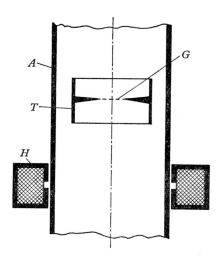

Fig. 7.12. Electron gauze filter of Boersch.

sufficient to block their way. After passing through the filter, the electron image is magnified and reprojected on to the photographic plate by the magnetic lens H in combination with the electrostatic lens formed by the tubular extension of T behind the gauze. A slight defocusing is sufficient to avoid a projection of the gauze structure into the picture.

Various other ingenious designs of lens filters have been employed. We mention only a five-electrode filter lens by Simpson and Marton (1961) in which two short-focus immersion lenses are placed back to back about a central electrode which contains the field saddle. On to this saddle an image of the entrance plane is projected by the first immersion lens, this image being reprojected by the second immersion lens on to the exit aperture. A special merit of this filter is its efficient energy resolution for relatively wide beams.

Another filter which should be mentioned here, is based on the electron mirror arrangement of Castaing and Henry (cf. fig. 4.14) which we have discussed in §4.7. Möllenstedt and Gruner (1968) used the mirror for sorting electron energies by applying the proper negative potential to the backing plate, so that only the slow electrons were reflected while the fast electrons landed on the backing plate where they were conducted away. The half width of a thermionic electron beam was reduced in this way from 0·5 to 0·2 eV.

§7.4. Relativistic aberration

A particular type of chromatic aberration is the relativistic aberration that is due to the variation of the mass of the electron with its velocity. This aberration is of a particular interest in the design of electron accelerators and the handling of the accelerated beams; the energy of which reaches millions of electron volts. Again, high voltage electron microscopes have been designed to energies up to 1 MeV where the mass of the electron has increased already by a factor of 2·9. The focal length of a lens for relatively slow electrons, say up to some keV, depends only on the ratio of voltages applied to the electrodes; it does not depend on the absolute values of these voltages. However, it has already been pointed out in §§4.2 and 4.6 that even for a fixed voltage ratio, the focal length of a lens starts to change when the electron velocities become so great that relativistic increases in electron mass can no longer be neglected. Now, this relativistic influence on the focal length of an electron lens is of great practical interest in connexion with the chromatic aberration of saddle-field lenses. In particular, the univoltage lenses described in §4.6, in which an intermediate electrode is connected to the potential of the cathode, have negligible ordinary chromatic aberration for electrons up to some keV travelling near enough to the axis. Since for relatively slow electrons the focal length of a univoltage lens is independent of the voltage of the external lens electrodes it has even been possible to project an image of the cathode by applying a.c. voltage to the external electrodes.

The great advantage of using such univoltage lenses in electron microscopes is based on their being fairly insensitive to voltage fluctuations. Somewhat earlier (§7.1) it has been pointed out that the accelerating voltage in magnetic electron microscopes has to be stabilized within a fraction of 10^{-4} or 10^{-5} in order to avoid a loss in definition due to chromatic aberration. The voltage in an electrostatic microscope using univoltage lenses need not be stabilized to such a high degree, but nevertheless it needs some stabilization, since the electron velocities are so great that relativistic effects can no longer be disregarded.

The relativistic aberration of einzel lenses (cf. §4.6) has been studied in detail by Ramberg (1942). He derived the following expression for the relative change in focal length f which applied for large image distances:

$$\Delta f = C_{rel} \Delta V, \qquad (7.19)$$

where ΔV is the voltage fluctuation and C_{rel} is a constant for a given lens. The coefficient C_{rel} is of the order of $7 \times 10^{-7} V_0 f$ for weak, initially decelerating einzel lenses, in which the potential V_i of the intermediate electrode is not much different from the potential V_0 of the two outer electrodes. If the power of the einzel lens is gradually increased by reduction of the potential V_i towards cathode potential, the relativistic aberration constant C_{rel} gradually decreases. Eventually C_{rel} reaches a value of about $2 \cdot 5 \times 10^{-7} V_0 f$ when V_i reaches cathode potential, i.e. for the univoltage lens.

Equation (7.19) fixes an upper limit for the voltage fluctuations ΔV that can be tolerated in an electrostatic electron microscope with a univoltage objective. Let the resolution of a certain distance Δr be required, Δr being the radius of the disc of confusion at the object side. For large image distances the permissible fluctuation in focal length is given by $\Delta f = \Delta r/\Theta$, where Θ is the semi-aperture of the focused rays. For example, a semi-aperture $\Theta = 10^{-3}$, a focal length $f = 4$ mm of the objective and a required resolution of 10 Å leads with (7.19) to a tolerance of 1 keV for the voltage fluctuations.

Some technical importance must be attributed to the fact that the relativistic and the ordinary chromatic aberration coefficients are of opposite sign. Le Rutte (1948) pointed out that the two aberrations can be made to cancel one another. For this purpose it is only necessary to apply a slight bias to the inner electrode of the einzel lens so that just the right amount of ordinary chromatic aberration is introduced.

The question, whether ordinary spherical aberration for a monochromatic electron beam could disappear at relativistic velocities, has been discussed by Rose (1967). He came to the conclusion that the aperture error of round lenses is unavoidable even in the case of very fast electrons. The aperture aberration of an electrostatic–magnetic quadrupole doublet, however, can change sign at certain accelerating voltages higher than 710 kV. At such high voltages the chromatic aberration of the above doublet is also, in principle, correctable.

§7.5. Space-charge error

The optics of a given electron lens is sometimes entirely changed by space-charge effects if it is traversed by an electron beam of great current density. In this way the cardinal points can be displaced and large third-order aberrations can be introduced. In principle it should be possible to distribute space charges in such a way that the properties of the lens are not altered. It should even be possible to correct geometrical lens errors by introducing space charges. Such theoretical possibilities still hold some hopes for future development. Practical electron optics, however, will have to develop still a long way before such beneficial effects of space charge could be realized. As a matter of fact, space charge seems at the moment only to upset correct image formation. The following empirical conclusions may be mentioned in this connexion:

(i) Spherical aberration in a beam of electron rays, focused by an electron lens, appears to increase with increasing beam current as a result of the space-charge carried by the beam itself (cf. §8.4).

(ii) Electron clouds circling about the lens axis have been injected into magnetic electron lenses in certain experiments which were started by Marton and Reverdin (1950) with the object of correcting aberration. Such clouds, however, seem to have on electron rays a similar effect to that which frosted glass has on light rays.

(iii) If an electron image from a thermionic cathode surface is to be projected by means of an electron lens, the emission current must be saturated. If, however, the emission density is raised, say by an increased cathode temperature, so that the emission current becomes space-charge limited, the image disappears, giving way to a diffuse brightness on the target.

Because of such disturbing effects it would be appropriate to deal with space charges here under the heading of electron-optical errors. However, a detailed discussion of 'electron optics and space-charge' is given in chapter 8, and it seems to be preferable to postpone the discussion of space-charge errors until the general principles and effects of space-charge have been expounded.

§7.6. Diffraction error

So far a treatment of electrons as charged corpuscles, moving through space according to ballistic laws and being deflected by fields according to electromagnetic laws, has been found to be quite adequate for the determination of electron trajectories. Alternatively the trajectories could be obtained by the geometric-optical approach, studying the refraction of rays at surfaces of given refractive indices. It may be pointed out now that the wave nature of the electron has to be taken into account as soon as electron rays are confined to apertures which are small enough to be comparable with the wavelength of the ray.

From light optics it is well known that two points of the object space will be projected as separate points into the image space only if their mutual distance exceeds a certain minimum which is a function mainly of the wavelength of the rays producing the image.

The wavelength λ of an electron ray is given by

$$\lambda = \frac{h}{p} = \frac{h}{e(Br)_e} = \left(\frac{h^2}{2em_0 V} \frac{1}{1 + \frac{e}{2m_0 c^2} V} \right)^{\frac{1}{2}} \qquad (7.20)$$

or $\qquad \lambda = \dfrac{0\cdot4135 \times 10^{-14}}{(Br)_e} = \dfrac{12\cdot26 \times 10^{-10}}{V_r^{\frac{1}{2}}},$

where h is Planck's constant, $p = mu$ is the momentum of the electron, $(Br)_e$ the electron momentum and V_r the relativistically corrected accelerating voltage (1.9).

If a train of waves passes through an array of narrow apertures, it will be diffracted into a sequence of intensity maxima and minima. According to elementary wave theory the first diffraction maximum is found at a deflexion angle α determined by the wavelength λ of the rays, by the refractive index N at the entrance relative to that at the exit side of the apertures, and by the mutual distance D of these apertures, namely, by

$$N \sin \alpha = \frac{\lambda}{D}. \qquad (7.21)$$

The diffracted rays emerging from the apertures can be used for the formation of an image if they are projected by a lens into the image points. Now practical electron lenses suffer from spherical aberration which grows with the lens aperture (cf. §6.3). To keep this aberration within practicable limits, the lens aperture has to be restricted by a stop. Abbe showed in 1874 that no image could be obtained so long as all the diffraction maxima were excluded from the image-forming beam by the use of too small an aperture stop. The first maximum at least must be admitted in order to give a recognizable image, and all the maxima are theoretically necessary for the formation of a perfect image. Therefore, if the aperture stop admits from the object to the lens a bundle of rays of semi-aperture α, the minimum distance of two object points which will be resolvable in the image should be of the order of D in (7.21), λ being the wavelength of the image-forming rays. The expression on the left-hand side of (7.21) which is characteristic of the resolving power is known as the 'numerical aperture'.

In electron-optical practice, the diffraction error is important only for electron microscopes of high magnification. There, the object is nearly in the focal plane of the objective lens and the refractive index is the same in the object space and in the image space ($N = 1$).

A scheme of rays as it occurs in the first stage of a highly magnifying electron microscope is shown in fig. 7.13. In this figure, the positions of the electron source, condenser lens, object and objective aperture stop are indicated. Bundles of rays leaving the source with small divergence are focused by the condenser on to the object. The semi-angle α_c of convergence being determined by the aperture at the condenser lens and by the mutual distances of source and object from the condenser lens. The rays passing through the object, are scattered into a bundle of semi-aperture α_o which is focused by the objective thus forming an image.

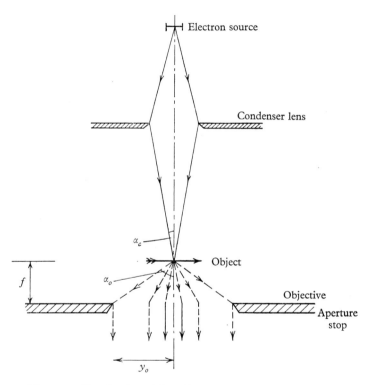

Fig. 7.13. Illumination of the object in an electron microscope.

The angular semi-apertures α_c and α_o are actually very small, only for reasons of clarity are they greatly exaggerated in fig. 7.13. The smallness of these angles must be stressed, typically they are $\alpha_o \approx 1/200$ and $\alpha_c \approx 1/1\,000$. However, the diffracted rays which are drawn in the figure in broken lines, enable the cone of electrons transmitted through the object to fill the objective aperture.

Since the angular semi-aperture α_o of rays transmitted through the object is so small, we can write

$$\sin \alpha_o \approx \alpha_o \approx y_o/f,$$

where y_o is the semi-aperture of the rays at the objective and f is the focal length of the objective lens. Hence, (7.21) may be written

$$D \approx \frac{\lambda}{\alpha_o} = \frac{\lambda f}{y_o}. \tag{7.22}$$

The exact form of (7.22) and thus the resolved distance D depends on the degree of coherence of the electron beam.

To understand this, it must be remembered that the derivation of (7.21) assumes that at any given instant the phase of the incident wave is constant over the whole array of apertures. Now if there is a fixed phase relationship between waves arriving at different points of the object within the field of view, the illumination is said to be coherent. This is the case, for instance, if a point source is used and the condenser aperture α_c is sufficiently small to make the resulting Airy disc in the object plane at least as large as the required field of view. There, for practical purposes, the illumination which arrives within the central maximum of the Airy pattern can be considered to be all in phase. An infinitesimally small point source is, however, an abstraction.

If the practical case of finite source size is considered, a high degree of coherence in the illumination can be retained, provided α_c is now small enough for the Airy disc, corresponding to a source point, to be large compared with the field of view at the object. This means that all the Airy maxima due to neighbouring points on the source will practically coincide in the object plane, resulting in only a small variation in phase over the illuminated area.

With increasing α_c, the diffraction maxima from each source become narrower until they cease to overlap, in which case the object is incoherently illuminated. This incoherent limit is reached when the diameter of the Airy disc at the object, due to a single point on the source, becomes comparable in smallness with the least resolved distance D. In this case, adjacent object points can be treated as independent self-luminous sources.

Hence we see that the coherence of the illumination increases if α_o is fixed and α_c decreases. However, if α_c is fixed and α_o decreases, the illumination becomes more incoherent, since D increases. It is thus that the ratio α_o/α_c of the angles subtended at the object by the condenser and objective apertures (see fig. 7.13), can be used as a measure of the coherence of the illumination. If $\alpha_o/\alpha_c \gg 1$, the illumination is highly coherent; whereas with decrease of this ratio, the degree of coherence decreases, the illumination being effectively incoherent for $\alpha_o/\alpha_c \approx 1$ (Fagot, Ferré and Fert, 1961). In practical electron mircoscopy it is more usual to work with quite high values of this ratio; a typical value is $\alpha_o/\alpha_c = 5$, and the illumination is therefore fairly coherent. The problem of microscope resolution with partially coherent illumination was analysed by Hopkins and Barham (1950), who showed that D decreases as α_o/α_c decreases from infinity, i.e. as the illumination becomes less coherent the resolving power improves. For $\alpha_o/\alpha_c = 5$, D in (7.22) is reduced by a factor 0·8.

If the illumination is incoherent, then the argument used to derive (7.22) is no longer valid. The object can be treated as self-luminous (Scherzer, 1939), the resolved distance for two self-luminous object points being simply the radius of the Airy disc, $0·61\lambda/\alpha_o$.

Referring to the question whether a coherent electron wave becomes incoherent due to scattering by the atoms of the specimen, it is expected that elastic scattering processes should give rise to a scattered wave coherent with the unscattered background wave, while inelastic processes, which result in a change of electron wavelength would produce an incoherent component. The relative probability of elastic and inelastic processes depends on specimen composition and thickness in a manner still not fully understood. It can be asserted, however, that the existence of phase contrast effects within the depth of field of the microscope (Fagot, Ferré and Fert, 1961) and of well-defined Fresnel fringes (cf. the end of this section), shows that an electron beam can largely retain its coherence while passing through a thin film specimen. Hence for all practical purposes, (7.22) can be regarded as giving a fairly close estimate for the radius of the disc of confusion due to diffraction.

The wavelength of the rays in a highly magnifying electron-microscope objective (60 kV electrons with 5×10^{-12} m wavelength) is by a factor 10^5 smaller than the wavelength of the visible light used in glass microscopy. However, the least resolved distance in the best electron microscopes is only by a factor 10^2 smaller than that resolved in a good oil-immersion light-microscope. The gain in resolving power obtained with the electron microscope is relatively small because (cf. 7.22) the numerical apertures used are minute as compared with the numerical apertures to which we are accustomed in light microscopy, since these are of the order of one unit (1·4 for oil-immersion objectives).

A further consequence of the small numerical aperture of the electron microscope is the large 'depth of field' of an electron microscope, as compared with the least resolved distance. In a good light microscope, these quantities are of the same order. The depth of field indicates the amount by which the objective lens may be out of focus before the blurring becomes comparable with the attained resolution. This is expressed as the distance between two planes, on either side of the correct focal plane, within which the above condition holds, and is given by $2S_{min}/\alpha_o$, S_{min} being the least resolved distance for the instrument (cf. §7.7). For an electron microscope of very high resolution $S_{min} \approx 5$ Å, $\alpha_o \approx 5 \times 10^{-3}$, giving a depth of field of 2000 Å.

The distance in final image space corresponding to the depth of field in object space is called the 'depth of focus'. In a high magnification (e.g. $\times 10^5$) instrument, this is of the order of a cm, and this is a very useful fact since it allows the photographic plate to be placed in a plane below that of the viewing screen, without loss of definition.

Another diffraction error may be mentioned in this section. Since the beginning of electron microscopy wide contour lines following the outlines of images have often been observed, especially when the picture was slightly out of focus. Boersch (1940, 1943) and Hillier (1940) interpreted these lines as Fresnel diffraction fringes due to the edges of the object. By their dependence on the focusing of the image these fringes can easily be distinguished from any real surface structures of the object.

In practice, these Fresnel fringes are of considerable assistance in adjusting a high resolution microscope. The exact focus may be found by adjusting the power of the objective so that the fringes which are visible for slight overfocusing and underfocusing disappear. The fringes also provide a very sensitive test for the presence of astigmatism, thus enabling the operator to judge the correct setting for the stigmator (cf. §6.2). The distance d between the zero and maximum of the first fringe in the diffraction pattern is, according to Haine and Mulvey (1954), approximately given by

$$d \approx \sqrt{\lambda z}, \qquad (7.23)$$

z being the distance off-focus of the plane of the fringes. Thus with 60 kV electrons ($\lambda = 0.05$ Å), defocusing by 2 μm should reveal a Fresnel fringe of width about 30 Å.

Astigmatism has the effect that the focal plane for edges oriented in one astigmatic plane are longitudinally displaced by a distance z_a from the focal plane for perpendicularly disposed edges. If, for example, the image of the edge of a hole in a carbon film is examined, the presence of a small amount of astigmatism will be revealed by a variation of fringe width with orientation. The astigmatic distance z_a can be deduced from the maximum and minimum fringe widths, viz.

$$z_a = \frac{1}{\lambda} (d_{max}^2 - d_{min}^2). \qquad (7.24)$$

For further details of the use of Fresnel fringes in electron microscopy, the reader is referred to Haine and Cosslett (1961).

In the projection of electron microscopic images of structures of the order of $\lesssim$ 10 Å by electron lenses of reasonably large aperture, scattering absorption is not likely to produce the observed contrast. Lenz was first to conclude that phase contrast here is the essential factor in producing the picture.

Near the resolution limit, the origin of contrast must be understood to be due to phase shifts of the electron wave in the potential field which surrounds the various parts of the object. These potential changes corresponding to changes in refractive index produce temporary changes in wavelength and thus phase shifts. The objective unites the diffracted and the undiffracted waves and interference between these waves produces the contrast in the image.

The potential field in the object can be considered as a super-position of sinusoidal potential fluctuations which can be represented as a Fourier series. Let the pitch of such a potential lattice be Λ, corresponding to a spatial frequency $1/\Lambda$. To every spatial frequency can be coordinated a certain angle of diffraction Θ where

$$\Theta = \lambda/\Lambda, \tag{7.25}$$

λ being the electron wavelength. Lenz (1965) assumed that $\frac{1}{2}\pi$ phase difference between diffracted and undiffracted waves would give a condition for maximum contrast. Now at a plane distant Δz from the object plane, the phase difference between the diffracted and undiffracted waves can be shown to be $(\pi/\lambda)\Delta z\,\Theta^2$. Hence from (7.25), there is a phase difference of $\frac{1}{2}\pi$, when

$$\Lambda = \sqrt{2\lambda\Delta z}. \tag{7.26}$$

Thus, maximum contrast will occur when the image is defocused from the Gaussian image plane by a distance Δz which is determined by the spatial frequency $1/\Lambda$ according to (7.26). In practice this simple expression is valid only for $\Lambda > 12$ Å. For smaller values of Λ, it is necessary to take account of the phase shifts associated with the spherical aberration of the objective; also, greater generality is obtained by assuming that maximum contrast occurs for phase shifts of $(2n-1)\frac{1}{2}\pi$. Thon (1965) quotes the following more accurate relationship between the degree of defocusing and the most favoured spatial frequency:

$$\Lambda = \lambda\left[\frac{\Delta z}{C_s} \pm \left(\frac{\Delta z^2}{C_s^2} + \frac{(2n-1)\lambda}{C_s}\right)^{\frac{1}{2}}\right]^{-\frac{1}{2}}, \tag{7.27}$$

where $n = \pm 1, 2, 3, \ldots$. It is seen that for a given Δz, Λ can now take many values, corresponding to different orders of diffraction.

According to experiments by Thon (1965, 1966), a series of electron micrographs of an object with sufficiently fine structure, e.g. of a carbon film, can be taken at several values z of defocus (e.g. -5000 Å $< \Delta z < +5000$ Å). In these micrographs the distribution of distances Λ between the various maxima of current density (e.g. 4 Å $< \Lambda < 18$ Å) is then determined. A series of average Λ-values is obtained, either by direct measurement of these distances or, more easily, from light-optical diffraction patterns of the micrographs (light-optical Fourier transformation: cf. Taylor, Hinde and Lipson, 1951). The diffraction patterns consist of a series of rings, and the radius of each of these rings corresponds to a Λ-value. By plotting Λ against Δz 'transfer functions' (cf. §7.7) can be obtained for the various orders of diffraction corresponding to different orders of n. These are in good agreement with the $\Lambda = f(\Delta z)$ curves constructed according to (7.27).

Much improved contrast can be obtained by introduction of a zone plate in the back focal plane of the objective. This was proposed first by Hoppe (1963) and successful experiments have been reported by Möllenstedt, Thon et al. (1969). We have already mentioned that such a zone plate can eliminate the spherical aberration of the objective. It also has the effect of limiting the image projection to certain spatial frequency bands passing only equiphase waves and obstructing the passage of waves of opposite phase. This can be understood by inspection of (7.25); the zone plate is in effect a series of annular stops which eliminate certain frequency bands by stopping off the corresponding range of diffraction angles. Pictures obtained in this way are not only of improved contrast but also more accessible to interpretation. Resolutions of 2·3 Å have been obtained. However, the difficulty of making the zone plate and adjusting it are very great. Such a plate may contain, e.g. five circular apertures and it must be adjusted accurately on an area of 0·1 mm radius.

A comparison of the mechanisms of phase contrast in electron microscopy and optical microscopy shows that the spherical aberration of the objective plays a vital role in the formation of phase contrast in the electron microscope image. In optical microscopy phase contrast is made visible by interposing a quarter wave plate in order to retard the phase of the axial, undiffracted wave by $\frac{1}{2}\pi$ with respect to the diffracted off-axis portions. By eventual superposition of the two-wave portions, the phase variations are transformed into amplitude variations in the image. The object thus becomes visible (see, e.g. Françon, 1963).

In electron microscopy, however, phase contrast is produced by defocusing; the spherical aberration, being a function of the zone radius, introduces the necessary phase difference between axial and off-axial portions of the transmitted waves. If the image is defocused by the correct amount, it is possible to locate an off-axial zone whose phase differs from that of the axial portion by approximately 90°.

A detailed consideration of the wave front aberration in the back-focal plane by Hanszen and Morgenstern (1965) has shown that for a particular distance from the Gaussian image plane, there can be found the off-axial zone whose phase relative to the axial portion is quite close to 90°.

§7.7. Combination of aberrations

The resolving power of an objective lens is determined by all the limiting aberrations which have been discussed in this and in the previous chapter. However, most of the aberrations can be kept small enough to be neglected in comparison with the three most important errors. These errors are:

   (i)   the diffraction error D given by (7.22);

   (ii) spherical aberration Δr_s given by (6.10), and

   (iii) chromatic aberration Δr_{cr} given by (7.3).

These three quantities are radii of discs of confusion in the Gaussian image plane: with the assumption that the intensities in these discs have bell-shaped distributions, they can be added as Gaussian functions so that the resulting error is given by

$$S = \sqrt{\{D^2 + (\Delta r_s)^2 + (\Delta r_{cr})^2\}}. \tag{7.28}$$

Now, the diffraction error D and the spherical error Δr_s vary in opposite senses with the aperture. Thus there will be an optimum angular semi-aperture Θ_{opt} for which the combined effect reaches the minimum confusion S_{min}. This minimum can be obtained by substituting $D = \lambda/\Theta$ (see 7.22) and $\Delta r_s = C_s\Theta^3$ (see (6.10)) into (7.28), differentiating with respect to Θ and taking $dS/d\Theta = 0$. Experimental results on the resolving power seem to indicate that the actual optimum aperture is slightly greater than that one calculated from (7.28).

For practical purposes, the estimation of the minimum combined confusion may be simplified. It seems quite adequate to neglect Δr_{cr} and to assume that S_{min} is obtained when D and Δr_s are of the same order of magnitude. Hence

$$D \approx \Delta r_s \approx S_{min}/\sqrt{2}.$$

Substituting again from (7.22) and (6.10) and introducing the equation for the wavelength

$$\lambda = 12 \cdot 2 \times 10^{-10}/\sqrt{V_r} \quad (\text{cf. (7.20)}),$$

one obtains for the optimum angular semi-aperture

$$\Theta_{opt} \approx \left(\frac{\lambda}{C_s}\right)^{\frac{1}{4}} = \frac{6 \times 10^{-3}}{C_s^{\frac{1}{4}} V^{\frac{1}{8}}}, \qquad (7.29)$$

where C_s is the spherical aberration coefficient (cf. §6.3). The minimum radius of the disc of combined confusions is

$$S_{min} = 1 \cdot 4 C_s^{\frac{1}{4}} \lambda^{\frac{3}{4}} = 3 \times 10^{-7} C_s^{\frac{1}{4}} V^{-\frac{3}{8}}. \qquad (7.30)$$

For example, an objective may be given with an aberration coefficient $C_s = 2$ mm. For electron rays of $V = 60$ kV, (7.29) and (7.30) yield $\Theta_{opt} = 7 \times 10^{-3}$ radian and $S_{min} = 10$ Å. Using this particular example, it may now be shown that the result of calculating the combined confusion is not much modified by consideration of the effect of chromatic aberration. A voltage fluctuation of 3 in 60000 and a focal length of the objective $f = 2$ mm may be given. Using the known value $C_{cr}/f = 0 \cdot 7$ given in §7.1 for the coefficient of chromatic aberration and with the above value $\Theta_{opt} = 7 \times 10^{-3}$, one obtains, according to (7.3), a radius $\Delta r_{cr} = 5$ Å for the chromatic confusion. If this value of Δr_{cr} is, according to (7.28), added quadratically to D and to Δr_s, one calculates $S_{min} = 11 \cdot 2$ Å instead of 10 Å for D and Δr_s alone.

The best dimensions of magnetic microscope lenses for minimizing spherical and chromatic aberrations have been discussed in §5.3 and §7.1 respectively. From the results quoted, it can be seen, however, that the two kinds of aberration reach a minimum at somewhat different gap-to-bore diameter S/D and at different excitations nI. As a compromise for good chromatic aberration without unduly large spherical aberration Mulvey and Wallington (1969) recommend a lens with $S/D = 2$ and with an excitation coefficient $nI/\sqrt{V_r} = 13$ (e.g. $nI = 4300$ A, $V_r = 110$ kV) resulting in $C_{cr} = 0 \cdot 076$ cm and $C_s = 0 \cdot 098$ cm.

According to (7.30) it might appear that the most effective method for improving the resolution is to raise the voltage V for accelerating the electron rays. However, Cosslett (1946) has pointed out that there should exist an optimum voltage since the aberration coefficient C_s is not independent of V. C_s is about inversely proportional to the focal length (cf. §6.5). Since in the available magnetic objectives the magnetic field cannot be raised beyond a certain value, the focal length is bound to increase with voltage once this saturation value for the magnetic field is approached. Already at voltages of about 60 kV, this increase in focal length is rapid enough to offset the parallel reduction in diffraction error, and with present-day best objectives hardly anything is gained by going to the higher voltages.

The optical quality of an electron lens will be the better, the more every possible aberration can be minimized. It is, however, of great interest if the quality can be judged without a separate measurement of all these errors. One relatively fast way of judging the lens is the inspection of the caustic which is the envelope of the focused rays. Caustic lines are obtained on any plane which intercepts the beam. If aberrations are present, these lines do not converge into a point when for any bundle of homocentric incident rays, this plane is moved from the lens to the Gaussian image distance.

Some discussion of caustic lines, as obtained on a fluorescent screen or on a photographic plate has been given by Leisegang (1953, 1954) and by Hanszen and Lauer (1969). The caustic lines obtained from an electron microscope condenser lens are discussed by Magnan (1961), who shows the form of the caustic in the presence of varying degrees of astigmatism.

A concept of great importance for judging the quality of a lens is the 'modulation transfer function'. The use of this function in glass optics has been reviewed by Rosenhauer and Rosenbruch (1967), it is, however, only beginning to be applied in electron optics. A plot of the transfer function contains as abscissa the 'spatial frequency' ν in cycles, say, per mm corresponding to a resolution of lines/mm. The resolution of the lines, however, depends on the clearness and the contrast of the object lines, hence a special definition of these is required. There, the object lines are produced by a sinusoidal variation of brightness of a given depth m of modulation. If now the depth of modulation m' occurs in the projected image, then the modulation transfer function is $T = \dfrac{m'}{m}$.

$T(\nu)$ is plotted, e.g. in per cent as the ordinate of the graph, frequently given as the 'response'. The graph may contain several curves for $T(\nu)$ the parameters of which are for instance the different apertures at which the lens under investigation has been tested. A discussion of the transfer function for electron lenses of different focal length and aberration constants has been given by Hanszen (1966).

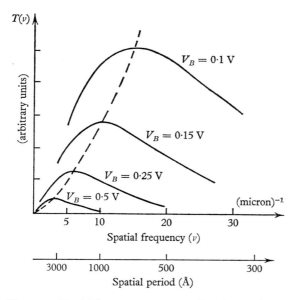

Fig. 7.14. Spatial frequency response of a mirror microscope.
Broken line: $T(\nu_{max}) \propto \nu_{max}^{\frac{3}{2}}$.

As an example, we show in fig. 7.14 a graph of the modulation transfer function which was deduced by Barnett and Nixon (1967) for their mirror electron microscope. Instruments of this kind have been developed by Mayer (1955, 1957) for examining the electrostatic field distribution at a solid surface. The electron beam is reflected at equipotential surfaces very close to the physical surface of the specimen which is held at a very small negative voltage with respect to the cathode from which the beam is originated (cf. §4.7). During the time that the beam is travelling slowly in the vicinity of the specimen, it is modulated by the surface microfields, this modulation giving rise to the image contrast on the final viewing screen. Parameters of the curves in fig. 7.14 are various bias voltages at the specimen. The response $T(\nu)$ could be given in arbitrary units only; the spatial frequency, say, of $\nu = 10\mu^{-1}$ denotes 10 line pairs per micron which corresponds to a least resolved distance of 1 000 Å.

The modulation transfer function for a photoelectric image converter in which a homogeneous electric field projects the photoelectrons on to the target (cf. §7.1) has been deduced by Grant (1966). Taking as an example, a 'proximity focus' tube in which the anode voltage $V_A = 5$ kV, the most probable emission energy is 1 volt, and the cathode to target spacing is 1 mm, his calculations indicate that at a spatial frequency of 7 line pairs/mm, $T = 0.5$ and at a spatial frequency of 18 line pairs/mm, $T = 0.07$.

CHAPTER 8

ELECTRON OPTICS AND SPACE-CHARGE

§8.1. Spreading of a homogeneous, homocentric beam

In a homogeneous, homocentric beam, all electrons have strictly the same velocity and they all aim at, or emerge from, a common geometrical centre. Homogeneous, homocentric bundles of electron rays can practically be realized only to some approximation. They are, however, of great interest because their space-charge effects

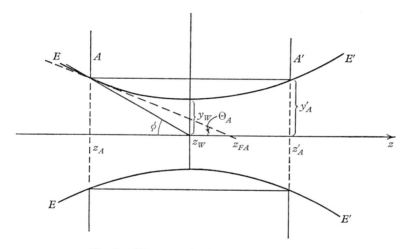

Fig. 8.1. Homocentric electron beam under the influence of its own space charge.

are easily accessible to electrodynamical calculation. It can be shown that mutual repulsion in homocentric beams deflects the rays as would an aberration-free optical lens. In fig. 8.1 is shown the cross-section of a homocentric beam of circular symmetry about the axis z, passing through a field-free space. The boundary surface of the beam is represented by the two curves EE'. At z_A the beam passes through a circular aperture A of radius y_A. There, the boundary curve EE' makes the slope angle Θ_A with a z-axis, aiming at the virtual focus z_{FA}. Under the influence of mutual repulsion, the slope angle Θ of the progressing rays decreases all the time until it becomes zero at z_W. There the beam aperture y reaches a minimum y_W, and the beam forms a waist. For further increase of z, the aperture y and the slope angle (now Θ') start to increase. They reach the values $y'_A = y_A$ and $\Theta'_A = \Theta_A$ in a position z'_A which is symmetrical to z_A with respect to z_W.

The spread of an initially parallel beam due to its own space charge was calculated first by McGregor-Morris and Mines (1925) by integration over the Coulomb forces between the electrons. Watson (1927) gave a more complete treatment by considering also the forces of magnetic attraction between the electrons. These magnetic forces, however, become noticeable for very fast rays only for which current density and space-charge effects are generally rather small. We shall follow here a straightforward approach given by Fowler and Gibson (1934) starting with a calculation of the field E at the surface of the beam which is assumed to be a cylinder filled uniformly with electric charge. From Gauss's law follows

$$-\int \rho \, dv = \int \epsilon E_y \, da, \qquad (8.1)$$

where $-\rho$ is the electron space-charge density,* dv the volume element, ϵE_y the normal flux in radial (y-)direction and da the surface element. Taking ρ as constant and integrating the volume over a cylinder of length Δz near the waist of radius y_W, we obtain

$$-\rho \pi y_W^2 \Delta z = \epsilon E_y^2 2\pi y \Delta z. \qquad (8.2)$$

For reasons of symmetry, no flux passes through the two flat end-faces of the cylinder, but the flux goes through all cylindrical surfaces of any radius y. The normal field component as given by (8.2) is

$$E_y = \frac{-\rho y_W^2}{2\epsilon y}. \qquad (8.3)$$

Now the current* I through the cross-section πy_W^2 is given by the linear velocity u_z of the electrons

$$\frac{I}{\pi y_W^2} = \rho u_z. \qquad (8.4)$$

---

* As in §1.2 (cf. footnote, p. 4) we have taken here $-e$ for the electronic charge and, analogously, we have used $-\rho$ for the electron space-charge density and $-I$ for the current carried by the electrons in the direction of the beam.

Moreover, from Newton's law together with (8.3) and (8.4) follows

$$\frac{d^2y}{dt^2} = -\frac{e}{m}E_y = \frac{e}{m}\frac{\rho y_W^2}{2\epsilon y} = \frac{e}{2mu_z}\frac{I}{\epsilon y\pi},$$

or

$$\frac{d^2y}{dt^2} = \frac{b^2}{2y}, \tag{8.5}$$

where

$$b^2 = \frac{eI}{mu_z\epsilon\pi} \tag{8.6}$$

is a constant for a beam of given electron velocity u_z and of given current I. According to (8.5) the radial acceleration of an electron at the surface will be inversely proportional to the radius y if the beam is assumed to have constant total current I. For constant current density $i = I/\pi y^2$, however, the acceleration of an electron in the beam is found to be directly proportional to its axial distance.

Multiplication of (8.5) by $2(dy/dt)dt$ and integration yields

$$\int \frac{d\left(\frac{dy}{dt}\right)}{dt} dt\, 2\frac{dy}{dt} = \int \frac{b^2}{y}dy,$$

or

$$\left(\frac{dy}{dt}\right)^2 = b^2 \log y + C. \tag{8.7}$$

Now take as the initial radius $y = y_W$, i.e. the coordinate origin may be chosen in the waist. There the rays travel parallel to the axis; thus $dy_W/dt = 0$ and the integration constant is

$$C = -b^2 \log y_W.$$

Hence (8.7) becomes

$$\frac{dy}{dt} = b\sqrt{\log\frac{y}{y_W}}. \tag{8.8}$$

With the abbreviation $\zeta = \sqrt{\log\dfrac{y}{y_W}}$,

or

$$y = y_W \exp \zeta^2. \tag{8.9}$$

Rewriting (8.8) with the help of (8.9) we find

$$\frac{dy}{d\zeta}\left(\frac{d\zeta}{dt}\right) = y_W 2\zeta \exp(\zeta^2)\left(\frac{d\zeta}{dt}\right) = b\zeta,$$

therefore
$$\int_{t_W}^{t} \frac{b}{2y_W}\, dt = \int_{0}^{\zeta} \exp(\zeta^2)\, d\zeta. \qquad (8.10)$$

The integral at the left-hand side is taken from the time t_W when the electron passes the waist, and the integral at the right-hand side is taken from $\zeta = 0$, since at the waist $y = y_W$, for which $\zeta = \sqrt{\log 1}$.

Integration of the left-hand side of (8.10) and replacement of $(t - t_W)$ by $\dfrac{z - z_W}{u_z}$ gives $\dfrac{b}{2y_W} \dfrac{z - z_W}{u_z}$.

Introducing b from (8.6) and using

$$u_z = \sqrt{\left(\frac{2e}{m} V\right)},$$

V being the energy in electron volts, we get

$$\frac{z - z_W}{2y_W} \left(\frac{I}{\pi \epsilon}\right)^{\frac{1}{2}} \frac{(e/m)^{-\frac{1}{4}}}{(2V)^{\frac{3}{4}}} = \int_{0}^{\sqrt{\log(y/y_W)}} \exp(\zeta^2)\, d\zeta,$$

or

$$\frac{z - z_W}{y_W} = \frac{4(\pi\epsilon)^{\frac{1}{2}}\left(\dfrac{e}{2m}\right)^{\frac{1}{4}}}{\sqrt{(I/V^{\frac{3}{2}})}}\, D(\zeta) = \frac{1 \cdot 15 \times 10^{-2} D[\sqrt{\log(y/y_W)}]}{\sqrt{(I/V^{\frac{3}{2}})}}, \qquad (8.11)$$

where

$$D(\zeta) = \int_{0}^{\zeta} \exp \zeta^2 d\zeta = \zeta + \frac{\zeta^3}{3} + \frac{\zeta^5}{5 \times 2!} + \frac{\zeta^7}{7 \times 3!} + \dots \qquad (8.12)$$

is the 'Dawson function' which first has been tabulated by Dawson (1898) and which has recently been calculated in great detail by Terrill and Sweeny (1949 a, b). Since ζ measures, according to (8.9), the beam spread y/y_W, (8.11) gives the relation for the radius y of the beam as a function of its distance $(z - z_W)$ from the waist, both quantities being measured in units of the waist radius y_W.

The beam-spread equation (8.11) is of a universal nature. It applies to all homocentric beams in a field-free space. The spread of a particular beam is seen to be completely controlled by the function $I/V^{\frac{3}{2}}$ of beam current I (ampere) and beam energy V

(electron volts); this function is called the space-charge factor or perveance of the beam. Again, (8.11) represents not only the paths of electrons along the boundary of the beam, it also represents the path of every electron inside the beam if we substitute for I the current I_1 that passes through a cross-section of axial symmetry containing the considered electron in its circular cylindrical boundary. This must be so since we can imagine the beam to be divided into an inner part containing the current I_1 passing through the cross-section of radius y_1 and an outer part with the current $I_2 = (I - I_1)$ passing through an annular ring aperture between the radii y and y_1. If we cut the outer part away, it will not affect the inner part, since according to Gauss's law the resultant of the forces exerted by the outer part in the region of the inner part must be zero.

Thus the paths of all electrons in the beam are geometrically similar to the paths of the electrons on the beam boundary, and the trajectories of individual electrons do not cross over one another. A beam with no trajectory crossings is called a 'laminar beam', laminar flow being a natural consequence of our hypothesis of homogeneity and homocentricity. Although, in practice, departures from laminar flow are inevitable (see §8.3), it is a useful idealization.

In fig. 8.2 which is due to Field, Spangenberg and Helm (1947), the function

$$Z = \frac{z - z_W}{y_W} \sqrt{\left(\frac{I}{V^{\frac{3}{2}}}\right)} \times 10^3 = 11 \cdot 5D \sqrt{\log (y/y_W)} \quad (8.13)$$

is plotted as abscissa against the ordinate $Y = y/y_W$. Thus fig. 8.2 represents the beam-spread equation (8.11). The centre of the beam waist is chosen as the coordinate origin, i.e. $z_W = 0$. The beam radius y measured in terms of waist radii y_W can be read directly as the ordinate Y of the graph. The axial beam co-ordinate z measured in waist radii y_W may be obtained by dividing the readings Z of the graph by 1000 times the value of the square root of the space-charge factor of the beam. For example, for $\sqrt{(I/V^{\frac{3}{2}})} = 10^{-3}$ (say, 100 V . 10^{-3} amp or say 10^4 V . 1 amp) and for a waist radius $y_W = 1$ mm, the figures of the graph could be read directly as millimetres. An initially parallel beam of this kind appears to spread out to $y = 5$ mm radius within a distance of $z = 28$ mm.

The spread curve of fig. 8.2 applies universally to any beam, since the abscissa can just be scaled up or down in inverse proportion with $\sqrt{(I/V^{\frac{3}{2}})}$. The curve is symmetrical about the value

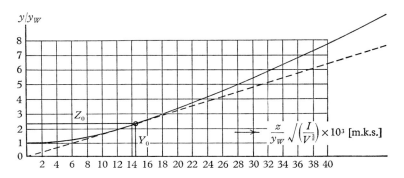

Fig. 8.2. Beam spread curve.

$z = 0$ and applies for electrons moving either to the right or to the left. All electrons, at a given instant, aim towards a virtual focus $z_{F\infty}$; however, when the bundle converges the current density grows, and each electron in the bundle follows a beam-spread curve as in fig. 8.2. The rays are gradually deflected as the current density increases, and they no longer aim towards $z_{F\infty}$, but towards a virtual focus z_F. As the electrons move forward, the focus z_F moves along the axis in the same direction. z_F will reach the coordinate origin which is placed at the beam waist when the slope of the beam envelope reaches the value

$$\left(\frac{\mathrm{d}y}{\mathrm{d}z}\right)_0 = \frac{y_0}{z_0} = \tan \Theta_0.$$

Substituting for z_0 from (8.11) yields the coordinates of the critical point

$$\left.\begin{aligned} y_0 &= 2\cdot35 y_W, \\ z_0 &= 14\cdot6 y_W/[10^3\sqrt{(I/V^{\frac{3}{2}})}]. \end{aligned}\right\} \tag{8.14}$$

The corresponding point (Y_0, Z_0) is plotted in fig. 8.2, where the critical tangent to the curve is shown as a broken line. Bedford

(1936) who first calculated the coordinates of the critical point, showed that the electron beam touches there an inscribed cone, the semi-vertical angle Θ_0 of which, according to (8.14), only depends upon the perveance, viz.:

$$I/V^{\frac{3}{2}} = 3\cdot9 \times 10^{-5} \tan^2 \Theta_0. \qquad (8.15)$$

With further advance of the converging electrons, the rate of change in slope of the rays becomes more rapid. The virtual focus moves farther along the axis and reaches infinity when the electrons reach the waist where the rays are parallel. Then as the beam gradually becomes divergent, the virtual focus moves from infinity in the direction from which the beam has come, it passes through the central position $z_F = 0$ and eventually approaches the extreme position $(-z_{F\infty})$.

Spreading of very fast beams. In addition to the electrostatic repulsion of the electrons there exists a magnetic attraction between the rays. This corresponds to the attraction between electric currents in two parallel conductors. The magnetic force is smaller than the electrostatic force by a factor $(u/c)^2$ where u is the electron velocity and c is the velocity of light. Thus the complete radial acceleration is obtained by multiplying (8.5) by the factor $(1 - (u/c)^2)$. The same expression for the radial acceleration is obtained if the field E_y and the space-charge density ρ in (8.3) are taken in a coordinate system which is fixed to a moving electron. If then the expression in (8.5) for the acceleration is transformed to a coordinate system fixed to the observer at rest, relativistic expressions must be introduced for the mass of the electron

$$m = m_0 \Big/ \left(1 - \frac{u^2}{c^2}\right)^{\frac{1}{2}},$$

and for the charge density

$$\rho = \rho_0 \Big/ \sqrt{\left\{1 - \left(\frac{u}{c}\right)^2\right\}},$$

where ρ is modified by the Lorentz contraction. The relativistically correct expression for the beam-spread curve (8.11) has been derived by Watson (1927), and we may write it here in the form

$$\frac{z - z_W}{y_W} = \frac{2e}{m} (\pi\epsilon)^{\frac{1}{2}} \frac{(Br)_e^{\frac{3}{2}}}{I^{\frac{1}{2}}} D \Big/ \sqrt{\left\{\log\left(\frac{y}{y_W}\right)\right\}}, \qquad (8.16)$$

where $(Br)_e$ is the electron momentum (cf. §5.1) and the constant factor is

$$\frac{2e}{m}(\pi\epsilon)^{\frac{1}{2}} = 1\cdot85 \times 10^6 \text{ m.k.s.}$$

Alternatively, the validity of equations (8.11) ff. can be extended into the region of great electron velocities simply by multiplying the beam voltage V everywhere by the relativistic correction factor (cf. (1.9)),

$$k_{rel} = V_r/V = (1 + 0\cdot97 \times 10^{-6} \text{ V}). \tag{8.17}$$

Thus for a 100 kV beam, say in a television projection tube, the relativistic reduction of the electrostatic repulsion, e.g. in (8.15), by nearly 5 per cent is quite noticeable. For great electron velocities, however, the practicable beam currents are nearly always so small that space-charge effects become negligible altogether.

§8.2. Application of laminar space-charge theory

A common problem in the design of high-power tubes is that of passing the maximum current at the minimum voltage through a cylinder of given dimensions. In fig. 8.1 a beam EE' is seen to enter a cylinder through an aperture A. It may be necessary that the whole beam leaves through the aperture A' which may have the same diameter as A. Obviously, at the entrance aperture A the beam should be convergent, so that it will come to a minimum diameter somewhere in the cylinder and then spread again until it just fills the exit aperture A'. From reasons of symmetry it follows that the waist should be formed exactly in the middle of the cylinder. In addition to these requirements, Spangenberg (1948) showed from the beam-spread curve (fig. 8.2) that the beam should enter the cylinder while aiming towards a point on the axis halfway between A and A'. The two circular edges defining the envelope of the beam as it passes the entrance A and exit aperture A' respectively should correspond to one point ($Y_0 Z_0$) of the beam-spread curve given by (8.14), where the tangent to this curve passes through the origin.

The slope angle of the tangent at $(X_0 Y_0)$ is identical with the semi-vertical angle of the cone tangent to the beam apertures A and A'. If radius and length of a given cylinder are R and $2L$

respectively, then $\tan \Theta_0 = R/L$. Moreover, the maximum current density at a given voltage that can be transmitted according to the first-order theory is determined by the following perveance factor:

$$I/V^{\frac{3}{2}} = 3 \cdot 9 \times 10^{-5} (R/L)^2 \text{ [amp/V}^{\frac{3}{2}}]. \qquad (8.18)$$

According to the theory, the value of Θ_0 for the maximum $I/V^{\frac{3}{2}}$ is not very critical. For example, the waist radius y_W may be varied from 0·25 to 0·6 of the cylinder radius R with an expected loss of only 10 per cent of the maximum current.

Another problem of practical interest arises in the design of cathode-ray tubes when the smallest possible spot is being focused on the fluorescent screen. Let the beam voltage and current of the beam and the coordinates z_A and y_A of the aperture (see fig. 8.1) of the final diaphragm be given. The spot size on the screen at a given distance $(z_0 - z_A)$ from the aperture A will now be only a function of the slope angle Θ_A of the beam envelope at z_A as controlled by the focusing lens. A minimum spot radius y_{min} would, however, not be obtained by moving the waist (y_W) to the screen, since shifting the waist by changing Θ_A also changes the size of the waist. Actually, the minimum spot radius y_{min} is obtained when $(z_W - z_A) < (z_0 - z_A)$, i.e. when the waist is formed at a small distance in front of the screen so that the beam has again slightly expanded when it reaches the screen. This can be derived from the theory outlined in §8.1. A comprehensive analysis of the problem was given by Weber (1967) and an empirical formula for the minimum spot radius has been obtained by Hollway (1962) namely

$$y_{min} = 5 \cdot 9 \times 10^4 y_A \left(\frac{z_W - z_A}{y_A} \frac{I^{\frac{1}{2}}}{V^{\frac{3}{4}}} \right)^{\frac{5}{2}}. \qquad (8.19)$$

For practical purposes it is often convenient to approximate the beam-spread curve by a simple parabola. For a small quantity $x \ll 1$, the approximation $\log (1+x) \approx x$ holds, and, according to (8.12), $D(x) \approx x$. Hence, the beam spread (8.11) may be written in the approximate form

$$\frac{y - y_W}{y_W} = \frac{I/V^{\frac{3}{2}}}{16\pi\epsilon (e/2m)^{\frac{1}{2}}} \left(\frac{z - z_W}{y_W} \right)^2, \qquad (8.20)$$

where $1/(16\pi\epsilon (e/2m)^{\frac{1}{2}}) = 0 \cdot 76 \times 10^4$ m.k.s. The ordinates of the parabola approximation deviate by less than 1 per cent from those of the real beam-spread curve, as long as the incremental spread $(y-y_W)$ is less than 15 per cent of the waist radius y_W.

Equation (8.20) may also be used to estimate the spreading of an initially very slightly converging or diverging beam. Taking Θ as the initial angle of convergence or divergence, the geometrical change in beam radius, $y-y_W = (z-z_W)\,\Theta$, may be a very small quantity. This geometrical change may, by independent super-imposition, simply be added to the very small increase by space charge given in (8.20).

Though (8.11) gives a complete representation of the beam spread it is generally inconvenient for practical purposes to relate the beam dimensions to the radius y_W of the beam waist, since y_W is rarely accessible to direct measurement. Hence the beam spread may be measured in units of an initial beam radius y_A which could be defined by a circular aperture A (see fig. 8.1) at an axial coordinate z_A. If this initial radius is introduced into (8.7), the value of the integration constant becomes

$$C = \left(\frac{\mathrm{d}y}{\mathrm{d}t}\right)_A^2 - b^2 \log y_A, \qquad (8.21)$$

where $(\mathrm{d}y/\mathrm{d}t)_A = (u_y)_A$ is the radial component of the electron velocity at the aperture A. With (8.21), (8.7) becomes

$$\frac{\mathrm{d}y}{\mathrm{d}t} = \sqrt{\left(b^2 \log \frac{y}{y_A} + (u_y)_A^2\right)}. \qquad (8.22)$$

Introducing $\mathrm{d}t = \mathrm{d}z/u_z$ and integrating yields

$$z - z_A = u_z \int_{y_A}^{y} \frac{\mathrm{d}y}{\sqrt{\{(u_y)_A^2 + b^2 \log (y/y_A)\}}}. \qquad (8.23)$$

Now the slope of the ray (see fig. 8.1) as it enters the aperture may be expressed either as the ratio of the radial to the axial velocity component or as the ratio of the beam radius to its subtangent

$$\tan \Theta_A = \left(\frac{u_y}{u_z}\right)_A = \frac{y_A}{z_{FA} - z_A}, \qquad (8.24)$$

where z_{FA} is the axial coordinate of the virtual focus at which the homocentric bundle is aiming while passing the aperture A.

If, as before, we use $u_z = \sqrt{\left(\dfrac{2e}{m}V\right)}$ and substitute b from (8.6), (8.22) with (8.23) and (8.24) becomes

$$z - z_A = (z_F - z_A) \int_1^{u/u_A} \frac{\mathrm{d}\left(\dfrac{y}{y_A}\right)}{\sqrt{\left\{1 + \dfrac{1}{\sigma}\log\left(\dfrac{y}{y_A}\right)\right\}}}, \qquad (8.25)$$

where $\qquad \sigma = 2\epsilon\pi\left(\dfrac{2e}{m}\right)^{\frac{1}{2}}\dfrac{\tan^2\Theta_A}{I/V^{\frac{3}{2}}} = 32\cdot95 \times 10^{-6}\dfrac{\tan^2\Theta_A}{I/V^{\frac{3}{2}}} \qquad (8.26)$

is called the 'beam-spread coefficient'.

The relativistically corrected beam spread coefficient becomes according to (8.17)

$$\sigma_{rel} = \sigma k_{rel}^{\frac{3}{2}}. \qquad (8.27)$$

The coordinates (y, z) of any beam electron may now be calculated everywhere from initial conditions which are accessible to direct measurement. Again, for convenience, the Dawson function, (8.12) could be introduced for the integral in (8.25). For this purpose take

$$\zeta^2 = \sigma + \log\left(\frac{y}{y_A}\right). \qquad (8.28)$$

The integration in (8.25) has then to be taken from $\zeta = \sqrt{\sigma}$ (cf. (8.28) with $y/y_A = 1$) up to ζ.

Moreover $\mathrm{d}(y/y_A)$ is found by differentiation of (8.28)

$$2\zeta\,\mathrm{d}\zeta = \mathrm{d}\left(\frac{y}{y_A}\right)\bigg/\left(\frac{y}{y_A}\right).$$

Hence (8.25) can be written

$$z - z_A = 2(z_F - z_A)\sigma^{\frac{1}{2}}\exp\left(-\sigma\right)\int_{\zeta=\sqrt{\sigma}}^{\zeta=\sqrt{\{\sigma+\log(y/y_A)\}}} \exp\left(\zeta^2\right)\mathrm{d}\zeta, \qquad (8.29)$$

where σ, the spread coefficient, is given by (8.26). The value of the integral in (8.29) is calculated as the difference of two Dawson functions (cf. (8.12)). Equation (8.29) has been fully discussed by Thompson and Headrick (1940).

Laminar flow in ribbon-shaped beams. Space-charge effects are much less pronounced in flat beams than in beams of circular cross-section. The simple theory of the spread of ribbon-shaped beams has been developed by Bouwers (1935) and by Thompson and Headrick (1940). This theory can be applied to all parts of the beam in which the velocity of the electrons is constant, and in which the direction of the electrons is homocentric. If calculations similar to those of §8.1 are made for two-dimensional beams, the spread curve can be shown to be a parabola. The parabolic path of each electron in the strip beam is then given by

$$y = y_A - \Theta z + (I/V^{\frac{3}{2}})\frac{z^2}{4k}, \qquad (8.30)$$

where y_A is the initial semi-aperture of the beam at $z = 0$ and Θ is the initial slope angle of its envelope with the z-axis. Moreover, V is the electron energy (eV), I is the beam current per unit beam width in the x-direction (A m^{-1}) (i.e. for a width of the strip beam $2x_A = 1$ m) and k is a constant of the value

$$k = 10{\cdot}4 \times 10^{-6}\,[\text{A}/\text{V}^{\frac{3}{2}}].$$

Differentiating (8.30) and taking $\mathrm{d}y/\mathrm{d}z = 0$ gives us the coordinates of the beam waist formed by convergent rays: location and aperture of the waist are seen to depend upon the initial angle Θ and upon the space-charge factor $I/V^{\frac{3}{2}}$ only.

With a decreasing ratio of $(I/V^{\frac{3}{2}})/\Theta^2$ the waist aperture gradually decreases. Very much in distinction with beams of circular cross-section, however, the beam waist of an intense strip beam can become of vanishingly small aperture ($y_W = 0$). For this critical case, the waist distance (z_W) from the initial aperture is found to be just twice as large as the distance (z_F) of the virtual focus. This critical case is presented in fig. 8.3, drawn to scale in the y-coordinate but scaled down by a factor 10 in the z-coordinate in order to make the essential features more visible. There, the initial convergence of the boundary of the beam (El) corresponds to $\Theta = 0{\cdot}1$ radian, the critical space-charge factor being

$$0{\cdot}1 \times 10^6\,[\text{A}/\text{V}^{\frac{3}{2}}].$$

For further decrease of $(I/V^{\frac{3}{2}})/\Theta^2$, i.e. beyond the critical value, the strip beam forms a real cross-over. This again cannot possibly occur with a quasi-continuous beam of circular cross-section.

The critical case represented by fig. 8.3 is of some practical interest; it defines theoretically the optimum-beam geometry if the maximum current has to be transmitted through a tunnel terminated by two equal slots. There, the waist has to be located in the middle of the tunnel.

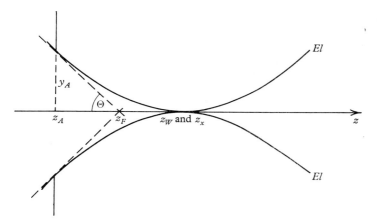

Fig. 8.3. Ribbon-shaped electron beam under the
influence of its space charge.

With increasing space charge, the beam geometry deviates from the predictions of the simple theory. A hump of negative potential is built up near the axial xz-plane. This hump slows down pre-ferentially the paraxial part of the beam, which eventually causes spherical aberration (cf. Klemperer, 1947 a). Space-charge errors are very similar to those discussed in §8.4; however, the disturb-ances of strip beams appear to be much smaller than those experienced with circular beams.

The influence of the termination of strip beams (i.e. of the final width in x-direction) upon the beam-spread has been discussed by Houtermans and Riewe (1941). The boundary of the finite beam in the yz-plane is given again by parabola, the beam edges having little influence upon the spread in the yz-plane as long as the cross-section of the beam in the xy-plane is a long rectangle with $x \gg y$. If, however, this cross-section approaches the shape of a square the expansion becomes rather similar to that of a round beam and the corners of the square beam do not play a great part.

§8.3. Focus spread and focus shift. Space-charge equivalent lens

From the electron-optical point of view, we are interested in the sharpest possible focus. Even if the space-charge factor $I/V^{\frac{3}{2}}$ is very small, the homocentric bundle of electrons will not come to a point focus such as is formed by a homocentric bundle of light rays. The radius of the disc of least confusion is the waist radius, which is easily calculated from (8.22). For the waist we have

$$\frac{dy}{dt} = 0.$$

Substituting for

$$u_y^2 = u_z^2 \tan^2 \Theta = \frac{2e}{m} V \tan^2 \Theta,$$

replacing b by (8.6) and using $u_z = \sqrt{\left(\frac{2e}{m} V\right)}$, one obtains

$$\frac{y_W}{y_A} = \exp(-\sigma), \qquad (8.31)$$

where σ is given by (8.26). It can be seen that the disc of least confusion decreases exponentially with decreasing space-charge factor $I/V^{\frac{3}{2}}$, and it can be made the smaller, the larger the semi-vertical angle Θ of the convergent beam.

Of further interest for electron-optical problems is the shift of focus due to space-charge effects. Assume the current of a beam of small space-charge density to be gradually increased. Initially the beam may be directed towards the virtual focus z_F while passing through the aperture A. The shift of the focus at z_W due to an increase in current can be calculated from (8.29). There, substitution of y_W of (8.31) for y yields

$$\frac{z_W - z_A}{z_F - z_A} = -\frac{2\sqrt{(\sigma)}D(\sqrt{\sigma})}{\exp \sigma}. \qquad (8.32)$$

This equation is represented by fig. 8.4, where $z_W - z_A/(z_F - z_A)$ is plotted against σ. Assume the practical case of a constant Θ_A, with $z_F = z_A = $ const. Auxiliary sketches (a)–(d) in this figure show four characteristic stages of the initially convergent electron beam. From (8.32) it follows that, for small values of σ, the entrance aperture should be nearer to the waist than to the virtual focus (see fig. 8.4(a)). Beams of $\sigma \approx 0.3$ are used in the design of some Klystron tubes. It is seen that the formation of a sharp focus in such a beam is quite out of the question. Its waist radius y_W is as large as three quarters of the aperture radius y_A.

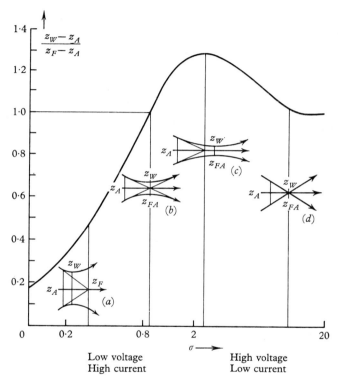

Fig. 8.4. Beam spread coefficient and shift of beam waist.

Now, when at a given constant slope angle Θ_A, an increase in voltage (or decrease in current) leads to an increase in σ, the virtual focal distance $z_{FA} - z_A$ remains constant but the waist distance $z_W - z_A$ varies. For $\sigma = 0.86$ (see fig. 8.4(b)) z_W and z_{FA} coincide. For $\sigma = \sigma_0 = 2.35$ (see fig. 8.4(c)), $(z_W - z_A)$ reaches a maximum. For large values of σ (see fig. 8.4(d)), the waist should again coincide with the virtual focus; in this region, however, the current density is generally low enough for the beam to cross over (cf. §8.4).

Since for every cross-section of a homogeneous, homocentric beam I is proportional to y^2 and hence to $\tan^2 \Theta$, the beam-spread coefficient σ is seen from (8.26) to be constant for all electrons in a given cross-section. Hence, whatever the value of σ, according to (8.32) a bundle of homocentric rays remains homocentric. Thus space charge effectively acts like an optical lens placed at the aperture A. Taking the actual focus as the waist of the beam, the focal length f of this 'space-charge equivalent lens' is given by

$$\frac{1}{z_W - z_A} - \frac{1}{z_F - z_A} = \frac{1}{f}. \qquad (8.33)$$

If (8.33) be multiplied by $(z_F - z_A)$, its first term can be obtained with the help of a graph (fig. 8.4). It appears that the space-charge equivalent lens is generally a diverging lens, which reaches its maximum power

$$-\frac{1}{f} = \frac{0.22}{z_F - z_A}$$

at $\sigma = 2.28$. Its power vanishes for $\sigma \to \infty$ and for $\sigma = 0.86$. On the other hand, for $\sigma < 0.86$ the space-charge equivalent lens may be compared with a converging lens in so far as it shortens the image distance. It is of particular interest that space-charge equivalent lenses are free from spherical aberration within the range of application of (8.32). Introducing the current density $i = I/\pi y_A^2$ into (8.26) and eliminating y_A by (8.24), the beam-spread coefficient may be expressed by

$$\sigma = \left(\frac{2e}{m}\right)^{\frac{1}{2}} \frac{2}{(z_F - z_A)^2} \frac{1}{i/V^{\frac{3}{2}}}. \qquad (8.34)$$

This equation again implies that for a homocentric, homogeneous bundle of rays, the focal length of the space-charge equivalent lens is independent of the aperture.

As an example, the focal length of a space-charge equivalent lens may be calculated for a beam of perveance

$$I/V^{\frac{3}{2}} = 10^{-6} \text{ amp volt}^{-\frac{3}{2}}$$

passing through an aperture stop of $y_A = 5$ mm radius and aiming towards a virtual focus $(z_F - z_A) = 20$ mm distant from the stop. There

$$\sigma = 2 \cdot 05, \quad \frac{z_W - z_A}{z_F - z_A} = 1 \cdot 28$$

and the focal length $f = -91$ mm. The focus z_F is shifted by $(z_F - z_W) = 3 \cdot 6$ mm in the direction of the electron motion.

§8.4. Aberrations of the space-charge equivalent lens. Non-laminar flow

The theory of the beam-spread developed so far is correct to a first approximation only. Maximum beam currents calculated from first-order theory are never reached in practice. Beams of great current density only approximately comply with the first-order theory since homogeneity of electron velocities and homocentricity of the rays can be maintained only for beams of relatively small current density. Borries and Dosse (1938) pointed out that the potential V_0 at the axis should always be more negative than the potential V_A at the envelope with the radial coordinate y_A. The potential difference is

$$V_0 - V_A = \int_0^{y_A} E_y \, dy = \frac{\rho}{2\epsilon} y_A^2, \quad (8.35)$$

where E_y is the normal component of the field as given by (8.3) which has to be integrated from $y = 0$ up to the full beam aperture y_A. Eliminating the space-charge density ρ by (8.4), etc., gives

$$V_0 - V_A = \frac{I/V_A^{\frac{1}{2}}}{(2e/m)^{\frac{1}{2}}} = 1 \cdot 52 \times 10^4 \, I/\sqrt{V_A}. \quad (8.36)$$

Due to this voltage difference, the paraxial electrons should be expected to have smaller velocities than those at the margin. Now, by the slowing down of the paraxial rays, the paraxial space-charge density is increased. This should slow down still more the paraxial part of the beam. As a consequence, the negative voltage in the paraxial region is raised further and so on. On the other hand, increased space charge will lead to increased beam-spread and consequently to reduced charge density. Eventually the competing processes set up an equilibrium. A theoretical investigation on the equilibrium potential distribution in an electron beam due to its space charge has been published by Smith and Hartman (1940), and approximate expressions have been derived for the beam-spread. According to the calculations of these authors a beam of high current density diverges a given amount in about one-half of the distance computed from the first-order theory equation, (8.25).

Klemperer (1947a) pointed out that the loss of homogeneity must lead to a loss of homocentricity and to a production of spherical aberration. The potential hump about the axis and the consequent reduction of electron velocities in the paraxial region will be seen to lead (cf. §8.9) to a variation of current density i along the radial coordinate y. Now, according to (8.34), the beam-spread coefficient σ and hence according to (8.32) the position z_W of the focus must be a function of the ray aperture y. In other words, the focal length of the space-charge equivalent lens (cf. (8.33)) will be a function of the aperture, i.e. the lens suffers from spherical aberration. The coefficient σ always decreases with increasing y, thus

$$\sigma_{Pa} > \sigma_{Ma},$$

where the suffix Pa stands for the paraxial region and Ma stands for the marginal region of the electron beam. Now, from the fact that the focus shift as shown in fig. 8.4 as a function of σ has a maximum at $\sigma_0 = 2\cdot35$, the following conclusion can be drawn: If $\sigma_{Ma} > \sigma_0$, then $(z_W - z_A)_{Pa} < (z_W - z_A)_{Ma}$ and the spherical aberration of the space-charge equivalent lens is negative. On the other hand, if $\sigma_0 > \sigma_{Pa}$, then $(z_W - z_A)_{Pa} > (z_W - z_A)_{Ma}$ and the spherical aberration of the space-charge equivalent lens is positive. To a certain extent, correction of the space-charge equivalent lens should be expected when $\sigma_{Pa} > \sigma_0 > \sigma_{Ma}$.

The experimental results of Dolder and Klemperer (1955) support the above arguments. These authors used a pepperpot diaphragm to study the spreading of beams over a wide range of σ, e.g. for a beam of 360 eV and $\Theta_A = 0\cdot084$ rad, beam currents in the range 10 μA to 2 mA were employed, the corresponding range of σ, averaged over the beam, being 160 to 0·8.

The case of small σ corresponding to relatively high current densities in which the aberration is positive is the more common one, and this is most troublesome in practice. However, the following examples show that in technical cathode-ray tubes σ is found on either side of the critical value σ_0. For the focused beam of a television receiver tube with 3 kV, 0·5 mA and $\Theta = 10^{-2}$, σ is found to be ≈ 1. For the focused beam in a projection tube with 50 kV, 3 mA and $\Theta = 0·03$, it is found that $\sigma \approx 100$. It will be shown later on (§9.7) that minimum cross-over size in an emission system is likely to occur when σ is of the order of σ_0.

Cross-over of low-intensity beams. Deviations from the beam-spread curve (8.11) should be expected for a beam containing so few electrons that its space charge can no longer be considered as a continuum. In particular, there will not be formed a waist of a radius given by (8.31), but some of the electrons will cross the axis. Not much information is available about low-intensity beams, but the following considerations will illustrate the problem. If two isolated electrons travel initially with equal axial coordinates, opposite and in equal distances from the axis they will, owing to their Coulomb repulsion, move along hyperbolic orbits which are somewhat similar to the beam-spread curve. If in another particular case initially homocentric rays again converge towards the axis but the relative positions of electrons travelling along these rays are not diametrically opposite, but interlaced, their mutual repulsion will increase less rapidly than with the inverse square of their distance from the axis. The orbits will form a waist only if the axial distances between single electrons are so small that the kinetic energy of the radial velocity component is outweighed by the potential on the axis. If this is not the case, the electrons will cross the axis. The electron orbits may still be curved near the axis, but with decreasing beam current they will straighten up until rectilinear motion of the electrons is reached. Only the cross-over of a very low-intensity beam completely corresponds to a light-optical focus as far as location and sharpness are concerned.

Now the question arises: When can the continuum theory of §8.1 be applied and when will the rays of a beam start to cross over? According to Barford and Klemperer (cf. Klemperer, 1953) a criterion can be developed from the assumption that the continuum theory is expected to fail when the waist radius y_W decreases below a certain critical value y_c which should be of the order of the mutual distance s of the electrons in the beam, i.e. when

$$y_W \lesssim y_c \approx s. \tag{8.37}$$

The mutual distance of the electrons in a homogeneous beam is given, according to (8.4), by

$$s = \left[\frac{\pi y^2 e}{I} \sqrt{\left(\frac{2e}{m} V \right)} \right]^{\frac{1}{3}}. \tag{8.38}$$

With the assumption of (8.37) we have $y = y_c = s$, hence (8.38) yields

$$y_c = \frac{\pi e}{I} \sqrt{\left(\frac{2e}{m} V \right)} = 3 \times 10^{-13} \frac{\sqrt{V}}{I}. \tag{8.39}$$

By substitution in (8.37) for y_c from (8.39) and for y_W from (8.31) with (8.26) one obtains the required criterion

$$\left(0 \cdot 3 \times 10^{12} y_A \frac{I}{V^{\frac{1}{2}}} \right) \exp \left(-33 \times 10^{-6} \frac{\Theta^2 V^{\frac{3}{2}}}{I} \right) \lessgtr 1. \tag{8.40}$$

If the condition (8.40) is fulfilled, the continuum theory of space charge will fail. It is interesting to notice that according to (8.40) a transition region should exist in which the outer-beam electrons still form a waist while the inner-beam electrons already cross over. This is easily seen. Since in beams of axial symmetry the inner parts (due to Gauss's law) are not influenced by the outer parts, a particular aperture y_{A1} can be derived from expression (8.40) inside which the continuum theory no longer holds. Hence, for rays of aperture y between y_A and y_{A1}, waist formation should be observed, while for all rays of an aperture $y < y_{A1}$ a true cross-over should be formed.

The coexistence of cross-over and waist in beams of low current density has been demonstrated by Dolder and Klemperer (1955) as part of the experimental investigation mentioned earlier. At very low currents, all electrons in a converging beam cross the axis. On increasing the beam current, the appearance of negative spherical aberration indicates the formation of an electron waist. The experiments indicated that over a certain range of beam current (beam voltage and initial convergence angle being fixed) a waist was formed by the marginal electrons, while the trajectories in the paraxial region still crossed over. The measurements of these authors also show that the Barford–Klemperer criterion (8.40) is in fair quantitative agreement with experiment. Thus (8.40) can be regarded as a useful guide as to whether to apply continuum space-charge theory or light-optical theory (i.e. space-charge negligible).

Non-laminar flow. The deficiencies of continuum theory at high current densities, and its breakdown at very low current densities, which we have discussed in this section show that the hypothesis of 'laminar flow' can be no more than an idealization for practical electron beams. Loss of homocentricity in an initially homocentric beam implies the occurrence of some trajectory crossings, so that beams suffering from spherical aberration cannot strictly be called laminar. Also, the formation of axial cross-overs in a low current density beam represents a very severe departure from laminarity.

The above considerations show that, in general, the flow, even in initially homogeneous homocentric beams, will not be laminar. In actual electron beams, conditions are further complicated by the fact that the rays are initially never strictly homocentric, owing to the imperfections of electron gun systems. Electron guns normally suffer from some degree of positive spherical aberration. Further-more, it is inevitable that radial velocity components are found in electron emission from the cathode. These give rise to 'trans-laminar' streams which profoundly modify the structure of high perveance beams (Brewer, 1959). It is possible for 'thermal' electrons to form axial cross-overs even in high power beams (Amboss, 1964), and their flow must be analysed, using light-optical rather than space-charge continuum theory, independently of the remainder of the electrons (Herrmann, 1958). The behaviour of 'thermal' electrons is of considerable practical importance in the design of high power beams, and we shall return to this point in §8.7.

The problem of determining the minimum spot size obtainable in a cathode-ray tube (cf. §8.2) is one in which the influence of thermal emission velocities as well as space charge forces must be taken into account. Weber (1967) has analysed the development of the beam in the field-free region of a cathode-ray tube, for condi-tions intermediate between the two extremes of space-charge

dominated laminar flow and thermal emission dominated optical
flow, and has presented his results in the form of universal curves
linking beam radius, spot size, lens strength, beam current and
geometrical parameters. Some indication of the relative importance
of space-charge repulsion and thermal emission velocities in determining the spot size can be obtained from the data of table 8.1,
calculated by Weber for a television display tube of beam voltage
18 kV, maximum current 1·6 mA, and effective source radius
0·3 mm. The last line of the table gives the spot radius which is
calculated in the absence of thermal velocities, i.e. as determined
only by space-charge. It is seen that at low currents, the spot size
is mainly determined by thermal velocities, and at maximum
current, while space-charge is the important factor, the effect of
thermal velocities is still not negligible.

TABLE 8.1. *Current I and radii y_s and y_0 of the focused spot in a
space-charge loaded beam, with and without consideration of thermal
agitation*

I (mA)	0	0·1	0·2	0·4	0·8	1·6
y_s (mm)	0·10	0·12	0·13	0·15	0·20	0·27
y_0 (mm)	0·0	0·0045	0·015	0·04	0·094	0·20

§8.5. Space-charge in interaction with electrodes and electron lenses

The discussions in this section on beam-spread were so far concerned with electrons moving through a completely field-free space.
No other fields than those set up by the space-charge have been
considered. In practice, however, the electrons may be accelerated
by axial fields or focused by electron lenses and they are emitted
and collected by electrodes. The mutual influence between space
charge and the fields set up by these electrodes cannot be neglected.

Borgnis (1943), for instance, investigated theoretically the
influence of space charge on initially parallel rays in a beam of
circular cross-section travelling in a space between two large, plane
parallel electrodes set up at right angles to the axis of the beam.
Both electrodes are at a common potential corresponding to the
kinetic energy of the electrons. The beam may, for instance, pass

these electrodes through apertures covered with wire mesh. Now, in addition to the mutual repulsion of the electrons, the forces between the electrons and their electric images in both electrodes have to be considered. The equipotentials about the beam are no longer co-axial cylinders, but now they are closed shells cutting the beam axis and the mid-plane (i.e. the plane parallel to and half-way between the conducting planes) at right angles. Along the axis is set up a purely longitudinal field which retards the electrons on their way to the mid-plane but accelerates them after passing this plane. The strength of these fields depends on the ratio y_A/L (= beam radius/conducting plane distance), the maximum possible field value for $y_A \gg L$ being $E = L\rho/8\pi e$, where ρ is the space-charge density and ϵ is the dielectric constant.

The beam periphery is cut by the equipotentials at angles which decrease from $90°$ near the conducting planes to $0°$ at the mid-plane. This accounts for a beam spread, which, however, is much smaller between the planes than it would be in free space. Again, the beam spread is found to depend very much on the relative distance (L/y_A) of the conducting planes. Details have been calculated by Goudet and Gratzmuller (1944). For instance, if the distance L is of the order of the radius of the entrance aperture $y_A = y_W$, the incremental spread is only a quarter of that calculated from (8.20). On the other hand, the values obtained from (8.20) would apply to the asymptotic case $L/y_W \to \infty$. For intermediate plate distances the following values may be quoted:

L/y_W	1	2	4	10	20
g	0·25	0·45	0·66	0·85	0·92

where g is a factor by which the right-hand side of (8.20) has to be multiplied in order to allow for the presence of the plane electrodes.

As a particular case of some practical interest, Goudet and Gratzmuller (1945) treated the spread of an initially parallel bundle of circular cross-section emitted from a plane, space-charge-limited cathode and intercepted by a plane anode. The beam spread again differs much from the result derived from (8.20). If the cathode is emitting under space-charge conditions, the spread is found to depend only on the ratio of the cathode–anode distance L to the initial beam radius y_c at the cathode. It is independent of the anode

voltage (cf. §9.7). The following numerical values may be sufficient to give an impression of the magnitude of the spread
$(y_A - y_c)/y_c$, (y_A = beam radius at the anode):

$\dfrac{L}{y_c}$	1	2	3
$\dfrac{y_A - y_c}{y_A}$	0·074	0·133	0·182

Moss (1945) has investigated theoretically the beam spread of
an initially convergent beam in an accelerating homogeneous
electrostatic field. He found that the beam-spread curve becomes
asymmetric, and that with a given initial beam-spread coefficient σ
the waist distance $(z_W - z_A)$ increases while the waist radius, y_W,
decreases with increasing field.

The beam spread in the accelerating field is of some interest
in connexion with the application of post-deflexion acceleration
(cf. §10.2) in cathode-ray tubes. In practice, generally the beam
current I and the ultimate target voltage V_T are given, and the
conditions leading to the least spot-size at the target have to be
studied. The two parallel, plane electrodes correspond to the anode
and to the target (screen) respectively, of the cathode-ray tube, and
the question arises as to what fraction of the available potential V_T,
should be applied to the anode in order to set up the post-acceleration field. As the result of some calculation, however, it follows
that the least beam-spread is obtained if full voltage is applied to
the anode so that the beam between anode and target travels in
field-free space. An example may illustrate the magnitude of the
effect. The target could be at 3 kV and only 1 kV is applied to the
anode which might be at 40 cm spacing from the target. This
implies a post-accelerating gradient of 50 V/cm. Under the given
conditions a minimum spot size is calculated as a waist radius of
0·25 mm. However, the waist radius can be reduced to 0·1 mm
simply by raising the anode voltage up to the target voltage.

Electron lenses. A frequently occurring problem is raised by space-
charge influence of a beam upon the position of the cardinal points
of an electron lens by which this beam is focused. With few
exceptions, the converging power of electron lenses decreases with

an increase in space-charge factor $(I/V^{\frac{3}{2}})$ at the focus. For a beam of constant current I, the space-charge factor changes with $V^{-\frac{3}{2}}$ while the beam passes through the lens. Thus, it is easily understood why the foci of decelerating lenses are found to be more sensitive to space charge than the foci of accelerating lenses.

A simple example of the effect of space-charge on the focal length of an electron lens is provided by the case of a space-charge limited diode (cf. §8.9) of spacing d, with a small hole in the anode. After leaving the anode aperture, the beam is assumed to enter a field-free space. In the absence of space-charge, the field about the anode hole would act as a diverging electron lens whose focal length is given from the Davisson–Calbick formula (cf. §4.4) by $f = -4d$ (since $E = V_A/d$, where V_A is the anode voltage). However, the potential in the diode region will be modified by space-charge. In the approximation of infinite flow (cf. §8.7) the potential variation is given by $V(z) = Az^{\frac{4}{3}}$. Hence, the field in the diode region to the left of the anode is given by

$$E(d) = -\frac{\mathrm{d}V(d)}{\mathrm{d}z} = -\tfrac{4}{3}Ad^{\frac{1}{3}},$$

and taking $E' = 0$ to the right of the anode the Davisson–Calbick formula now shows that the anode hole is equivalent to a diverging lens of focal length $f_s = -3d$. Thus the space-charge modifies the field in such a way as to increase the divergence of the beam through the anode aperture lens.

Initially accelerating saddle-field lenses are little affected by space-charge. Initially decelerating saddle-field lenses, however, which are practically more important, sometimes suffer so much from space-charge aberrations that they can be employed for focusing currents of the order of microamperes only. For instance, the three-tube einzel lens with long intermediate tube is greatly affected. Klemperer (1941) investigated the mid-image distance of such a lens as a function of the angular aperture of the rays. In fig. 8.5 there are plotted two aberration curves which refer to a three-tube lens in which the outer electrodes are maintained at 3 kV and the intermediate electrode is at 1 kV. $(\Theta')^2$ is plotted as the ordinate. The continuous line refers to the case in which the beam carries very small currents of the order of only a microampere. This line is approximately straight, and the slope indicates positive aberration throughout (cf. §6.4). If, however, the beam current is gradually increased, the aberration curve gradually changes its shape and position. When the current reaches the order of a milliampere, the aberration curve is given by the broken line

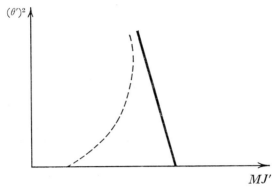

Fig. 8.5. Spherical aberration curves for an initially decelerating three-tube einzel lens with 1 microamp (——) and 1 milliamp (– – – –) beam current.

in fig. 8.5. In this case the aberration is seen to be highly negative for rays near the axis, but it gradually changes with increasing ray aperture into positive values for marginal rays.

Three-diaphragm saddle-field lenses consisting of large, plane widely spaced electrodes with fairly small apertures are relatively free from space-charge trouble even for initially decelerated electrons. This has been shown by Klemperer (1942), who, in addition, described three-tube saddle-field lenses which are relatively free from space-charge effects. Such tube lenses, however, have to contain some small diaphragm arrangements with the help of which the potential distribution in the tubes is modified to approach closely the potential distribution of the high-quality large-diaphragm lenses. Such tubular lenses are preferable to the plain diaphragm lenses if the space available for the electrode structure is restricted.

Referring to fig. 8.5, it is seen that over a zone near the margin of the beam, the positive aberration of the lens is approximately cancelled by the negative aberration due to space-charge, which leads naturally to the speculation that space-charge might be of some use in correcting the spherical aberration of electron lenses. However, Ash (1955) has shown that space-charge can be of little practical use for aberration correction. Referring to the important example of a high-resolution electron microscope objective, he showed that the beneficial effects of a space-charge suitable for aberration correction would be much vitiated by the disturbing effects of collisions between the beam and the electrons in the space-charge cloud which result in a blurring of the projected image.

The analysis of Kanaya, Kawakatsu and Yamazaki (1965) gives some insight into the effect of the beam-space charge on the operation of a high resolution magnetic objective lens. It appears that the defocusing effect of beam-space charge leads to a deterioration of resolving power at higher beam currents. A criterion is given which can be expressed as follows: If a resolution limit of δ is desired, then the beam-spread coefficient (cf. (8.26)) must be such that

$$\sigma \geqslant 0 \cdot 8 r_0 / \delta, \tag{8.41}$$

where r_0 is the radius of the objective aperture. This leads to the conclusion that if a resolution limit of a few ångström units is to be attained at 50 kV, then the beam current at the specimen must not exceed about 1 μA.

§8.6. Space-charge and positive ions

According to experimental evidence, the beam-spread is often appreciably reduced by the presence of positive ions. It is quite unavoidable that electron beams above a critical velocity corresponding to the ionization potential (about 12 V) should ionize the residual gas. The highest electron densities now obtainable are of the order of 10^{16} electron m^{-3}, but even in a vacuum of 10^{-8} torr, the number of molecules of residual gas is still of the order of 10^{14} molecules m^{-3}. In the usual electron-ray tubes the density of residual gas molecules far exceeds the density of electrons.

Owing to their far greater mass, the positive ions generally travel much more slowly than the light electrons, consequently they are very effective in setting up a field to counteract the negative space charge. Field, Spangenberg and Helm (1947) found, even at pressures of 10^{-7} torr, that the beam-spread was reduced—due to the presence of positive ions—to nearly half the value predicted by the simple beam-spread theory.

Wadia (1958) has made a detailed theoretical investigation of the dependence of beam-spread on pressure, expressing the distribution of positive ions created by an electron beam in the form of a universal curve. The influence of the ion distribution on the beam profile was estimated by a numerical method of step-by-step integration. Fig. 8.6 shows the results obtained for an initially converging 5 kV beam of semi-angle $7.5°$ and perveance 1.5×10^{-6}.

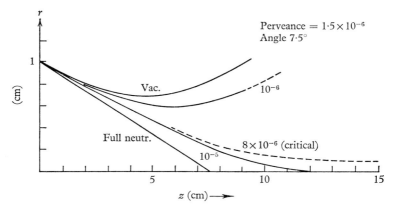

Fig. 8.6. Beam profiles as a function of gas pressure.

As the pressure increases the beam waist decreases and moves in the positive z-direction. At a critical pressure (8×10^{-6} torr in this case), the beam reaches its minimum radius and then travels parallel to the axis. Above this pressure, the beam forms an axial cross-over. It is interesting to note that according to these calculations, pressures of 10^{-8} torr and below are equivalent to total vacuum as far as ion production is concerned. The beam profiles of fig. 8.6 are consistent with the observations of Klemperer and Mayo (1948), who found a marked improvement in the current efficiency of a cathode-ray tube gun on raising the pressure from 10^{-7} to 10^{-4} torr.

The mechanism of space-charge neutralization may be understood as follows. The positive ions produced by the beam form an ion gas which spreads out by diffusion under the influence of electric forces. If the electron beam travels in accelerating fields, the positive ions will drift along the beam from the positive to the negative electrodes. For instance, the ions may be captured eventually by the cathode. If, however, the electron beam travels through a field-free space, a potential depression will be set up due to the electronic space charge (cf. (8.35)). The depression extends along the axis but decreases when a grid or an apertured diaphragm is approached. For instance, the beam may enter and leave a field-free space through such diaphragms, these being at the potential of the conducting tube which surrounds the beam or—for increasing the above depression—they may be positively biased with respect to the tube. Thus there is formed a potential valley surrounded by higher potential regions on all sides. Such a potential valley acts as a trap by which the positive ions are attracted and retained. On the other hand, low-velocity electrons, formed by the ionization process, are forced away from the centre of the beam.

The resulting accumulation of ions in the beam causes a progressive neutralization of the electron charge and consequently the depression decreases. Eventually the ion trapping becomes ineffective enough to be counterbalanced by the diffusion of the ion gas, and a state of equilibrium is attained. Linder and Hernquist (1950), who discussed this process in detail, derived a quantitative relation between residual gas pressure P and the equilibrium depth of the potential depression V, namely

$$P = \frac{R}{\pi^{\frac{1}{2}}r^2} \frac{{}^+\rho_0}{{}^-\rho S} \left(\frac{m}{M}\frac{{}^+V_w}{V_{el}}\right)^{\frac{1}{2}} \exp\left(\frac{\Delta V}{{}^+V_w}\right), \qquad (8.42)$$

where R and r are the radii of the tube and of the electron beam respectively. ${}^+\rho_0$ and $-{}^-\rho_0$ are the charge densities of ions at the axis and of electrons in the beam respectively. m/M is the mass ratio of electron to ion. ${}^+V_w$ is the kinetic energy of the ion near the wall as calculated from the kinetic theory, while V_{el} is the energy of the beam electrons. Moreover, S is the number of ions formed per unit path length of one beam electron per unit pressure.

S is known to be a function of the electron velocity; it rises to a maximum near 100 eV (e.g. 10^3 ions per metre in air of 1 torr pressure) and then drops slowly as V_{el} increases. For instance, taking $R = 5$ mm, $r = 1$ mm, $M/m = 52000$, $V_{el} = 300$ V and $^+V_w = 0.026$ V, and with $^+\rho_0 \approx {}^-\rho$, one obtains from (8.42) the following table for the voltage depression as a function of the gas pressure:

P (torr)	10^{-4}	10^{-5}	10^{-6}	10^{-7}	10^{-8}
V (V)	-0.01	-0.07	-0.13	-0.19	-0.25

The time to build up the state of equilibrium resulting in the above voltage depression has been calculated by Linder and Hernquist (1950) to be

$$t = (m/2e)^{\frac{1}{2}}/SPV_{el}. \tag{8.43}$$

For instance, for a 300 eV beam, build-up times t of the order of 1μs are required at $P = 10^{-5}$ torr pressure. As t is seen to increase inversely proportional to P, build-up times of the order of milliseconds are needed in the best obtainable vacua.

Linder and Hernquist confirmed their theory by experiments with convergent electron beams emitted from a large spherical concave cathode and accelerated by a concentric grid into a field-free space. The relatively small aperture of a collector cage was arranged at the centre towards which the beam converged. The fraction of the beam current which passed the aperture was found to increase as the beam-spread was counteracted by space-charge neutralization due to positive ions formed in the residual gas. Maximum increases in current density from 2 to 60 [mA mm^{-2}] could be observed due to space-charge neutralization in 300 eV beams. The build-up time of the positive-ion gas could be determined from oscillographic studies. Unfortunately, another practical drawback is linked with the relatively long build-up time. When the initial potential depression in the beam is being neutralized by the ion gas which is rapidly increasing in density, an inertia-like overshooting of the equilibrium point and a swing back towards it seems to occur. In this way oscillations are set up. These are observed as noise which appears abruptly at a time when the space-charge neutralization process just appears to be completed.

Ginzton and Wadia (1954) discussed the design of an effective ion barrier in the form of an einzel lens with positively charged intermediate diaphragm to counteract axial drainage of ions to the cathode in high-power beam tubes. In this way, not only damage of the cathode by the impact of the ions could be avoided, but also the beam diameter could be kept very small owing to the neutralizing of space-charge effects by the ions retained in the beam. Using a suitable design of ion-retaining electrodes it was possible to effect an improvement of transmission through a drift space. A 330 eV beam of perveance 2×10^{-6} and minimum diameter 4 mm was injected into a drift tube of 12·5 mm diameter and 15 cm length. Operating without the ion-retaining electrodes the transmission was 12 per cent, but with these electrodes 80 per cent of the current passed through the tube.

Barford (1957) derived an integro-differential equation for space-charge flow in the presence of ions. The influence of ions is represented by a single parameter ϕ, defined as follows:

$$\phi = \frac{6·45 \times 10^4 p V_0^{\frac{3}{4}}}{(V_0 + 267) I^{\frac{1}{2}}}, \qquad (8.44)$$

where p is the gas pressure in torr, I (A m^{-2}) is the beam current density, V_0 the anode voltage. Solutions of the flow equation show that for $\phi \geqslant 0·1$ ions are produced at such a rate that they can neutralize the electron space-charge even when there exist appreciable fields to draw them off, whereas for $\phi \leqslant 0·01$, neutralization can only occur if the ions are trapped. Application of these results to a practical low voltage (300 V) reflex klystron oscillator shows that at 10^{-3} torr the potential distribution is determined almost completely by the positive ion space-charge, whereas at 10^{-6} torr the effect of ions is negligible. In the case of a high voltage (18 kV) klystron amplifier it is concluded that appreciable space-charge neutralization would occur at 10^{-5} torr, while below 10^{-7} torr the ions would have a negligible effect.

§8.7. Balancing of space-charge pressure by external fields. Confinement of high-density beams

The flow of electrons, even in beams of very great space-charge density, can be kept rectilinear by arranging for a suitable potential distribution around the beam. Potential distributions which fulfil the condition of compensating the mutual repulsion can be found from the laws of electron flow between certain infinitely extended electrodes, the cathode being space-charge limited.

The flow between parallel plane electrodes, between coaxial cylinders and between concentric spheres is well known from the investigations of Langmuir (1913), cf. Langmuir and Compton (1931). In all three cases the flow of the electrons is rectilinear and the potential along any ray changes in proportion with the 4/3 power of the distance from the cathode. Take, for instance, the flow between two infinite parallel planes and imagine a beam of circular cross-section with its axis at right angles to the planes cut out of the flow. There the potential within the beam and at its boundary must change according to the 4/3 power of the axial coordinate. Therefore, an isolated beam of initially parallel electrons could be kept parallel and prevented from spreading by establishing an external potential distribution according to the 4/3 power law. The same reasoning, applied to the flow between concentric spheres, would provide the correct external potential distribution for rectilinear, homocentric, convergent or divergent bundles of rays.

The method of finding the correct potential distribution has been worked out by Pierce (1940). Within the beam, Poisson's equation must hold, namely,

$$\frac{\partial^2 V}{\partial z^2} = \frac{\rho}{\epsilon} = \frac{i}{\epsilon}\left(\frac{m}{2eV}\right)^{\frac{1}{2}}, \qquad (8.45)$$

where $V = V(z)$ is the potential, $^-\rho$ the electron space-charge density and $-i$ the electron-current density in the beam. If no spreading by mutual repulsion occurs, the radial potential gradient vanishes, i.e.

$$\frac{\partial V}{\partial r} = 0. \qquad (8.46)$$

Moreover, if the beam is space-charge limited, the gradient

$(\partial V/\partial z)_{z=0}$ at the cathode must vanish. With this condition (8.45) twice integrated yields at the edge of the beam the potential

$$V = Az^{\frac{4}{3}}, \qquad (8.47)$$

with $\qquad A = \left(\dfrac{9i}{2\epsilon}\right)^{\frac{2}{3}} \dfrac{1}{2}\left(\dfrac{e}{m}\right)^{-\frac{1}{3}} = 5 \cdot 7 \times 10^3 . i^{\frac{2}{3}} \quad [\text{Vm}^{-\frac{4}{3}}]$

being independent of z. Equations (8.46) and (8.47) must hold inside the beam as well as in the space-charge-free region surrounding it. Analytical methods have been developed which permit the determination of electrode shapes required in order to fulfil (8.46) and (8.47) simultaneously. These methods are beyond the scope of this book but are fully discussed by Kirstein, Kino and Waters (1967). The problem can also be attacked by analogue methods (cf. §§8.8 and 9.8).

Brillouin flow. It is also possible to use a uniform magnetic field in order to balance space-charge pressure. Electron motion under the influence of space-charges in longitudinal homogeneous magnetic fields has been discussed by Haeff (1939), by Smith and Hartman (1940) and by others. The theoretical study of beams of constant diameter in magnetic fields is due to Brillouin (1945), to Samuel (1949) and to Pierce (1949). Electrons moving in the z-direction along a field $B = B_z$ but having a small transverse velocity component will spiral about the z-axis with an angular frequency given by (5.6), namely

$$\omega = \frac{e}{2m} B. \qquad (8.48)$$

If the radial distance of the electron from the z-axis of the beam is not to change ($dr/dt = 0$), the centrifugal force and the electrostatic force must be balanced by the magnetic force, i.e.

$$mr\omega^2 + e\frac{\partial V}{\partial r} = eBr\omega. \qquad (8.49)$$

Substituting for ω the value of (8.48), one obtains

$$\frac{\partial V}{\partial r} = \frac{e}{4m} B^2 r, \qquad (8.50)$$

and by integration $\qquad V_1 - V_0 = \dfrac{e}{8m}(Br)^2. \qquad (8.51)$

Now, according to Poisson's equation, the space-charge density is given by

$$\frac{\rho}{\epsilon} = \frac{\partial^2 V}{\partial r^2} + \frac{1}{r}\frac{\partial V}{\partial r}. \tag{8.52}$$

Substitution of (8.50) for $\partial V/\partial r$ yields

$$\frac{\rho}{\epsilon} = \frac{e}{2m}B^2. \tag{8.53}$$

It is convenient to express (8.53) in terms of the cyclotron frequency $\omega_c = (e/m)B$ and the 'plasma frequency' ω_p defined by

$$\omega_p^2 = \frac{e\rho}{m\epsilon}. \tag{8.54}$$

In this case (8.53) becomes

$$2\omega_p^2 = \omega_c^2. \tag{8.55}$$

In order to introduce into (8.53) the current I (cf. (8.4)) for the space-charge ρ, u_z has to be known. u_z is obtained from the energy equation

$$\frac{m}{2}u_z^2 + \frac{m}{2}(r\omega)^2 = eV, \tag{8.56}$$

which, with (8.48) and (8.51), yields

$$u_z = \sqrt{\left(\frac{2e}{m}V_0\right)}. \tag{8.57}$$

Hence it follows that all electrons have the same z-component of velocity depending upon the potential V_0 on the axis only. The current within a cylinder of radius r_1 is then

$$I = \pi r_1^2 \rho u_z$$

$$= \frac{\pi\epsilon}{2}\left(\frac{e}{m}\right)^{\frac{3}{2}} r_1^2 B^2 \sqrt{V_0}$$

$$= 1\cdot45 \times 10^6 r_1^2 B^2 \sqrt{V_0}. \tag{8.58}$$

This relation allows us to calculate the flux density B which constrains the space-charge of a beam of current I, and of axial electron voltage V_0 within a cylinder of radius r_1. The outer electrons whirl about the axis with an energy $V_1 > V_0$ given by (8.51).

This type of flow is called Brillouin flow. If the beam is not parallel on injection, or if the magnetic field does not have the correct value B_{Br} for Brillouin flow, given from (8.58) by

$$B_{Br} = \frac{IV^{-\frac{1}{4}}}{1 \cdot 2 \times 10^3 r_1} \qquad (8.59)$$

or from (8.54) by $\omega_c/\omega_p = \sqrt{2}$, then the beam will not stay parallel, but will be periodically broad and narrow along its length. If $B < B_{Br}$ ($\omega_c/\omega_p < \sqrt{2}$), the beam should scallop outwards in a periodic manner from its equilibrium value and r_1 will represent its minimum radius. If $B > B_{Br}$ ($\omega_c/\omega_p > \sqrt{2}$), the beam should scallop inwards so that r_1 will be its maximum radius.

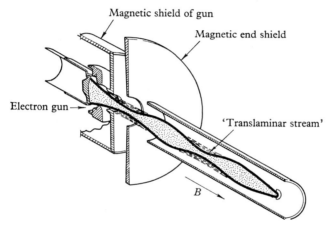

Fig. 8.7. Shielded gun and translaminar stream in Brillouin flow.

It is found, however, that in practice the critical field for obtaining a constant diameter beam is somewhat higher than the value B_{Br} predicted from (8.59). This discrepancy is due to the fact that in practical beams, the flow is not laminar (cf. §8.4), and (8.59) is strictly correct only for laminar beams. In particular, the presence of transverse thermal emission velocities has been shown by Herrmann (1958) to lead to an increase in the magnetic field needed to confine the beam within the equilibrium radius r_1. Brewer (1959) showed that nature of the variation with axial distance of the current density of the beam indicated the existence of 'translaminar streams' due to the spherical aberration of the gun, and that if the spherical aberration of the gun is sufficiently bad, appreciable scalloping of the beam occurs at all values of magnetic field. This is shown in fig. 8.7.

Of some practical interest is the problem of injecting a parallel electron beam into the homogeneous magnetic field in which the Brillouin flow can take place. The electron gun can, for instance, be magnetically shielded, the cathode being in a magnetically field-free space, as indicated in fig. 8.7. In this case, the electrons must be accelerated into the magnetic field with the proper energy given by (8.59). The magnetic field can, however, not rise suddenly from zero to the value B_{Br}, and the transition region of B must be shaped in such a manner that in combination with the accelerating electrostatic field, the electron orbits are as parallel as possible when they reach the onset of the Brillouin field.

In the other extreme case, the electron gun is entirely immersed in the uniform Brillouin field. In tubes with long beams, e.g. in travelling wave tubes, both methods have been employed. Calculations of the electron motion in the combined electrostatic and magnetic field have been presented by Wang (1950), and the starting of beams of constant diameter in the magnetic field has been discussed by Pierce (1949), but the problem is essentially a practical one. Thus, the injection of electrons into the Brillouin field has been studied experimentally by Mathias and King (1957), while Bevc, Palmer and Susskind (1958) have obtained the design for the transition region by analogue computation.

Immersed flow. Frequently in practice, magnetic fields several times the Brillouin field have been employed. It is of course possible to increase the flux density of the magnetic field to such an extent that the electrons are forced to move about the flux lines in extremely narrow spirals. In this way, for instance, a beam drawn from a cathode of radius r_c by an accelerating voltage V can be constrained by an axial magnetic field B within a certain radius r_1. The beam expansion is given according to Pierce (1949) by

$$\frac{r_1}{r_c} = [(k^2 + 1)^{\frac{1}{2}} + k]^{\frac{1}{2}}, \tag{8.60}$$

where
$$k = 3 \cdot 5 \times 10^{-7} \frac{I}{V^{\frac{1}{2}}(r_c B)^2}. \qquad (8.61)$$

(8.60) holds for the limiting case of immersed flow in which B is so high that the beam emerging from the gun is parallel to the axis; this case is sometimes called magnetically 'confined flow'. At lower values of B (and it should be remembered that in practice it is desirable to economize on the power requirements and weight of solenoids), the diverging effect of the anode aperture of the gun (cf. §8.5) must be taken into account. Brewer (1957 a) has shown that the fluctuations in beam width in immersed flow are strongly dependent on the beam divergence at the anode, and whereas the magnitude of the radial fluctuations can in principle be reduced to a low level by increasing B, the magnetic field required to confine a beam of given perveance within a given radius is about twice as great as that required in Brillouin flow. Thus in terms of magnetic field intensity per unit beam perveance, immersed flow is less efficient than Brillouin flow. Furthermore, the beam has usually about the same diameter as the cathode, which effectively limits the beam current density to that obtainable from the cathode, whereas in Brillouin flow (see fig. 8.7) the cathode radius can be larger than the beam radius. This explains why in spite of the simplicity of the immersed flow configuration, Brillouin flow is often preferred for high-power beam applications.

Periodic focusing. Although Brillouin flow and immersed flow are the most widely applied methods for containing high-density beams, there are other methods which find application in certain special cases.

Periodic magnetic focusing (see e.g. Mendel, Quate and Yocom, 1954) is used extensively in travelling wave tubes, its advantage being that it is very efficient in terms of power consumption (or, in the case of permanent magnets, weight) per unit beam perveance. The magnetic focusing field is provided by a series of pole pieces, alternately N and S arranged along the length of the beam, and the magnetic field variation with axial distance is approximately sinusoidal. Since the focusing effect of a magnetic field depends on the square of the field (cf. §5.2), there is always an inward focusing effect to balance space-charge pressure, although the field periodically reverses in direction. The condition for equilibrium between magnetic and space-charge forces is identical with that for Brillouin flow (8.58), except that B must be replaced by the root mean square value of the magnetic field. Pierce (1953) has discussed the use of spatially alternating magnetic fields produced by a series of permanent ring magnets for focusing low-voltage electron beams. For example, a beam of 20 mA at 1·7 kV was confined within 2 mm diameter over a length of 15 cm. The alternating field has the advantage of dying off at a much faster rate than the uniform field and far less ampere turns are required for its production. If permanent magnets are used, much weight can be saved. In the above example, the weight of magnet required was 1·4 lb instead of 38 lb for the corresponding uniform field.

It is also possible to use a periodically varying electrostatic field to contain a long beam of charged particles. Clogston and Heffner (1954) have analysed periodic focusing of high perveance beams for a number of simple configurations. As an example of beam confinement by an axially symmetric periodic electric field, they consider the case of a series of annular discs held at alternately higher and lower potentials $V_0 + V_1$ and $V_0 - V_1$, V_0 being the initial beam voltage. The axial potential can be approximated by

$$V(z) = V_0 + V_1 \left(1 - \cos \frac{2\pi z}{\lambda} \right), \qquad (8.62)$$

where $\frac{1}{2}\lambda$ is the distance between the electrodes. For an initially parallel beam of radius r_0, they find that the condition for space-charge forces to be balanced by the strong-focusing action is

$$\frac{V_1}{V_0} = \frac{4}{\sqrt{3}} \left(\frac{\lambda \omega_p}{2\pi v_z} \right), \qquad (8.63)$$

where ω_p is the plasma frequency, as defined by (8.54) and v_z is the initial axial beam velocity. If (8.63) is satisfied then in the paraxial approximation, the equation for the trajectory of an electron at the beam edge is

$$\frac{r}{r_0} = 1 - \frac{V_1}{4V_0} \left(1 - \cos \frac{2\pi z}{\lambda} \right). \qquad (8.64)$$

If a 10 mA beam with $V_0 = 1000$ volts is considered and λ is $\frac{1}{4}$ inch (6·35 mm), then it is found from (8.63) that the optimum V_1 is 26·5 volts and that the depth of the beam ripple is about $2\frac{1}{2}$ per cent of the beam radius.

The periodic focusing of sheet beams has been analysed by Waters (1960) in connexion with 'tape ladder' lines. These structures are used in certain types of travelling wave tube oscillators; they serve for focusing the beam and also as the radio-frequency circuit.

§8.8. Electrode design, electrostatic fields and electron trajectories in presence of space-charges

A particular potential distribution which simultaneously satisfies both (8.45) and (8.46) can be found with the help of an electrolytic field-plotting trough. In the wedge-shaped trough as described in §3.3, all electrodes are represented by metal strips in the usual way, while the edge of the beam has to be represented by an insulating strip. This guarantees the fulfilment of (8.46), since the normal field component must vanish at the beam edge and no current can flow across the insulator. Then the electrode strips representing cathode, anode and other field-shaping electrodes are altered in shape, position and potential until probe measurements show that the potential along the insulating strip varies with the 4/3 power of the distance from the cathode in order to fulfil (8.46). Fig. 8.8 shows such a field plotting trough seen from above; its four corners are marked T_1, T_2, T_3, T_4. Its bottom is slightly tilted with respect to the plane of the drawing. The liquid in the trough forms a wedge with the thin end indicated by the edge $W_1 W_2$. A spherical cathode surface is represented by C. The rectilinear convergent electron beam is represented by the bakelite strip B standing on the bottom of the trough in the liquid. Five very fine probes P_1, P_2, P_3, P_4, P_5 are permanently fixed to the bakelite, these being spaced to give equal increment of voltage rather than of distance. In this case the distances of the probes from the cathode are given, according to (8.47), by the ratios $1:2\cdot52:4\cdot33:6\cdot35:8\cdot5$, etc. Now the electrode strips S_1 and S_2 (or in an alternative case S_1 and S_3) connected to cathode and anode potential respectively are shifted and shaped empirically until the voltages at the probes are found to increase in the ratios 1:2:3:4:.... Calculation shows that the electrode S_1 should make an angle of 67·5° with the beam edge simulated by the bakelite. The shapes of electrodes to produce the right field are unique. However, there is a great variety of dissimilar electrode shapes which form the required field to a good approximation. In this way one is at liberty to choose electrode shapes to meet the exigencies of construction.

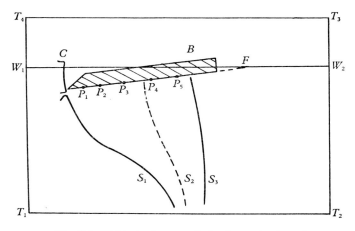

Fig. 8.8. Field plotting trough for determination of electrode shapes in presence of a beam.

The electron beam in fig. 8.8 passes through an aperture in the anode S_3 into the field-free space, and in the drawing it simply is extended to reach a focus at F. In the practical gun design, however, difficulties always arise with respect to the location of the actual beam focus. The field between S_1 and S_3 penetrates through the aperture in S_3 as discussed in §4.4 and §8.5, and a diverging electron lens is set up; moreover, the beam will diverge outside S_3 by mutual repulsion. Both effects can be estimated to some extent from the formulae and data given in these sections. To compensate for the later divergence, the beam may be initially overfocused. But in practice (cf. §8.4) it is found that homocentricity and homogeneity of the beam is to some extent lost before it reaches the focus.

The method discussed above starts from the desired beam shape, and finds the electrode structure necessary to maintain it. An alternative approach to design is to postulate the electrode shape and derive the field distribution, and hence the electron trajectories by solving Poisson's equation. Owing to the fact that the shape of the beam influences the potential distribution, the solution of the field plotting and ray tracing problem is more complicated than in the Laplacian problem discussed in chapter 3, where it was possible to separate the operations of field plotting and ray tracing. A method of a successive approximation must be used, the steps being as follows:

(a) Solve Laplace's equation (space-charge zero) for the electrode structure.

(b) Trace representative electron trajectories in the Laplacian field.

(c) Deduce the beam current density distribution $i(r, z)$ from the electron trajectories.

(d) Use the current density distribution to substitute for the charge density term in Poisson's equation, which can be written, in the case of circular symmetry,

$$\frac{\partial^2 V}{\partial r^2} + \frac{1}{r}\frac{\partial V}{\partial r} + \frac{\partial^2 V}{\partial z^2} - \frac{i(r, z)}{\epsilon\sqrt{\left(\frac{2e}{m}V\right)}} = 0. \qquad (8.65)$$

(e) Recalculate the potential distribution by solving Poisson's equation.

(f) Redetermine the electron trajectories in the Poisson field.

(g) Repeat steps (c) to (f), until a self-consistent solution is obtained.

Steps (a), (b), (e) and (f) may be carried out using methods discussed in chapter 3. In practice, the most frequently used methods are the electrolytic tank (see e.g. Hollway, 1955; Barber and Sander, 1958; Van Duzer and Brewer, 1959), the resistance network (e.g. Hechtel and Seeger, 1961; Archard, 1959) and the digital computer (Kirstein and Hornsby, 1964; Kulsrud, 1967; Weber, 1967).

In step (c), the approximation is made that all electron paths from one section of the cathode are similar, and that only one

central trajectory need be considered from each section. Thus the
beam is regarded as being split up into a number of beamlets each
centred on its representative trajectory. The current density in
each beamlet may be calculated from a knowledge of the potential
at a small distance from the cathode along a central ray, and from
the laws of space-charge limited emission (see (8.66) and (8.67)).

Once the function $i(r, z)$ has been determined, the solution of
Poisson's equation may proceed either by analogue or digital
methods. If an analogue method is being used, some means of
simulating the space-charge field is required. In the case of the
electrolytic trough, current can be injected into the bottom of the
trough through a network of conducting studs, the current at each
stud being proportional to the value of the space-charge term in
(8.65). Similarly, space-charge can be simulated in a resistance network analogue by injecting appropriate currents at each mesh
point.

It has already been noted in §3.6 that, for the purpose of
trajectory plotting, it is possible to use a special purpose on-line
digital or analogue computer in conjunction with the field analogues. This is an expensive procedure, but it may well be thought
to be worth while in space-charge problems, where owing to the
iterative method of solution, many trajectory plots must be made
in the course of a single problem. It is of course possible to feed a
general purpose off-line computer with data from the field analogue, though if many cycles of iteration are required, this is clearly
a cumbersome procedure. It might be thought preferable, where a
large high-speed computer is available, to dispense with field
analogues altogether, and to programme it to solve the whole
problem, carrying out all the steps (a)–(g).

The computing techniques are fully discussed in the references
quoted. The basic procedures differ little in principle from those
discussed in chapter 3, Poisson's equation being solved by
relaxation methods. Some attention must be given to the overall
rate of convergence of the iteration process comprising successive
performance of the steps (c) to (f). It appears that it is possible in
some cases to accelerate this convergence by the technique of
under-relaxation (Kino and Taylor, 1962). For example, in the
design of a magnetron injection gun, it was found possible to

reduce the number of cycles needed to achieve a given accuracy from ten to five.

It should be remarked that although increased employment of digital computers in the solution of space-charge problems may be foreseen, large-scale field analogues are very helpful for obtaining first information, and they are still in use currently in major industrial laboratories.

Finally, we mention here a relatively simple, though only approximate, method for the tracing of space-charge loaded beams through electrostatic fields due to Walcher (1951). The given field is divided into a finite number of steps as shown in §3.4. At the boundary of each step, a ray is refracted according to Snell's law. In distinction to the procedure of §3.4, however, the ray is not drawn as a straight line in the field-free space between two boundaries, but, due to its space charge, it has to follow a beam-spread curve. The method is simple for beams of small incremental spread when the beam-spread curve can be approximated by the parabolic law (8.20). For every step, however, the parameter of the parabolic arc representing the ray has to be changed according to the change in potential.

§8.9. Minimum in the potential distribution along the beam. The virtual cathode

The influence of space-charge on the potential distribution along the axis of a beam has been considered here so far for one case only. This particular case was concerned with a beam entering the space with zero velocity from a space-charge-limited cathode, and the potential along the beam was found to change according to the 4/3 power law of (8.47). We now want to investigate this longitudinal potential distribution for a beam entering with a given velocity into a field-free space or into a homogeneous retarding field. To avoid edge effects and a lateral potential distribution a flow of infinite lateral extent may be considered first. This ideal case is of some importance for the study of triodes, of tetrode valves and of beam tubes; and it has been investigated theoretically by Gill (1925), by Salzberg and Haeff (1937), by Birdsall and Bridges (1961, 1963) and by others. The results are essentially derived by integration of Poisson's law.

In the simple triode arrangement shown in the lower part of fig. 8.9, electrons may be emitted from a plane cathode c and may be accelerated by a plane grid g into a space which is limited by a plane anode a. A z-axis may be assumed at right angles to the three parallel electrodes, the origin ($z = 0$) being in the cathode plane. The potential distribution $V(z)$ between c and g is known from

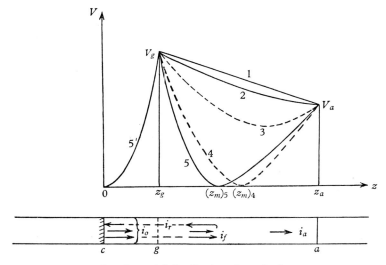

Fig. 8.9. Potential distribution along the beam.

(8.47), while the current density in this space is given by the following well-known equation which was first derived by Child (1911), i.e.

$$i = G_0 V_g^{\frac{3}{2}}, \qquad (8.66)$$

with
$$G_0 = \frac{4\epsilon}{9} \sqrt{\left(\frac{2e}{m}\right)} \frac{1}{s^2} = \frac{2 \cdot 33 \times 10^{-6}}{s^2} \text{ [m.k.s.]}, \qquad (8.67)$$

where s (here $= z_g$) is the distance between the electrodes.

We now assume the ideal case of a grid electrode which passes all electrons with the full energy of V_g eV exactly in the z-direction into the anode space. Some possible potential distributions in the anode space are plotted in the upper part of fig. 8.9. They are essentially determined by the value of a certain constant which appears after a second integration of Poisson's equation. This constant is given by the potentials V_g and V_a and by a current density i_0 which passes through the grid electrode g. Assuming both potentials V_g and V_a to be positive with respect to the cathode and $V_g \geq V_a$, the electrons are either entering a field-free space or they may run into a retarding field.

There are two ways of altering the current density i_0:

(i) A negatively biased control grid can be inserted between the cathode and the positive grid g. Then i_0 can be controlled by changing the bias just as it is done in a screen-grid tetrode valve.

(ii) The distance between the space-charge limited cathode c and the grid g is varied. Then i_0 is changed in proportion with $1/z_g^2$ according to (8.66) and (8.67).

For $i_0 = 0$, the potential distribution between the two planes g and a is represented by the straight line (1) in fig. 8.9. If the current i_0 begins to flow, the potential between g and a is depressed as shown by curve (2). Further increase in i_0 leads to a potential distribution as curve (3) in fig. 8.9. There a minimum of potential $V = V_m > 0$ is formed at a plane $z = z_m$, where the field $(\mathrm{d}V/\mathrm{d}z)_m$ vanishes. This potential minimum is initially formed at the anode, but with increasing current it gradually recedes towards the grid. For the special case $V_a = V_g$, however, the potential minimum always stays half-way between the electrodes.

With increasing i_0, the potential minimum gradually descends until it reaches a critical value. A further increase in injected current starts the following cycle of action: Increase in space-charge leads to reduction of potential which by slowing down the electron current leads to further increase in space-charge which leads to further reduction of potential and so on. The potential curve can drop only as far as zero, at which point a virtual cathode is formed. When the potential minimum reaches its critical value, a theoretical maximum current density

$$(i_{max})_{crit} = G_0(V_g^{\frac{1}{2}} + V_a^{\frac{1}{2}})^3 \qquad (8.68)$$

can be calculated to pass to the anode. The constant G_0 of (8.68) is given again by (8.67) with $s = (z_a - z_g)$ being the distance between anode and grid. A comparison of (8.68) with (8.66) shows the remarkable result that the maximum current density $(i_{max})_{crit}$ which can be transmitted through an originally field-free space

$(V_g = V_a)$ is eight times greater than the current density that is transmitted from a space-charge-limited cathode, provided that anode potentials V_a and electrode distances s are the same in both cases.

In order to force the maximum possible current $(i_{max})_{crit}$ through the anode space of the simple triode arrangement of fig. 8.9, a cathode-to-grid distance

$$z_g = \frac{z_a - z_g}{\sqrt{8}} \qquad (8.69)$$

has to be adjusted according to (8.66). If now the cathode-to-grid distance z_g is still further reduced and additional electrons are injected into the anode space, the potential minimum is found to drop abruptly down to zero. This is the formation of the virtual cathode mentioned above. The unstable distribution (4) is obtained. It implies the reflexion of some fraction i_r of the current density i_f which moves forwards in the $(+z)$-direction. The total current density i_0 in the space between cathode and grid and the current density i_a which proceeds to the anode are thus given by

$$i_0 = |i_f| + |i_r|, \qquad (8.70)$$

and $$i_a = |i_f| - |i_r| = |i_0| - |2i_r|. \qquad (8.71)$$

The current density i_r which is reflected by the virtual cathode passes back through the grid g and eventually reaches again the cathode c. The resultant redistribution of space-charge changes the potential distribution of fig. 8.9, curve (4), into the distribution curve (5) with a corresponding shift of the virtual cathode from $(z_m)_4$ to $(z_m)_5$ towards the grid. The potential distribution (5) is stable again.

The left branch of curve (5) between z_m and z_g must be symmetrical about the plane $z = z_g$ to the potential distribution curve (5′) between o and z_g, since the total current density i_0 is the same at both sides. Hence $(z_m - z_g) = z_g$, and the position z_m of the virtual cathode can no longer depend upon V_a but only upon V_g and z_g.

With still further increase in current i_0, say by further reduction of the cathode-to-grid distance z_g, the virtual cathode moves nearer and nearer towards the grid g, and the total current density

i_0 between grid and virtual cathode increases correspondingly. Simultaneously, the fraction i_a that reaches the anode, gradually decreases. This can be seen, for instance, by applying Child's law equation (8.66) to the right branch of the potential distribution between z_m and z_a, where with decreasing $(z_m - z_g)$, the complementary $(z_a - z_m)$ must increase. Hence, the largest current densities which can be transmitted from the grid to the anode in the presence of a virtual cathode are much smaller than the maximum current densities that are transmitted in the presence of the potential minimum of positive voltage.

If starting from large current densities the space current i_0 is gradually decreased, the anode current i_a will be found to increase. The operation now, however, cannot be performed in just the reverse of the way that has been described for gradually increasing current. This is due to the fact that originally the operation began with a potential minimum of finite voltage while now it begins with a virtual cathode. Now with decreasing i_0, the virtual cathode will move back towards the anode. This behaviour will continue until the space-current density reaches a critical value which is calculated to be

$$(i_{min})_{crit} = G_0(V_g^{\frac{3}{4}} + V_a^{\frac{3}{4}})^2, \qquad (8.72)$$

where G_0 is again given by (8.67), with s being the grid to anode distance. The slightest further reduction of space current causes the virtual cathode to disappear. Now, however, the jump in distribution into stage (3) produces no change in current, though it will be observed that the highest current reached on the retrace of the cycle is less than that obtained when the injected current i_0 was increased. From stage (3) onwards the operation with decreasing current is again the inverse operation to that with increasing current.

The whole cycle for the transmitted current i_a as a function of the injected current i_f is shown in fig. 8.10. Along the points (1), (2), (3) and (4) which mark the corresponding stages shown in fig. 8.9, the curve is a straight line through the origin, since in these stages $i_a = i_0$. At (4) the change from the positive potential minimum to the virtual cathode occurs, and the distribution changes abruptly into that of stage (5). The dashed line indicates that the intermediate states between the critical potential minimum distribution and the stable virtual cathode distribution (5) are unstable. With further increase of the injected current, the transmitted current only decreases. If by decreasing the injected current the cycle is reversed, the transmitted current will gradually increase along the lower branch of the loop shown in fig. 8.10; then, after reaching the stage (3), i_a will decrease again with decreasing i_0.

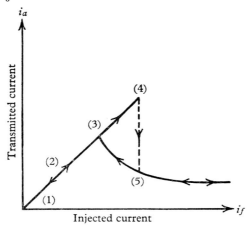

Fig. 8.10. Transmission of current through a space charge.

We must emphasize at this point that the simple theory of transmission of current through a diode region which we have presented is idealized and approximate. Various factors limit its validity. Birdsall and Bridges (1961, 1963) have predicted, by means of a time-dependent analysis, that the transition initiated at the critical point (4) should give rise to oscillations in which the position and value of the potential minimum are subject to cyclic variation, and according to these authors, the potential minimum can go negative during these oscillations. The time average position and value of the potential minimum correspond only approximately to distribution (5). Likewise, the time-averaged value of the transmitted current in this dynamic analysis does not correspond very exactly to the current predicted by the simple theory.

A further uncertainty concerns the role of positive ions, which are not taken account of in our discussion. According to Czarczynski, Ryley and Gambling (1966) the size of such hysteresis loops is strongly dependent on ion concentration, the loops practically disappearing in the absence of ions. It is clear then that the relationship between theory and experiment in this area is still not entirely satisfactory.

Magnetically contained beams. Potential distributions in quasi-parallel, magnetically confined beams which are long in com-

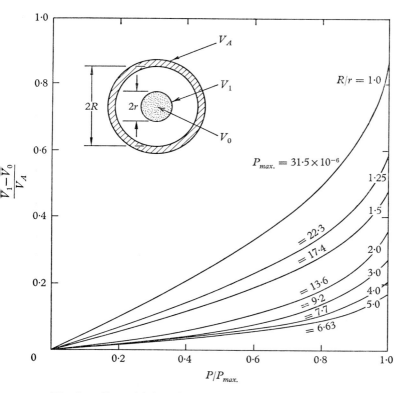

Fig. 8.11. Potential drop from edge to axis of a solid cylindrical beam in immersed flow.

parison with their radius have been discussed by Haeff (1939), Smith and Hartman (1940) and by Brewer (1967). Let such a beam pass co-axially through a conducting tube at anode potential V_A, and let space-charge spreading be constrained by a very strong longitudinal magnetic field. As the beam current I is increased, the potential V_0 at the beam axis falls progressively with respect to the potential V_1 at the edge until the ratio $V_0/V_1 = 0.17$. At this potential distribution the maximum possible current I_{max} passes along the beam, and the perveance $I/V_A^{\frac{3}{2}}$ reaches its maximum. If

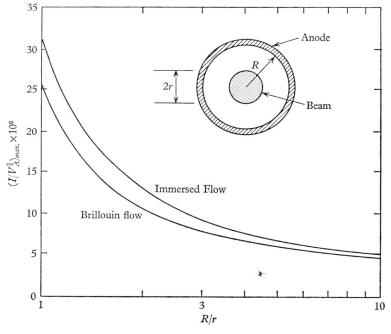

Fig. 8.12. Maximum perveance of a solid cylindrical beam as a function of R/r, the ratio of the radius of a surrounding conductor R to the beam radius r.

further current is injected into the beam, some electrons are turned back, and by a sudden transition through an unstable region, a virtual cathode is formed in the beam. Now only a reduced current I_0 will be able to pass, resulting in a reduced perveance $(I/V_A^{\frac{3}{2}})_0$. Critical perveances $P_{max} = (I/V_A^{\frac{3}{2}})_{max}$ and $P_0 = (I/V_A^{\frac{3}{2}})_0$ can be found for every ratio of radii of beam and surrounding electrode. Both these factors are the greater, the nearer the beam radius approaches the tube radius, i.e. the nearer the potential V_1 at the edge of the beam comes to the potential V_A of the tube. Fig. 8.11

shows some curves calculated by Brewer (1955) for the potential drop $V_1 - V_0$ from the edge to the axis of a solid beam as a function of the perveance P. The beam, of radius r, passes co-axially through an outer conductor of radius R and at anode potential V_A, and it is immersed in a strong longitudinal homogeneous magnetic field. If the beam completely fills the tube, $V_1 = V_A$, and for this particular case one obtains $P_{max} = 31 \cdot 5 \times 10^{-6}$, and for the flow with virtual cathode $P_0 = 29 \cdot 3 \times 10^{-6}$, while the Brillouin type of flow (cf. (8.58)) in the properly adjusted magnetic flux density yields a maximum

$$(I/V_A^{\frac{3}{2}})_{max} = 25 \cdot 4 \times 10^{-6}.$$

Fig. 8.12 shows the relationship between maximum perveance P_{max} and the ratio of radii R of the outer tube to r of the beam for the two cases of immersed flow and Brillouin flow.

CHAPTER 9

EMISSION SYSTEMS AND
ELECTRON GUNS

§9.1. The acceleration of the emitted electrons. Cathodes in emission systems

The light sources in light optics correspond to the cathodes in electron optics. While, however, a light-emitting body, for instance an incandescent filament, can be used as a source of light-rays immediately, a simple electron emitter such as a thermionic cathode is not sufficient for the production of electron rays in practical electron-optical gear. When passing through the refracting elements of glass optics, the speed (phase velocity) of light emitted from the filament is changed by a relatively small fraction only. On the other hand, electrons leaving a filament have to be accelerated first; their kinetic energy must be increased many times, in some cases even many thousand times, before they can enter a particular electron optical device.

The electrons from the cathode have to be injected into a relatively field-free region, i.e. they have to enter a space which is surrounded by an electrode of positive potential with respect to the cathode. This electrode is the anode of the emission system, hence electric lines of force starting at the cathode will end on the anode. The emission will first follow approximately the direction of the lines of force, since within some distance from the cathode, the kinetic energy of all electrons is still small in comparison with the potential drop along its path. On their way to the anode, the electrons are more and more accelerated. Eventually their accumulated kinetic energy will be much greater than the potential drop along a given element of the path. Hence along this path element the electron can deviate considerably from the direction of the field. For this reason it is possible to accelerate the emission of a relatively large cathode surface through a small aperture of the anode into a field-free space which is surrounded by the anode.

In good emission systems, the whole emission from the cathode passes the aperture in the anode, no electrons being intercepted. In this case, the electric field at the cathode is directed initially towards the anode aperture, the deviation of the field from the initial direction occurring rather close to the anode only. For the production of such a field an auxiliary electrode (see for instance G in figures 9.1 and 9.2) between cathode and anode is essential. This electrode has always to be substantially negative with respect to the anode. Generally, it is kept at cathode potential or slightly negative with respect to the cathode. As early as in 1903, Wehnelt recommended such auxiliary electrodes for focusing the electron emission towards the anode. The auxiliary focusing electrodes often serve the additional purpose of controlling the intensity of the emission. In this case, 'intensity modulation' is effected by changing the potential of the auxiliary electrode. Hence, in analogy to the terminology in valve technique, this electrode is called the 'grid', the 'screen' or the 'modulator' of the emission system. The simplest emission systems are triodes containing a cathode, a grid electrode and an anode, but a complicated system might contain many electrodes which sometimes are of complicated shape.

Before discussing the properties of complete assemblies, we shall give a survey of only the most important types of cathodes, which are suitable for use in emission systems, introducing the reader to the specialized literature.

We mention first the metallic thermionic cathodes (cf. Nottingham, 1956). These are useful if the vacuum is not of the best quality and if relatively high emission velocities can be tolerated. Flat or concave metal cathodes can be made from 0·02 to 0·1 mm sheets of tungsten, tantalum or molybdenum. These cathodes can be heated from a hot tungsten spiral or by electron bombardment. Heating by transition resistance from a carbon point touching the back of the cathode (Soa, 1959) has special advantages for emission microscopes (§9.2). Saturation densities of emission of a few 10^4 A m^{-2} can be reached at sufficiently high temperatures, e.g. 3 000° C for tungsten or 2 500° C for tantalum. The life time of the cathode, however, decreases with its temperature. A scheme for direct heating by running the current through the cathode has been given by Nicoll (1955).

Very robust and suitable e.g. for electron beam welding (§§9.9

and 9.10) are cathodes made from a massive, indirectly heated tungsten bolt (Bas, Cremosnik and Lerch, 1962). Up to 10^5 A m^{-2} can be drawn for a few hours from these bolt cathodes.

In electron microscopes the thermionic cathode is usually made from tungsten wire bent in the shape of a hairpin (cf. §9.9) and emitting at the tip only. From these cathodes current densities of 10^5 A m^{-2} can be drawn from an area of $\frac{1}{10}$ mm diameter. The emitting area can be reduced further by grinding the tip (Bradley, 1961). This source size is still too large for various purposes (cf. §9.9) and considerable interest has been shown in 'point cathodes' in which thermionic current is drawn from a sharply pointed tungsten wire welded to the top of the hairpin cathode. These point cathodes are etched down to a point of a few μm diameter with a cone angle of 0·1 radian (Sakaki and Möllenstedt, 1956; Swift and Nixon, 1962). Special jigs have been developed for the etching of these points (cf. Niemeck and Ruppin, 1954; Schreck and Placius, 1956).

Thoriated tungsten which, owing to its relatively low work function, is conveniently used for some thermionic cathodes is not recommended for electron optical work as the electron emission from it is very unequal over the surface.

Very common in cathode-ray tubes are oxide cathodes consisting of a layer of alkaline earth oxides (e.g. 60 per cent Ba; 40 per cent Sr). The layer of the carbonates is deposited, say, by dipping, spraying or cataphoresis on to a metal core (Ni, Pt) which is indirectly heated in the way explained above. The carbonates are changed into oxides by heat in the 'activation' process. At 1 000–1 100° K, an emission of 10^3 A m^{-2} is drawn while a saturation density of 10^4–10^5 A m^{-2} can be reached in short pulses (cf. Herrmann and Wagner, 1951; Beck, 1966). Of some interest in electron optics are oxide cathodes of very small emitting area. Ando et al. (1959) have used oxide cathodes of 30 μm diameter in the bore of platinum capillaries of 0·3 mm outer diameter.

As an alternative to the alkaline earths, layers of thoria (Hanley, 1948) or lanthanum hexaboride (Lafferty, 1951; Broers, 1967, 1969) have been used. The latter (LaB$_6$) cathodes are particularly recommended for demountable work as they may be exposed to the atmosphere when cold.

Finally, we mention the matrix—or dispenser cathodes. They are produced, for instance, from a mixture of nickel-powder (80 per cent) with strontium and barium carbonates (20 per cent) and with an activator (1 per cent), e.g. zirconium. The mixture is sintered under pressure and heat (Fane, 1958). Another useful matrix cathode is made from tungsten powder sintered and impregnated with barium aluminate (Levi, 1953, 1955). This was developed from the so-called 'L-cathode' in which the barium was introduced to the tungsten matrix as a carbonate (Lemmens, Jansen and Loosjes, 1950). It is a great merit of these dispenser cathodes that they can readily be machined to shape, and that they are very robust and not easily spoiled when heavy currents are drawn from them.

The conditions under which the cathodes are operated in the emission systems are of great importance. Some cathodes, for instance oxide cathodes in electron guns (§9.2), must be used under space-charge conditions. On the other hand, an investigation of these oxide cathodes in an electron emission microscope (§9.2) can be carried out only when the emission is temperature limited. Again, the emission of the tungsten hairpin cathodes in the highly magnifying electron microscopes is always practically temperature limited.

Under space-charge conditions, the cathode is heated sufficiently so that a space-charge cloud is formed in front of it. The emission current then increases with the 3/2 power of the applied anode voltage, but it does not depend on small changes of the cathode temperature. On the other hand, when the emission is temperature limited all available electrons are withdrawn by the applied field. The emitted current cannot be increased by an increase in anode voltage; however, it will grow exponentially with increasing cathode temperature.

§9.2. Triode emission systems with plane cathode and their use in emission microscopes and in television guns

Of some technical importance are the 'two-diaphragm emission systems' shown in fig. 9.1, and the emission system with open anode in fig. 9.2. In all the drawings, the cathode is marked C, the grid electrode G and the anode A.

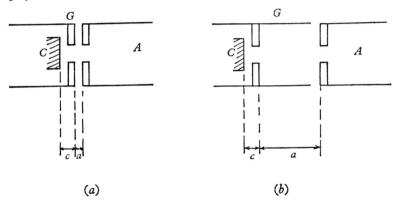

Fig. 9.1. Two-diaphragm emission systems.

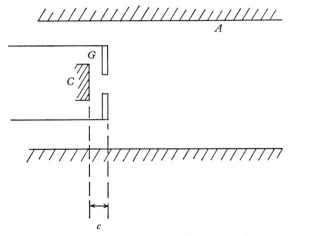

Fig. 9.2. Emission system with open anode.

These systems have flat equipotential cathodes which are heated
either by thermal conduction or radiation from a hot filament
directly behind the cathode, or by electron bombardment from a
filament behind the cathode and at a few hundred volts negative
with respect to it. The grid electrodes are flat diaphragms with a
circular aperture. They are fixed in tubular sleeves which also
serve for shielding the electron rays from external electrostatic
disturbances.

The anodes in the system of fig. 9.1 are also apertured flat diaphragms. In the system of fig. 9.1 (*a*) electrostatic shielding of the space between grid electrode and anode is provided automatically, since the radius of the diaphragms is large in comparison with their spacing. Otherwise, the shielding may be effected by surrounding the grid–anode space with a cylinder that is either connected to the grid electrode as in fig. 9.1 (*b*) or to the anode. The two-diaphragm systems of fig. 9.1 have found wide application. Their use in electron microscopy was first studied by Brüche and

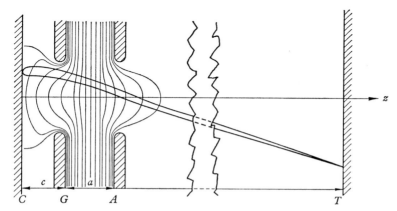

Fig. 9.3. Emission microscope of Brüche and Johannson.

Johansson (1932). Their system with its equipotentials is shown in fig. 9.3. The potential difference between grid electrode *G* and cathode *C* is very small and by a slight change of it, the image can be focused. Both spacings *c* and *a* are of the order of one grid aperture radius $r_g \approx 1$ mm. A magnified image of the cathode surface is projected on to a fluorescent target *T* at about 20 cm distance from the cathode. Johannson (1933, 1934) studied the magnification, and resolving power as a function of the distances *c* and *a* and electrode voltages. He achieved, with 750 volts at the anode *A*, an image of 180× magnification with a resolution of 10 μm. Later observers improved the resolution by reducing the unevenness of the cathode surface, by increasing the field in front of the cathode (cf. §9.7, (9.18)) and by optimizing the geometry. We mention only the study by Mecklenburg (1942) who obtained a resolution of 300–500 Å. In a very detailed investigation, Soa (1959) measured the optical properties of the emission system as a function of the thickness of grid and anode, of the electrode spacing a and c, their aperture ratio r_g/r_a and their voltage ratio V_g/V_a. We shall report some of his results in §9.4.

In the emission microscope, the electrons originating from the cathode may have been liberated by photo-effect, by secondary emission or by thermionic emission. In the instrument employing secondary emission, the cathode is bombarded with a primary electron beam or with positive ions. In both cases, the secondary electrons emitted from the cathode have energies extending over several eV. Hence it is usual to arrange an aperture stop in the back focal plane of the immersion objective in order to exclude the electrons of high emission energy, especially those with large transverse velocity components (cf. Möllenstedt and Lenz, 1963). In the instrument employing thermionic emission, the cathode temperature must be kept below a certain limit, so that the emission is saturated and space-charge formation is avoided (cf. §9.1).

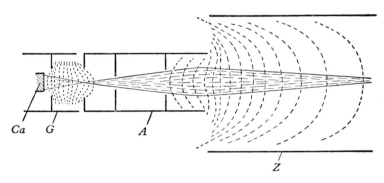

Fig. 9.4. Electron gun of Zworykin.

On the other hand, in electron guns such as shown in fig. 9.4, the emission is usually space-charge limited and the field from the anode is shielded to some extent from the cathode by keeping the grid electrode negative or at zero potential with respect to the cathode. The electrostatic field then converges the electron rays into a disc of least confusion which is called the 'cross-over'.

In the electron gun, an image of this cross-over is projected by an electron lens on to target, e.g. a fluorescent screen. This is shown in fig. 9.4 for the example of one of the first television guns (Zworykin, 1933). There, a beam from a two-diaphragm emission system forms a cross-over near the aperture of the anode A. An electrostatic focusing lens between A and a second anode Z serves to project an image of the cross-over on to the fluorescent target. With this gun, currents of a few hundred microamps could be projected into a spot of the order of 1 mm² at electron energies of a few keV. The first anode in a two-diaphragm emission system is usually kept at a few hundred volts, with respect to the cathode, hence the electrostatic focusing lens must accelerate the beam at a voltage ratio $V(Z)/V(A)$ of about 10 to 20. It will be seen later on (§9.6) that if an open system, such as in fig. 9.2, is used in an electron gun, it requires much higher voltage at the anode A to draw a current comparable with that of the two-diaphragm system in fig. 9.1. The cross-over, which is formed just behind the grid aperture, is conveniently projected by a magnetic lens, when the open system is used in a gun.

Guns in television viewing tubes. The prototype of such guns, shown in fig. 9.4, would still serve well in picture tubes, and all later guns have been developed from it (cf. Moss, 1968). The developments consist substantially in the addition of further electrodes and of an ion trap.

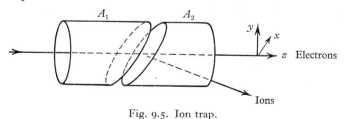

Fig. 9.5. Ion trap.

A frequently used trap is shown in fig. 9.5. There, the anode of the gun consists of two parts A_1 and A_2 terminated by planes which are inclined at about 45° to the axis (cf. Morton, 1946). A slightly accelerating potential difference is applied between A_1 and A_2 setting up an electrostatic field E. The acceleration of the electrons in the $(-y)$-direction is, however, compensated by a magnetic field B in the $(-x)$-direction so that the electron beam continues to travel in the z-direction. Negative ions which are always emitted from oxide cathodes, suffer—owing to their relatively large mass over charge ratio—sufficiently large overall deflexion

$$(y = (m/e)\,(E/B^2))$$

in the combined field that they are filtered out of the electron beam. In this way, the fluorescent screen of the tube is saved from the damaging effect of these ions. As to the addition of electrodes to the original triode (fig. 9.1) we mention first a plane diaphragm, the 'accelerator' A_c, which is inserted between the grid electrode G and the anode A. The brilliance of the spot produced, say by the magnetically focused beam of a triode, on the fluorescent screen depends on $V_a I_s$ the product of anode voltage and screen current. With the modulation of the emission current I_{em} by the grid bias V_g, adequate stabilization of V_a (≈ 15 kV) which takes a substantial part I_a of the current I_{em}, becomes difficult and expensive. After the introduction of the accelerator, however, much of I_a is intercepted by it and since its voltage is low ($V_{ac} \approx 300$ V) it can be stabilized much more easily.

A further improvement of the gun for picture tubes was obtained by inserting a pre-focusing electrode P between A_c and A. The pre-focusing of the beam by the electron lens set up between P and A serves to reduce the beam diameter at the scanning coils (§ 10.3). Deflexion defocusing (§ 10.7) which increases rapidly with the beam diameter, can thus be kept very small. According to paraxial theory, it is immaterial where pre-focusing takes place. However, the closer the pre-focusing lens to the emission system, the greater will be the power of the lens in order to obtain the same width of the beam. Moreover, the beam width in the pre-focusing lens will still be small so that aberrations are minimized (§ 6.3) and space-charge effects in the region of lower velocity will be reduced. A gun has been constructed according to these principles by Francken, Gier and Nienhuis (1956) and is shown here in fig. 9.6. In this gun, the voltage of the pre-focusing electrode P is close to cathode potential $V_p \approx V_c = 0$ but can be regulated in order to minimize the beam diameter. P draws no current. The accelerator

is at $V_{ac} = 300$ V, the anode at $V_A = 15$ kV and the beam is
focused on to the screen by a magnetic lens. As can be seen in the
drawing, the axis of the gun is curved after P and a transverse
magnetic field is superimposed over this region so that the combination of electrostatic and magnetic fields can serve as an ion trap
in the manner explained above.

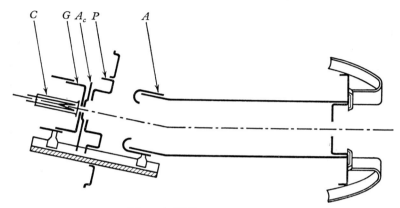

Fig. 9.6. Philips pentode gun.

Other developments of the gun seem to be less important. We
mention here the triple gun of the R.C.A. colour picture tube
(Morrell and Hardy, 1964) which is still representative of the
latest designs. This gun contains a flat grid electrode at -200 to
-400 volts, a flat accelerator at 1 kV, a tubular pre-focusing
electrode at 6 kV and a tubular anode at 27 kV. Three guns of this
kind are assembled into a unit, carrying three radially converging
pole pieces with magnetic shields between them. The magnetic
fields direct the three emerging electron beams on to the three
associated types of phosphors to be excited in the three basic
colours.

Some recent guns for picture tubes have been developed in a
somewhat different way. This was based upon some work done
by Law in the late 1930s on the use of grid electrodes at positive
potential with respect to the cathode. In some constructions, this
positive gride electrode which modulates the beam intensity is
preceded by a positive screen which is arranged in close proximity

to the cathode. We mention here the 'Rank Monocon' gun and the 'GEC Laminar Beam' gun. These guns are capable of producing very fine foci of high current intensity but they require relatively large 'drive' voltages to modulate the beam. This is a definite disadvantage as compared with the usual gun construction employing negative modulating potentials.

The guns in picture tubes are required to produce the greatest possible current into the smallest possible spot from a beam converging into the smallest possible angle. The upper limit of performance is clearly indicated by Langmuir's equation (7.8) in §7.1. Hence the final voltage of the beam should be increased as much as compatible with insulation and safety requirements; cathode surfaces have to be used from which highest current densities can be drawn, and focusing lenses have to be employed which are able to handle beams of widest angle without aberration. In practice, the Langmuir limit is never reached owing to the various unavoidable aberrations. Hasker and Groendijk (1962) defined accordingly a figure of merit

$$Q = D_s D_b / L,$$

where L is the distance from the deflection plane to the screen, D_s is the diameter at the focused spot and D_b is the diameter of the beam at the scanning field, these diameters being measured at a circle at which the current density has dropped to $1/e = 0.37$ of its maximum value at the centre (cf. §7.1). For good picture quality Q must be small. E.g. with their gun, Hasker and Groendijk (1962) measured $Q = 6.9 \times 10^{-3}$ mm radians at a focused current $I = 0.85$ mA. Q was reduced to 1.8×10^{-3} when the current was biased off to 0.015 mA.

§9.3. Fields and electron trajectories in emission systems. Cross-over formation

In the great majority of all practical emission systems, electrostatic fields converge the electron rays into a disc of least confusion which is called the 'cross-over'. The cross-over formation is explained by fig. 9.7. The electrons leave the cathode surface at different angles and with different speeds which are due to the thermal velocity distribution. Let electrons of a homogeneous velocity be considered first. Rays marked in the figure as (1), (2) and (3) leave the cathode surface C at right angles; they intersect at a common point on the axis (z) forming the centre of the cross-over X. These rays are the chief trajectories of electron bundles,

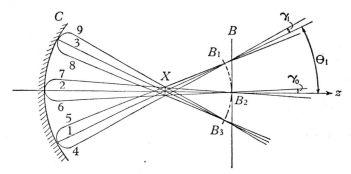

Fig. 9.7. Formation of cross-over and of cathode image.

each bundle emerging from a common point of the cathode surface. The peripheral rays of these bundles leave the cathode surface at grazing angles; they are marked in fig. 9.7 by numbers (4)–(9). The widths of the bundles at their mutual intersection is responsible for the finite extension of the cross-over X. Since the field tends to converge each ray of the bundle towards its chief trajectory, all bundles eventually come to focus at the surface $B_1 B_2 B_3$ where an optical image of the cathode is formed.

The size of the cross-over has been estimated by Ruska (1933), who assumed the field between cathode and centre of the cross-over to be a central field with strictly spherical equipotentials. He started from the equation of motion for the electrons under these idealized conditions, and derived a radius of cross-over

$$y_X = \frac{y_c}{\sin \Theta} \sqrt{\frac{V_{em}}{V_X + V_{em}}}, \qquad (9.1)$$

where y_c is the semi-aperture of the cathode surface, V_{em} is the thermal emission energy (electron-volts) of the electron, V_X the potential at the cross-over and Θ the angular semi-aperture between chief ray and axis. (A derivation of (9.1) from optical principles will be given in the next section.) Law (1937) generalized (9.1) for any non-uniform field of rotational symmetry by introducing a 'field-form factor' L which is a function of the potential distribution in the system. He expressed the cross-over radius as

$$y_X = \frac{z_X}{L} \sqrt{\frac{V_{em}}{V_X + V_{em}}}, \qquad (9.2)$$

where z_X is the distance between cathode and cross-over. Law's factor $L \approx z_X \Theta/y_c$ would be unity for the field with strictly spherical equipotentials. For the fields of practical systems L is found to be of the order of a few tenths. Disregarding aberrations, Dosse (1940) took the field-form factor as the ratio of the cross-over distance from the cathode to the focal length of the emission system, i.e.

$$L = z_X/f'. \qquad (9.3)$$

Cross-over size and location depend very much on the potential distribution in the system.

Einstein and Jacob (1948) found that the empirical potential distributions along the axis of triode systems could be expressed by

$$V(z) = A[\exp(kz) - \exp(-kz)], \qquad (9.4)$$

where A and k are characteristic constants for a given system. A is a 'voltage scale factor' measuring a certain fraction of the cross-over voltage V_X, while k is a 'geometrical scale factor' which is inversely proportional to the cross-over distance (z_X) from the cathode. The values of the parameters A and k depend mainly upon the grid to cathode distance c; however, k is always of the order of a reciprocal grid-aperture diameter and, for zero grid bias, A is of the order of a small fraction (≈ 3 per cent) of the 'black-out voltage' (cf. §9.5). Series development of (9.4) allows us to visualize the change of potential at various distances from the cathode. Close to the cathode the field will be approximately constant, since for $z \ll k$,

$$V(z) = 2Akz. \qquad (9.5)$$

Farther away from the cathode, there follows a simple expotential region which includes the cross-over.

Accurate electron trajectories can be obtained by ray tracing

(§3.4 ff.) through the actual field plots. These traced paths, however, are of little practical value, since in emission systems they generally are only a rough approximation to the actual paths. It will be shown in §9.4 that all properties of the emission system are mainly controlled by the concentration of space charge. The problems of ray tracing and field plotting are therefore interlinked, as explained in §8.8, since it is necessary to solve Poisson's equation rather than Laplace's equation. Thus the methods of trajectory determination discussed in §8.8 must be used in the theoretical study of many emission systems; details of computational procedures for particular types of electron guns are to be found in the references quoted in that section.

Experimental investigation of the cross-over. This has been attempted in two different ways. Usually a magnified electron-optical image of the cross-over is projected by an electron lens on to a fluorescent target (e.g. Klemperer and Mayo, 1948). The use of high-quality electric or magnetic lenses of relatively large diameter is essential so that unavoidable lens errors are sufficiently small and do not modify the projected image.

In an alternative method (Klemperer and Mayo, 1948; Klemperer and Klinger, 1951) pencils emerging from the cross-over are selected by a pepperpot diaphragm and traced by a sliding fluorescent target. In fig. 9.8, X represents the cross-over, P a part of the pepperpot diaphragm and T the target. The diaphragm P contains a large number of circular holes; only two of them are shown in the figure. Through one of these holes, which is the axial hole, there passes the chief ray O which coincides with the axis z of the system. The two outermost peripheral rays passing through the same hole are marked O_a and O_b. All three rays intersect at the vertex S_0; they represent the axis and, respectively, the axial section through the envelope of a small conical pencil with the semi-vertical angle γ_0. Analogously, the rays through a hole off the axis are the chief rays I and the peripheral rays $1a$ and $1b$, forming a cone with the vertex S, and semi-vertical angle γ_1. In each case, the chief rays are traced by following up the centres of the fluorescent spots on a target T which is arranged perpendicularly to z and can be moved along this axis.

The extrapolated intersection of the chief rays with the axis allows
a location of the cross-over position z_X. If the chief rays of
different divergences Θ have no common point of intersection,
their tracing will provide information on the spherical aberration
of the cross-over.

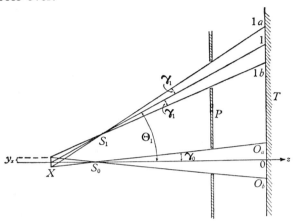

Fig. 9.8. Selection of pencils by a pepperpot diaphragm.

The divergences γ of the pencils are obtained by tracing the
diameters of the fluorescent spots.* The cross-over diameter as
traced by the pencil o can be derived from the figure to be

$$(y_X)_0 = (z_p - z_X)\gamma_0 - r_p, \qquad (9.6)$$

where z_p is the distance of the pepperpot diaphragm from the
cathode and r_p is the radius of the pepperpot hole which can be
measured directly under a microscope. A pencil intersecting the
axis at an angle Θ_1 leads to a cross-over radius $(y_X)_1$, which, in
principle, can be different from $(y_X)_0$, since it is formed by a
pencil emitted from another zone of the cathode and hence has
travelled through a different field. Moreover, the radii $(y_X)_0$,
$(y_X)_1$, $(y_X)_2$ belong to a virtual cross-over which is extrapolated

---

* A traced spot diameter does not necessarily correspond to the diameter of a
  disc with sharp boundaries. It is shown below (§ 9.7) that the current density
  in the cross-over fades away as an exponential function of the radius, and that
  a cross-over radius can be fixed only by special definitions. The visually
  determined radius, however, seems roughly to coincide with definitions given
  below by (9.25).

from straight rays. Actually after crossing over, the rays are often subject to some deflexion by a field which, still belonging to the emission system, can extend some distance behind the cross-over. Hence the real cross-over does not need to coincide with the virtual.

§9.4. Optical approach to emission systems

It has been pointed out that an accelerating electron lens is necessary for the projection of the electron emission from the cathode into a field-free space. This lens should be of positive focal length and located approximately at its focal distance from the cathode. A single aperture in front of the cathode would be insufficient. This follows from the optical properties of apertured diaphragms given in §4.4. Electrons drawn from such a system into a field-free space always form a divergent beam, since its focal length—according to the Davisson–Calbick equation (4.8)—is negative, the field E in front of the diaphragm always being greater than the field E' behind it. However, by provision of a second diaphragm of a relatively high positive potential it is easy to make $E' > E$. Hence a system of positive focal length results. (4.8) can be used to calculate this focal length to some approximation as long as space charge is negligible (cf. Klemperer, 1947 b).

The application of some general optical principles to emission systems may be explained after Dosse (1940) by comparing the electron emission system of fig. 9.9 (b) with its light-optical analogue in fig. 9.9 (a). In both drawings, there are shown three chief rays o, 1 and 2 which all leave the emitting surface C at right angles. The rays 1a and 1b form the envelope of the elementary bundle 1 which emerges from the point P.

In the light-optical case, in which the refractive index in object and image space is of the same order of magnitude, the limiting rays 1a and 1b are given by the aperture stop S of the lens L. In the electron-optical case, however, the refractive index along the electron path increases very rapidly. Hence the radius of an elementary bundle of electrons is decreasing so rapidly that it is not limited by an aperture in a focusing electrode, but 1a and 1b are rays which are emitted at a grazing angle. At the cathode the elementary bundle then fills an entire hemisphere.

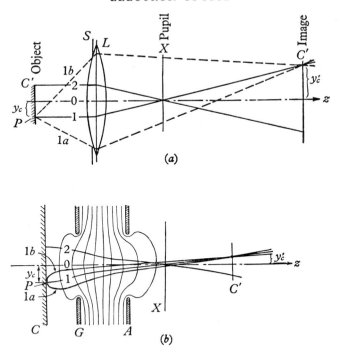

Fig. 9.9. Optical interpretation of an emission system.

The cross-section of the beam in the plane X forms the cross-over. This corresponds to the exit pupil of the optical system which is defined as the cross-section of the beam in a plane perpendicular to the axis at the point of intersection of the chief rays with the axis (cf. Drude, 1925, p. 73). Again, it can be seen that the size of the pupil of the light-optical system is determined by the limiting physical aperture stop S, while in the electron-emitting system the exit pupil is given by the emission velocity of the peripheral electrons which left the cathode at grazing angle.

The chief rays from a plane cathode surface as shown in fig. 9.9 are initially parallel. Hence, in this case, the cross-over X coincides with the focus of the emission system. On the other hand, the image C' of the cathode surface C is formed by the foci of the elementary bundles. The magnification y'_c/y_c is, according to (2.2), given by the ratio of the distance XC' to the focal length f' of the emission system. This relation and (2.8) provide a practical means of determining the focal lengths which are given by

$$f' = XC'y_c/y_c'$$ (9.7)

and

$$f = f' \left(\frac{V_{em}}{V_{em} + V}\right)^{\frac{1}{2}},$$ (9.8)

where V_{em} is the energy with which the electron is emitted (e.g. 0·1 eV for an oxide cathode) and V is the electron energy in the image space. Magnifications y_c'/y_c have been measured by Johannson (1933) with the help of a pattern of lines scratched in the cathode surface. The distances of the lines can be measured with a microscope, first on the cathode and later on a fluorescent target, to which the image is projected. Magnifications of the cathode surface of the order of 100× can be obtained easily with two-diaphragm emission systems. In this way, the system has been used as we have shown in §9.2 as an electron microscope, where the object is immersed in a region of high field. Accordingly, the system is called an immersion system, and it can be compared with the immersion objective of a light-microscope, the object (cathode) being immersed in the field of the two-diaphragm lens. Focal lengths and magnifications of the objective are controlled by the distance c of the cathode from the grid electrode and by the grid bias V_g.

Bas and Preuss (1964) have published data on cardinal points of an immersion objective which they obtained experimentally and by computation. Soa (1959) made detailed experimental investigations on various immersion systems and we present in fig. 9.10 some of his results which refer to a two-diaphragm system with equal aperture radii ($r_a = r_g$) as shown in fig. 9.1(a).

The spacings are $c = 1\cdot9r_g$ and $a = 4\cdot4r_g$, and the thickness of the grid diaphragm is $t_g = r_g$. The thickness of the anode diaphragm has no influence on the performance and hence does not need to be specified. A cross-section through the system is shown on the right hand side of the figure. The abscissa of the graph, representing the voltage ratio (V_g/V_a in per cent) between grid and anode, is drawn along the inside of the anode. The ordinate, which is drawn along the axis of the system, applies to the various

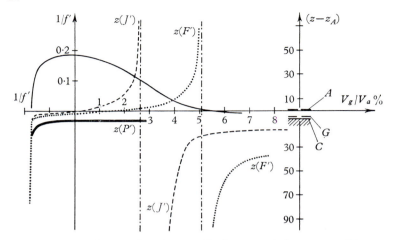

Fig. 9.10. Optical properties of a two-diaphragm emission system. Distances $(z - z_A)$ from the anode A: (i) of the focus $z(F')$ . . . ., (ii) of the principal plane $z(P')$ ——, (iii) of the image of the cathode surface $z(J')$ – – – –, and the reciprocal focal length $1/f'$ ——. f' and z are measured in units of a grid aperture radius (r_g) and plotted against the ratio of grid voltage to anode voltage (V_G/V_A).

distances z from the location z_a of the anode, namely: that of the image-side focus F', of the principal plane P' and of the image J' of the cathode surface. Also, on the far left hand side there is drawn an ordinate referring to the power $1/f'$ of the immersion objective. Both f' and z are measured in terms of grid aperture radii (r_g). It can be seen that P' is practically coincident with the cathode for a wide range of voltage ratios. On the other hand, with increasing V_g/V_a, the focus F' passes through the plane of the anode, moving up to infinity at $V_g \simeq$ 0·05 V_a. For $V_g >$ 0·05 V_a, the immersion objective behaves like a concave lens. Correspondingly, the power $1/f'$ reaches its maximum near $V_g \simeq$ 0 and disappears for a V_g between 0·05 and 0·06 V_a. Moreover, the image J' of the cathode lies exactly in the plane of the anode when $1/f'$ reaches its maximum value. The image distance increases rapidly with increasing positive grid bias and for $V_g/V_a >$ 2·6 per cent the image becomes virtual.

Two-diaphragm systems with plane cathode but with different geometrical dimensions of c, a and t_g behave optically in a somewhat analogous way. Thus, for all investigated systems, the principal plane P' is at all voltage ratios always quite close to the plane of the cathode, and at the minimum of f' the image appears always to be close to the plane of the anode. The thickness of the grid t_G and its spacing c from the cathode have the greatest influence on the value of f'. The focal length f' passes through a distinct minimum when the grid bias (V_g/V_a) is changed monotonically. $(V_g/V_a)_0$ at f'_{min} is the smaller, i.e. more negative, and the minimum gets the more pronounced, the smaller the spacing c.

Thus, to obtain highest magnification y'/y and short focal lengths, one has to chose thin electrodes and small spacings. On the other hand, to obtain highest field strength at the cathode one must take thick grid electrodes and smallest c-spacing. Variation of the shape of the grid and of the anode have according to Soa's (1959) experiments very little influence on the optical properties of the system.

If an emission system is used in an electron gun, the electrons emitted from the system have to be focused on a fluorescent screen by a focusing lens of a given aperture. There, one attempts to concentrate the maximum current into the smallest spot. Hence it is usually necessary to project the largest possible current density i_X at the cross-over into the smallest possible solid angle Ω. In other words, the directional intensity of the beam, as given in §1.6 should be as large as possible.

In the absence of aberrations and if the beam is not cut down by aperture stops, the brightness of the emitting area of the cathode and the voltage at the target will alone determine the brightness of the focused spot as shown in §1.6 (1.50). If the aperture of the beam is reduced by stops, which generally is necessary in practical guns in order to avoid excessive aperture errors of the focusing lens, the best obtainable current density i_{max} in the spot is given by the 'Langmuir limitation' of (7.8) in §7.1. This applies in the absence of geometric aberrations only. Even in the best practical guns the focused current density is substantially smaller than i_{max}.

§9.5. Beam intensity modulation

The current I_{em} drawn from the cathode of a triode emission system of given geometry depends on the potential V_g of the grid and on the anode potential V_a, with respect to the cathode potential. A plot of I_{em} versus V_g is shown in fig. 9.11. At $V_g < 0$, the grid takes practically no current. With increasing grid bias, i.e. negative grid potential, I_{em} decreases, but owing to the Maxwellian distribution of emission energies it only asymptotically converges to zero. It has been found of technical interest, however, to define an

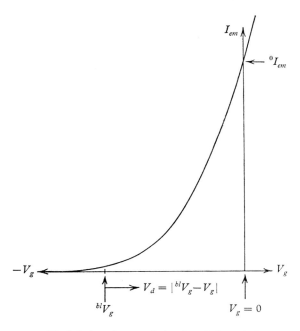

Fig. 9.11. Modulation characteristic of a triode emission system.

approximate value for the grid bias at which the emission current I_{em} is 'practically' zero. This is the so-called 'cut-off voltage' or 'black-out voltage' ${}^{bl}V_g$, at which the spot on the fluorescent screen of a cathode-ray tube incorporating the emission system seems to disappear in a 'dark' room. If the grid bias is increased beyond ${}^{bl}V_g$ the resulting current I_{em} will be due to the Maxwellian tail of the energy distribution of the emitted electrons. For grid voltages more positive than ${}^{bl}V_g$ the notion of the 'grid drive' $V_d = |{}^{bl}V_g - V_g|$ is frequently met in the literature. As indicated in fig. 9.11, V_d increases when the bias V_g is reduced.

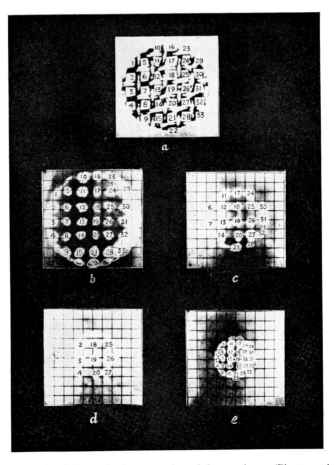

Fig. 9.12. Emitting cathode area and modulator voltage. (Photograph.)

(*Facing p.* 341)

To calculate the space-charge limited current of an emission system, Poisson's equation must be solved. As will be shown, there is no simple solution of this problem. In principle, a solution can be obtained with the aid of a computer but in this manner, it is not possible to obtain any quick insight into the working of the system. Hence approximate and experimental methods have been developed.

We deal first with the investigation of the emitting area of the cathode surface. The results demonstrated by fig. 9.12 are due to Moss (1946). Fig. 9.12(a) shows a light optical image of the cathode as seen through the grid aperture. A wire gauze is arranged in contact with the cathode surface so that the different parts of it can be located in the image. Fig. 9.12(b)–(d) show electron optical images (cf. fig. 9.7) taken with various values of bias V_g. It can be seen that the magnification y_c'/y_c remains constant, but that the radius $(y_c)_0$ of the emitting cathode area decreases. According to a semi-empirical approximate formula given by Moss (1946) this radius is

$$(y_c)_0 = r_g \frac{V_d}{{}^{bl}V_g},$$ (9.9)

where r_g is the radius of the grid aperture.

The potential at the cathode surface may be obtained by super-imposition of the potentials due to the anode and due to the grid. It can be checked by field plotting that this combined potential becomes zero along a circle of radius $(y_c)_0$ given by (9.9). With increasing grid bias $(-V_g)$, the circle contracts like an 'iris' reaching the axis when V_g reaches the value

$$^{bl}V_g = DV_a,$$ (9.10)

where V_a is the anode voltage and D is the penetration factor which indicates the extent to which the anode field penetrates through the grid aperture to the cathode. The penetration factor is the reciprocal of the 'voltage-amplification factor' in valve technique. The penetration factor of an emission system changes substantially along the modulation characteristics due to the above 'iris effect'; however, D of (9.10) is its largest value and this is reached at cut-off; it corresponds to the field penetration at the axis.

Cathode loading. Due to the different field penetration at different distances from the axis, the field at the cathode surface is inhomogeneous, it grows from o at $y_c = (y_c)_0$ to a maximum value at $y_c = $ o on the axis. For this reason the emission density is inhomogeneous. A semi-empirical formula given by Moss (1946, 1968) allows one to estimate the peak loading at the centre of the cathode:

$$i_0 = 14 \cdot 5 \, V_d^{\frac{3}{2}}/r_g^2. \tag{9.11}$$

According to experimental results (9.11) has no discontinuity at zero grid bias. This implies, for instance, that a system with $^{bl}V_g = -20$ V operating at $V_g = +10$ has the same peak loading as a system with $^{bl}V_g = -50$ operating at $V_g = -20$, provided the aperture radius r_g is the same in both cases. The maximum peak loading that can be afforded depends on the quality of the emissive surface and on the duration of the emission (see next section).

A useful empirical formula for the total current emerging from the grid hole is, according to Moss (1946), given by

$$I_{em} = 3 \times 10^{-6} \, V_d^{\frac{7}{2}} \, {}^{bl}V_g^{-2}. \tag{9.12}$$

This equation has been discussed by Jacob (1952) and it was found to apply with good approximation for relatively large anode voltages ($V_a \gg V_{em}$) and small grid-cathode spacings ($c \lesssim r_g$). Alternatively, the current obtained from an emission system under space-charge conditions can be estimated from Child's law (8.66), the emission system being treated as a diode. In this 'equivalent diode', the flow is somewhat different from that specified in (8.66); hence the proportionality constant between I_{em} and $V^{\frac{3}{2}}$ cannot be expected to have exactly the value G_0 of the plane parallel diode. The potential difference at the equivalent diode which results from a superimposition of the potentials of the electrodes of the emission system is, according to Schottky (1919), given by

$$V = \frac{V_g + DV_a}{1 + D}. \tag{9.13}$$

This equation essentially applies to the paraxial region where the field penetration is represented exactly by D. However, (9.13) substituted into Child's equation, (8.66), allows us, for technical purposes, to obtain a fair estimate of the total emission current over the whole cathode surface as a function of V_g and V_a. For the particular case of emission with zero grid bias ($V_g = 0$) and with small penetration factor ($D \ll 1$) one obtains by substituting (9.10) into (9.13) the familiar approximation

$$I_{em} = G^{\,bl}V_a^{\frac{3}{2}}, \qquad (9.14)$$

which also follows from (9.12). (9.14) is of much practical use for the design of cathode-ray tubes. There, the perveance of the system would, according to (9.10) be given by

$$P = I_{em}(DV_a)^{-\frac{3}{2}}. \qquad (9.15)$$

Emission velocities and grid modulation. In the discussions of this section no account has been taken so far of the Maxwellian distribution of emission velocities. Though the fluorescent spot of a cathode-ray tube appears to be blacked out at a fairly well-defined grid voltage $^{bl}V_g$, it appears to be impossible completely to cut-off the emission current of a thermionic cathode if sufficiently sensitive current meters are used for its measurement. The emission current I_{em} can be shown (cf. (7.5) in §7.1) to decrease with grid bias according to the following exponential law:

$$\log_{10}(I_{em}/I_s) = 0.4\,\frac{eV}{kT}, \qquad (9.16)$$

where V is the equivalent diode voltage (see (9.13)), and I_s is the ideal saturation current that can be drawn from the cathode which is, for example, of the order of 10 mA for an oxide cathode of 1 mm² surface. Again, $e/k = 11\,600°$ V^{-1} and T is the absolute temperature of the cathode. If, for instance, by visual measurements, black-out is found to occur when the emission is suppressed to 10^{-7} A, a retarding voltage V of about -1 V would be calculated from (9.16). Now, if $^{bl}V_g$ is plotted as a function of V_a, the straight-line relation through the origin implied by (9.10) should be expected to hold only if the above retarding voltage $|V| = 1$ were subtracted as a correcting term from every measured value of black-out voltage.

In practice the correcting term is found by measuring a few $^{bl}V_g/V_a$ ratios for different V_a and by subtracting an empirical term from $^{bl}V_g$ such as to make D approximately independent of the anode voltage. The empirical term is found to be of the order of a volt. Besides the correction for the emission velocities it actually includes a correction for contact potential difference between cathode and grid and another term which is due to the curvature of the modulating equipotential (cf. Langmuir and Compton, 1931, p. 231), but this cannot be discussed here. The correction term cannot be neglected for emission systems with large grid-cathode spacings or with very small cathodes (e.g. hair-pin cathode, §9.9), but in technical cathode-ray tubes which have black-out voltages of the order of 50 V it is quite unimportant.

Beam divergence. Of practical interest is the relatively large change of angular divergence of the emitted cone of rays observed for any change of grid bias. Such a change is demonstrated by the pictures shown in fig. 9.12. According to Jacob (1948) a linear relation holds, namely

$$\Theta = \frac{\Theta_0}{{}^{bl}V_g} V_d, \qquad (9.17)$$

where ${}^{bl}V_g$ is the black-out voltage of the emission system and Θ is the semi-divergence of the bundle of rays when a grid drive V_d is applied, while Θ_0 is the corresponding semi-vertical angle for zero grid drive. Θ may refer to a cone comprising all rays about the axis from the maximum value of the angular intensity distribution of emitted rays, down to any conventional value of it as defined, for example, by (9.25). Jacob (1944) found (9.17) to hold accurately enough to utilize the variation of beam angle for the design of an electron-optical voltmeter. According to measurements of Moss (1946), however, the relation is fulfilled to a good approximation only if the angles are not too large.

Cross-over changes. Important effects of grid modulation concern a change of cross-over size and a cross-over shift. Both are mainly space-charge effects and can be understood qualitatively by applying the conclusions drawn in §8.3. These modulation effects can be very disturbing in high-definition television tubes for instance.The highlights in the picture tend to be blurred, and a simultaneous sharp focusing of the highlights and shadows of the scanned picture is impossible. Both the change of cross-over size and the cross-over shift are generally the larger, the greater the perveances $I/V^{\frac{3}{2}}$ involved. However, both effects are reduced to a relative minimum when the beam-spread factor (cf. (8.31) and (8.32)) reaches its maximum shown in fig. 8.4. This apparently happens for the preference spacings discussed in the later §9.7.

§9.6. Design of simple triode systems with plane cathode

For the practical application of emission systems in electron-optical gear certain qualities of performance are required. Homogeneous cathode loading, high-emission density, great homocentricity but small divergence of the emitted cone of rays, great slope of the modulation characteristics and various other properties may be desirable. From the study of the previous sections, however, it appears that not all requirements can be fulfilled simultaneously. A gain in one of the desirable properties might necessarily mean a loss in others. Practical compromises have to be effected in operational performance, and for this purpose some knowledge of the dependence of the various properties on the geometry of the electrodes is desirable.

Let the well-established empirical systems of figs. 9.1 and 9.2 be discussed in the first instance. Their electrodes consist of plane diaphragms and of cylindrical tubes only. For a given aperture radius r_g all essential geometrical information is contained in the two spacings c and a indicated in figs. 9.1 and 9.2. It should be emphasized that both c and a are defined as the distances of cathode and anode respectively from the top surface of the grid electrode which is facing the anode. If these 'grid-top' spacings are given, the thickness of the grid diaphragm can be disregarded, since its effect is unimportant for the electron optics of the system. Both spacings c and a may be measured in terms of grid-aperture radii r_g so that the given information can be applied to any size of emission system. Generally, the scale of the system does not affect its performance excepting for a limitation of the peak loading of emission density at the cathode (cf. 9.11).

Most important for the electrical performance of the systems are two parameters, namely, the penetration factor D defined by (9.10) and the geometric factor G of Child's equation (9.14). Now, nearly all experimental values for G of systems of the pattern of figs. 9.1 and 9.2 under full space-charge conditions are found to lie between

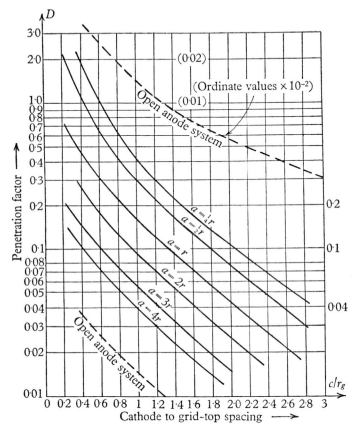

Fig. 9.13. Penetration factors D of symmetrical two-diaphragm systems (fig. 9.1) and of diaphragm-tube systems (fig. 9.2).

the extreme values 3×10^{-6} and 5×10^{-6} [amp volt$^{-\frac{3}{2}}$]. The higher values of G are reached only at extremely small values of c/a.

The penetration factor D, on the other hand, changes greatly even for small changes in the spacings of the emission system. Experimental values of D for the systems of figs. 9.1 and 9.2 are plotted in fig. 9.13 as a function of the spacing c from the cathode to the top of the grid (Klemperer, 1953). In case of the two-diaphragm systems, various spacings a from the top of the grid to the anode are taken as constant parameters of the curves. Both c and a are measured in grid-aperture radii r_g. The six solid curves belong to the symmetrical two-diaphragm system (fig. 9.1). The broken curves belong to the diaphragm-tube system (fig. 9.2). The penetration factors of these systems hardly depend upon a spacing between grid electrode and anode as long as the tube radius is many times greater than the grid aperture.

A knowledge of the G and D values is most useful when the design of a system of given emission current I_{em} is required for a given anode voltage V_a. For instance, let a system be required which emits about $550\,\mu\mathrm{A}$ at 1 kV. The necessary black-out voltage is calculated from (9.14) to be

$$^{bl}V_g = \left(\frac{I_{em}}{G}\right)^{\frac{3}{2}} = \left(\frac{550\times10^{-6}}{3\cdot5\times10^{-6}}\right)^{\frac{3}{2}} \approx 30 \text{ volts.}$$

The necessary penetration factor is calculated from (9.10) to be

$$D = \frac{^{bl}V_g}{V_a} = \frac{30}{1\,000} = 3 \text{ per cent.}$$

A glance at the curve of fig. 9.13 shows that the necessary D can be realized either by an open anode system with $c = 0\cdot5r_g$ or by a series of symmetrical two-diaphragm systems, e.g. with $a = 4r_g$ and $c = 1\cdot2r_g$, or with $a = r_g$ and $c = 2\cdot3r_g$, or with any other combination of c and a which may be taken from the given curves or interpolated for intermediate values.

For the electron-optical performance of emission systems in electron guns, a knowledge of the angular intensity distribution of emission is of great interest. Some measurements of such distributions have been made by Jacob (1939) who applied a scanning method in which the beam was moved periodically over the entrance slit of a Faraday cage by means of an alternating deflecting field. The current distribution was amplified and registered by an oscillograph.

Klemperer (1953) used for the measurement of this angular current distribution a Faraday cage with an iris diaphragm such as is commonly used in photographic cameras. The iris could be opened gradually; in this way the current in the cage could be measured as a function of the aperture. By differentiation of the measured curves, zonal current distributions were obtained. These zonal distributions, divided by the areas of the zones

$$(a \simeq 2\pi\Theta\,d\Theta),$$

yield the required angular density distributions such as shown in fig. 9.14. Abscissae are the semi-vertical angles Θ of co-axial cones

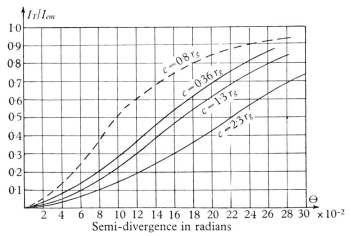

Fig. 9.14. Angular current distributions of some symmetrical two-diaphragm emission systems.

into which the current I_T is emitted. Ordinates are the fractions (I_T/I_{em}) of the emission into the cone divided by the total emission into the hemisphere. All curves apply to two-diaphragm systems with equal grid and anode aperture ($r_g = r_a$) in which the grid to anode spacing is equal to the grid-aperture radius ($a = r_g$). The three solid curves have been taken with grid-cathode spacings $c = 0.36r_g$, $1.3r_g$ and $2.3r_g$ respectively. They seem to suggest that the beam spread increases monotonically with c. Actually, however, a pronounced minimum for the beam spread is found at about $c = 0.8$. It is also found that the distribution for such small c-values is very sensitive to any change in the quality of the vacuum and it must be concluded that the beam divergence depends greatly on the space-charge density in the cross-over.

It has already been pointed out in §8.3 at the discussion of a space-charge loaded beam, that if starting from a steep initial angle of convergence this angle is decreased, the distance $(z_W - z_A)$ of the waist from the electron source passes through a maximum (cf. fig. 8.4). For a given angle of convergence, this distance will also pass through a maximum, when the factor $I/V^{\frac{3}{2}}$ is increased. Now in the emission system, a decrease in initial angle of convergence and an increase of $I/V^{\frac{3}{2}}$ take place simultaneously when the grid to cathode spacing c is decreased. Hence the distance of the waist from the cathode should pass through a maximum when the grid spacing c is varied. This is illustrated in fig. 9.15 where the

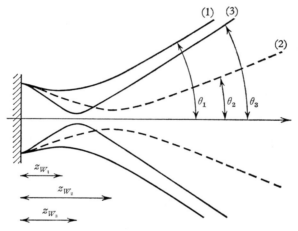

Fig. 9.15. Waist formation for various grid-cathode spacings.

envelopes of beams emitted from the cathode c are shown for three different grid spacings c. The beam 3 corresponds to the largest c, the beam 1 to the smallest c. It is seen that for an intermediate c of the beam 2, the waist reaches a maximum elongation z_{W_2}. Simultaneously with the largest z_W occurs the minimum divergence Θ_2 behind the waist. In this way we can understand the result shown in fig. 9.16 that the maximum I_T/I_{em} into any given cone of rays occurs at a certain optimum grid spacing c. We should also understand that this maximum depends upon the quality of the vacuum, since the mutual repulsion of the electrons in the waist is greatly reduced by the presence of only very few positive ions. The optimum grid spacing c must be expected to be a function of the anode spacing a. This is confirmed by the current efficiency curves in fig. 9.16 which are based on measurements by Klemperer

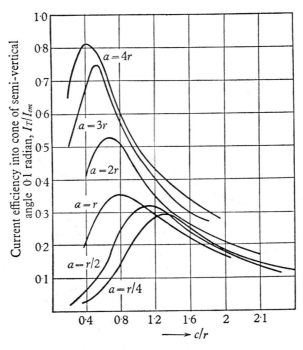

Fig. 9.16. Current efficiencies of two-diaphragm systems for various grid-anode spacings a and grid-cathode spacings c.

and Mayo (1948) and show the ratio I_T/I_{em} for the fixed semi-vertical angle $\Theta = $ o·1 radian as a function of c/r for six different grid-anode spacings a. Each curve exhibits a distinct maximum at certain conjugate a and c values which are called the 'preference spacings' for the two-diaphragm emission system.

Similar curves are obtained for open anode systems (fig. 9.2). There, however, the angular divergences are much smaller. However, the preference spacing c_p of the open anode system is so small that it can hardly be realized experimentally with grid electrodes of practicable thickness. The measurements shown in figs. 9.14 and 9.16 have been taken at a vacuum of 10^{-5}–10^{-6} torr but the maxima of current efficiency have been found to persist—at slightly changed preference spacings and at a somewhat reduced height—down to the best vacua obtained in these experiments (10^{-8} torr). There are further slight changes in the values of the preference spacings when the emission current I_{em} is changed; these were distinctly observed when I_{em} was increased from 1 to 6 mA.

§9.7. Aberrations in emission systems

Gaussian optics cannot present anything but a very rough qualitative description of emission systems. However, it applies at least to paraxial rays of systems which are operated near to current saturation conditions.

Field curvature. This is certainly the most disturbing error when a plane cathode is used in a low magnification emission microscope (§9.2) or in an image converter (§9.8). There, the marginal regions of the images are always quite out of focus. The curvature of the image plane is as bad as could be expected from the experience with lenses reported in §6.7. For a given system with given voltages at the electrodes, the centre of the image will be in focus at a given target distance. Any other zone of the image, however, can be focused by varying the grid voltage.

The curvature of the field has been found to depend upon the spacings and on the shape of the electrodes. For example, for a nearly symmetrical two-diaphragm system Johannson (1933) measured the diameter of the sharply projected paraxial area of the cathode as a function of the grid to cathode spacing c and of the ratio of grid voltage V_g to anode voltage V_a. He found that the radius of this sharp area reaches a maximum of about one-third of the grid aperture radius r_g, when

$$c/r_g \approx 1 \quad \text{and} \quad V_g/V_a \approx +0 \cdot 1.$$

Under these conditions magnifications of about 50× have been reached within an image distance of about $330 r_g$. Such an image can be seen, for example, in fig. 9.17, which shows the surface of an oxide cathode scratched with a pattern of straight lines. The sharply projected area is marked in the picture by a circle. Johannson (1933) also found that the diameter of this sharply projected area could be increased by about 70 per cent when the grid electrode was provided with a funnel-shaped aperture with its narrow end extending towards the cathode.

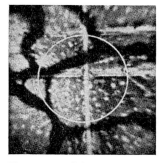

Fig. 9.17. Electron optical image of the cathode of a two-diaphragm emission system. (Photograph.)

Spherical and chromatic aberration. Current distribution in the cross-over. From the good optical definition of the central part of the image (fig. 9.17), the conclusion can be drawn that spherical aberration in the emission microscope is practically negligible. This is because the aperture of rays reaching the image is small. Chromatic aberration appears to be negligible too, for the electron energy in the image space is of the order of several hundred, sometimes many thousand volts, which is very large in comparison with the thermionic energy distribution which covers not more than some tenths of a volt.

According to a theory by Recknagel (1941) the radius of the disc of confusion at the Gaussian image plane caused by aberration in the homogeneous field E in front of the cathode can be calculated to be

$$\Delta r = \frac{y'}{y} \frac{2V_{em}}{E} \qquad (9.18)$$

where V_{em} (volts) corresponds to the emission energy of the electrons and y'/y is the magnification. Thus best resolution could be obtained by increasing E as much as, with respect to the insulation of the system, could be tolerated. Septier (1955) pointed out that (9.18) predicts an aberration disc several times greater than that obtained in practical emission microscopes. By a detailed consideration of the caustic cross-sections in the region of focus, he showed that the disc of confusion in the Gaussian image plane is several times larger than the minimum disc of confusion which, just as in the case of spherical aberration (cf. §6.4), is found in a plane close to, but not coincident with, the Gaussian plane. A review of some numerical methods for the calculation of the properties of immersion objectives, in particular their potential distribution and the resulting spherical and chromatic aberration can be found in an article by Hahn (1958) who also has given results of the calculation for a few systems with specified spacings.

The conditions for an emission system are quite different when it is used in an electron gun. First, the cathode is now used under space-charge conditions. Secondly, not an image of the cathode but an image of the cross-over is projected onto the fluorescent target. Both spherical and chromatic aberration of the emission system form a serious limitation for the sharpness of focus of the electron gun. It has been shown in §7.1 that in the absence of spherical aberration, the highest obtainable current density in a focused spot (cf. (7.8)) is limited only by the chromatic aberration which is due to the given band width of emission velocities. Law (1937), Langmuir (1937b) and Dosse (1940) have discussed the influence of the thermionic energy distribution of the emission on the current density distribution in the cross-over. Each emission velocity group forms a pupil whose radius is given by Abbe's sine law as

$$y_X = \left(\frac{V_{em}}{V_X + V_{em}}\right)^{\frac{1}{2}} \frac{y_c \sin \Theta_c}{\sin \Theta_X}, \qquad (9.19)$$

where V_{em} is the kinetic energy of emission of the electron and Θ_c is the initial slope angle, while V_X and Θ_X are the voltage and slope angle respectively at the cross-over. By differentiating (9.19) one obtains

$$\Delta y_X = \left(\frac{V_{em}}{V_X + V_{em}}\right)^{\frac{1}{2}} \frac{y_c}{\sin \Theta_X} \cos \Theta_c \Delta \Theta_c, \qquad (9.20)$$

which means that a small variation $\Delta \Theta_c$ of emission angle at the cathode gives rise to a variation Δy_X in cross-over radius y_X. Therefore, since the number of electrons in the energy range

$$(V_{em}, \quad V_{em} + \Delta V_{em})$$

emitted between angles Θ_c and $\Theta_c + \Delta \Theta_c$ due to (1.38) of §1.6 is of the form

$$f(V_{em}) \sin \Theta_c \cos \Theta_c \Delta \Theta_c \Delta V_{em},$$

the number of electrons of initial energy $(V_{em}, V_{em} + \Delta V_{em})$ passing through an annulus $(y_X, y_X + \Delta y_X)$ will be of the form

$$I_X = f(V_{em}) \sin \Theta_c \cos \Theta_c\, g(y_X)\, \Delta y_X\, \Delta V_{em}. \qquad (9.21)$$

To find the total current passing through a circle of radius y_X of the cross-over expression (9.21) has then to be integrated over V_{em} and over y_X. Hence, there follows a distribution of current density as a function of the radial coordinate y_X of the cross-over which originally is due to Law (1937), namely,

$$i(y_X) = \frac{I_{em}}{\alpha\pi} \exp\left[-\left(\frac{y_X}{\alpha}\right)^2\right], \qquad (9.22)$$

where I_{em} is the total emission current from the cathode,

$$\alpha = \frac{kT}{eV_X}\left(\frac{z_X}{L}\right)^2, \qquad (9.23)$$

and e/k = electron charge/Boltzmann constant = 11 600 (degree/volt), T = absolute temperature of the cathode, V_X = voltage at the cross-over, and L = field form factor introduced by (9.2).

The Maxwellian distribution of emission energies spreads to very large values which, however, are very rare. Hence the current density distribution (9.22) must spread up to very large radii y_X. There, however, it contains very few electrons indeed.

The maximum current density is found in the centre of the cross-over $(y_X = 0)$ and is, according to (9.22) and (9.3), given by

$$i_{max} = \frac{I_{em}}{\alpha\pi} \approx i_{em}\frac{eV_X}{kT}\left(\frac{y_c}{f'}\right)^2. \qquad (9.24)$$

With increasing distance from the axis, the current density i decreases according to (9.22) at a rate given by a Gaussian error function. For instance, i decays to $1/n$th of its maximum value at a radius

$$y_X = (\alpha \log n)^{\frac{1}{2}}, \qquad (9.25)$$

and, in particular, the half-maximum width of the distribution is given by

$$2y_X = 1\cdot66\sqrt{\alpha}.$$

By definition and with (9.25) Jacob (1939) introduced an 'effective cross-over radius' given by $1/n = 0\cdot2$, while Dosse (1940) preferred a conventional cross-over size with a limiting $1/n = 0\cdot05$.

The current passing outside Dosse's cross-over just amounts to 5 per cent of the total current, and the emission energies passing through this cross-over do not exceed $3V_\alpha$, where $V_\alpha = (k/e)\,T$ is the most frequent emission energy.

The average directional intensity i/Ω, i.e. the current density per solid angle (cf. §1.6) depends upon $i(y_X)$ as given in (9.22). This is of great practical importance, since i can be increased by reducing y_X. Hence the intensity emitted by a given system can be increased by cutting off marginal zones either from the cross-over area or from the cone of emitted rays. The reduction of the cone of rays by aperture stops in the anode is quite usual in the practice of cathode-ray tubes. On the other hand, the beam reduction by a stop in the cross-over plane has been successfully performed in the gun design by Law (1937).

Experimental investigations of the current distribution in the cross-over have essentially confirmed the Gaussian distribution of (9.22) though, due to the presence of various aberrations, the experimental values for the current density are generally by a factor two or more, below the theoretical optimum given by Langmuir's equation (7.8). Various experimental methods have been used for measuring the current distribution in the cross-over. Law (1937), for instance, inserted aperture stops of different sizes into the plane of the cross-over, measuring for each size the ratio of the intercepted to the transmitted current. Measurements of the cross-over size have been performed by Jacob (1939) and by Dosse (1940) using the 'slit scanning' method which has been described in §9.6.

The measured angular semi-aperture, say β, generally results from superimposition of all elementary bundles with angular semi-aperture γ, the axis of each elementary bundle having the slope angle Θ (cf. figs. 9.7 and 9.8). Since also the distributions in the elementary bundles are error functions, one obtains by super-imposition

$$\beta^2 = \Theta^2 + \gamma^2. \tag{9.26}$$

But in most cases $\Theta \gg \gamma$, so that $\beta \approx \Theta$.

Influence of space-charge on the cross-over. The cross-over size determined by one of the direct experimental methods generally appears to be far greater than one could expect from the results of ray tracing through the field plot. The 'swollen' cross-over size is quite clearly caused by space-charge repulsion.

The conclusion, that space-charge could be the determining factor for the cross-over size, has been reached by Klemperer and Mayo (1948), who investigated the properties of a symmetrical two-diaphragm system as a function of the diaphragm spacings. For large grid-cathode spacings, the curvature of equipotentials near the cathode is large, hence the electrons reach the cross-over at a steep angle Θ. Simultaneously the perveance $I_{em}/V_X^{\frac{3}{2}}$ is relatively small because the large spacing leads to small fields at the cathode and to small emission currents I_{em}. Under these conditions the beam-spread coefficient $\sigma \sim \Theta^2/(I/V^{\frac{3}{2}})$ (cf. (8.26)) is large. On the other hand, decreasing of the grid-cathode distance flattens the equipotentials near the cathode, decreases and increases $I_{em}/V_X^{\frac{3}{2}}$. Now with decreasing σ the distance of the waist of a convergent beam under space-charge conditions has been shown (cf. fig. 8.4) to increase first and then to decrease again. Precisely this effect has been observed by Klemperer and Mayo to happen with the cross-over diameter of a space-charge-limited emission system.

The maximum displacement of the cross-over from the cathode has been found to occur simultaneously with minimum beam divergence and with least spherical aberration. Again, least aberration is—according to §8.3—expected for a 'space-charge equivalent lens' in the region in which the beam-spread coefficient σ reaches its maximum. Hence it appears that the spherical aberration of emission systems must be to a great extent an effect of space-charge. Since for practical purposes systems with least errors are required, the grid-cathode spacings at which maximum σ and minimum aberration occurs are of technical importance. These spacings are the preference spacings, which we have described in §9.6 (cf. fig. 9.16).

It has so far not been possible to reduce the spherical aberration of a triode system by geometrical modification of the electrodes. Hence it seems very likely that the spherical aberration observed in the cross-over is caused by hardly anything else but by space-charge, and that the amount of spherical aberration added by the

Laplacian field is negligible. Due to the small quantities involved, it is difficult to measure exact values of the aberrations. For symmetrical two-diaphragm systems with small grid-cathode distances Klemperer and Mayo (1948) found the following values of longitudinal aberration Δz in terms of grid radii r_g as a function of the slope angle Θ of the ray, expressed in radians:

Δz	$2r_g$	$9r_g$	$20r_g$
Θ	0·02	0·05	0·07

Here, Δz increases roughly in proportion to Θ^2. The magnitude of the aberrations in a similar system with larger cathode-grid distance, producing a bundle of rays of greater divergence, is indicated by the following values measured by Klemperer (1953)

Δz	$1r_g$	$3r_g$	$6r_g$
Θ	0·03	0·1	0·3

Here, Δz increases obviously less than in proportion with Θ^2, i.e. the higher aberration coefficients (cf. (6.5)) are negative.

Modification of the potential distribution in an emission system by space-charge. This modification is very substantial and has been determined by Maloff and Epstein (1934, 1938) in the following way:

(1) Equipotentials $V(y, z)$ are obtained by measurements in the electrolytic trough as shown in §3.3.

(2) The electron beam is traced through this potential plot $V(y, z)$.

(3) The charge density is calculated throughout the beam from the values of current and of electron velocities.

(4) From the charge density and the shape of the beam a potential distribution $V_1(y, z)$ due only to the beam is calculated.

(5) The charges induced on the electrodes are calculated and a corresponding potential distribution $V_2(y, z)$ is obtained.

(6) The true potential distribution results by superimposition of the potentials $V + V_1 + V_2 = V'$ at every point (y, z).

(7) A better approximation is then obtained by repeating the above procedure starting with the modified distribution V' instead of the original V.

A field plot obtained in this way is shown in fig. 9.18. The modified equipotentials (V′) are indicated as broken lines; the original Laplacian distribution (V) is drawn in solid lines. Near the cathode, the V′-lines are of greater curvature and they are more widely spaced than the V-lines, though the greatest differences (V′ − V) are found in the cross-over region. A maximum space-charge density occurs at the potential minimum close to the

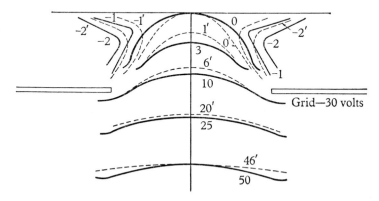

Fig. 9.18. Effect of space-charge on potential
distribution in an emission system.

cathode. In the example of fig. 9.18. this amounts to 10^{17} electrons m^{-3} The space-charge density in the cross-over may be of the same order of magnitude or smaller on account of the greater velocity of the electrons. The greatest space-charge densities obtained in modern electron guns are estimated to reach 10^{19} electrons m^{-3}. The results of Maloff and Epstein (1938) have been confirmed and extended by Plocke (1952) and the following important effects of space-charge in emission systems may be enumerated:

(1) Decrease of the emitting area of the cathode.

(2) Decrease in potential gradient along the axis in the region between cathode and grid.

(3) Increases and decreases in curvature of equipotentials in the grid-cathode and grid-anode region respectively.

(4) Large increases in both focal lengths (f and f′) of the emission system, e.g. in the example of fig. 9.18 by a factor 4.

(5) Increase in cross-over size and decrease in beam divergence as discussed earlier in this section.

A beneficial electron optical effect of space-charge should be mentioned. It is generally difficult to obtain a perfectly smooth and plane surface at the cathode of a thermionic emitter. Any unevenness, however, results in anomalously large radial velocities of the emitted electrons. These velocities are easily detected and measured with a 'pinhole spectrometer'. This is a pinhole set up in the anode diaphragm of the emission system, so that transverse velocities show up on a fluorescent target as an increase in size of the spot which is a 'shadow' cast by the pinhole. Now, pinhole experiments by Moss (1961) with a plane oxide cathode in a two-diaphragm emission system showed that sometimes transverse emission energies of about 10 eV were obtained from a cathode with temperature limited emission. These energies were reduced to the negligible proportions expected from a Maxwellian distribution belonging to the particular temperature, when the cathode temperature was increased only slightly, so that a space-charge was built up in front of it. Apparently, the electrons emitted from the side of a valley in a rough surface will initially experience a retarding field in the space-charge cloud against which they will dissipate their ejection kinetic energy. The fact that these ejection energies can be caused by the fields arising from uneven textures, was demonstrated by Moss (1961) who found that he could reduce such energies from 10 to 0·5 eV by physically smoothing the cathode surface.

Aberrations in emission systems can be predicted by the ray-tracing methods which we have described in §3.6 and §8.8. We should mention here also a ray-tracing technique due to Vine (1966) which is particularly suited to large emission systems in which space-charge is negligible, as for instance, in the image tubes which will be described in §9.8. The method is basically a circle method (§3.4) but computations are performed by a digital computer. The accuracy in computing aberrations is obtained by referring the trajectory of an electron in a bundle to the principal ray of this bundle as (curvilinear) axis, rather than measure the electron position relative to the axis of symmetry of the system.

Errors due to thermal velocity effects can be estimated according to a theoretical study by Hamza (1966). His results apply in particular to guns with concave cathodes (Pierce guns, cf. §9.8) which are working under space-charge conditions. Using a digital computer for solving Poisson's equation, he found that a useful parameter in discussing such effects is given by PV_A/T_c, where P is the perveance, V_A the anode voltage and T_c the cathode temperature. If this parameter is small, e.g. $PV_A/T = 10^{-7}$ or less, serious beam spreading owing to thermal velocities must be expected. If it is large, e.g. 10^{-6} or greater then thermal spreading is negligible compared with space-charge repulsion.

The various focusing defects in electron guns do not combine in any simple manner and their net effect is best evaluated experimentally by determining the current density distribution in various planes downstream from the anode aperture of the gun. This technique of examining the caustic (cf. §7.7) is of some interest in the design of electron guns used in microwave beam tubes (cf. §8.7, §9.8) where the properties of a particular gun are frequently characterized pictorially by a series of plots of the current density against radius at various planes along the axis of the beam. The simplest method of determining such current density profiles is to measure the current passing through a small pinhole aperture which is traversed across the beam (see e.g. Brewer, 1957b). This is, however, a rather laborious procedure and Ashkin (1957) has shown how more rapid results can be obtained by automating the method. By means of electrostatic deflexion plates, the beam is scanned across a small fixed pinhole on axis. A signal, derived from the current transmitted through the aperture, is fed onto the y-plates of an oscilloscope whose time base is synchronized with the beam deflexion. Thus the CRO trace gives a direct measurement of the current density distribution in the beam.

An exact experimental method for the determination of the combined aberrations in emission systems has been developed by Hanszen (1964). The underlying principle is that of the shadow method described in §2.4 and §6.4. Now, however, the shadow casting wire gauze is replaced by a diaphragm with an extremely fine aperture of 5 μm. This aperture is moved in the y-direction across the z-direction of the emitted beam. A photographic plate which intercepts the 'shadow'-image of the aperture is moved simultaneously and automatically in the x-direction. In this way, a trace is obtained on the photographic plate, the evaluation of which yields the location and slope of the rays across the entire caustic (cf. §7.7).

§9.8. Emission systems with concave cathodes

The Pierce system. It has been pointed out (cf. §1.6 and §9.7) that the available current, as well as, for instance brightness, intensity and current density focused in a spot, can be increased by increasing the emission density at the cathode. The available emission density, however, is strictly limited and it is a property of the emitting surface. In order to preserve the cathode for a reasonable lifetime, say of a few thousand hours, its temperature must not be raised too much.

The maximum practical emission current densities for most emitting cathodes lies in the range $5 \times 10^3 – 5 \times 10^4$ amp m^{-2}. However, the current density required for instance in the beams of high power microwave tubes is of the order of 10^5–10^6 A m^{-2}. Thus for such applications, the electron gun must compress the current so that the emergent beam has a smaller cross-section than that of emitting area of the cathode.

An ideal arrangement for converging the total emission of a large cathode surface into a small point focus would be provided by a field with strictly spherical equipotentials. Such a scheme could be realized by a spherical concave cathode with an extremely small anode in the centre of the sphere. The emission system of fig. 9.19 due to Klemperer and Wright (1939) has a concave cathode and a funnel shaped anode and it represents an early practical example based on this idea. The approximately radial lines of force between cathode and anode ensure a fairly constant field at the cathode surface, and thus some degree of homogeneity of emission density. This is important since in order to reach optimum performance of an emission system the maximum available emission density should be drawn everywhere from the emitting surface. If the emission density is not uniform as for instance in the plane cathode systems of figs. 9.1 and 9.2, then only a fraction of the cathode gives maximum emission and the average emission density is substantially below the peak value.

The gun of fig. 9.19 was developed empirically and the electrode geometry is not ideal. For obtaining high perveances, in excess of, say 5×10^{-7} amp volts$^{-\frac{3}{2}}$ a more systematic approach has to be applied.

A general procedure for the design of emission systems with a homogeneous field at the cathode surface has been put forward by Pierce (1940). He argued that the electrodes of the emission

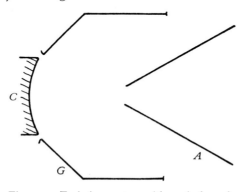

Fig. 9.19. Emission system with conical anode.

system should be designed to create an electric field configuration *external* to the beam which ensures rectilinear flow between cathode and anode. If the beam from the cathode can be considered to be cut from the unbounded, radially symmetric flow which would occur in a *complete* spherical diode, then the emission density will be homogeneous. Hence the problem becomes that of finding electrode arrangements which produce such fields and potentials at the boundaries of the actual beam as the 'removed' charge would have produced there. Now, the rectilinear flow of space-charge has already been treated in §8.7, for the plane case where it was shown that in a space-charge limited flow the voltage increases with the 4/3 power of the distance from the cathode. Moreover, detailed explanations have been given in §8.8 for the use of the field-plotting trough for finding experimentally the right electrode arrangement which produces a potential increase at the edge of a limited beam according to the 4/3 power law.

Now in the case of spherical flow, the 4/3 power law is, in general, only obeyed along the portion of the beam edge in the region of the cathode. The actual potential variation in the spherical diode has been deduced by Langmuir and Blodgett (1924). Unfortunately, this potential distribution takes a non-analytical form, being expressed in terms of the 'Langmuir α-function', which the authors have tabulated for a wide range of ratios of cathode to anode curvature. However, satisfactory electrode structures for guns of perveance up to 5×10^{-7} A V$^{-\frac{3}{2}}$ can be

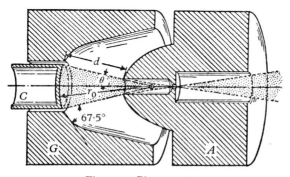

Fig. 9.20. Pierce system.

found using the electrolytic trough as explained in §8.8, the potential along the beam edge being specified by the Langmuir–Blodgett distribution.

An emission system according to Pierce is shown in fig. 9.20. The surface of the grid electrode G meets the beam envelope under the universal angle 67·5° as postulated in §8.8. Samuel (1945) has published some detailed material about the theoretical relationship between perveance ($I/V_A^{\frac{3}{2}}$), semi-vertical angle (Θ) of the conical beam and anode-to-cathode spacing (d) measured in terms of the radius of curvature (r_c) of the cathode. For any required $I/V_A^{\frac{3}{2}}$, Θ can be presented as a unique function of d/r_c. However, in high perveance systems (e.g. $> 5 \times 10^{-7}$ A V$^{-\frac{3}{2}}$), the experimentally obtained perveances fall well short of theoretical expectations, because the anode aperture in such systems must be made comparable in size to the anode to cathode spacing. This results in some non-uniformity and in a decrease of the fields at the cathode surface.

The perturbing effect of the anode aperture on the field at the cathode surface is illustrated in fig. 9.21 (*a*), which shows equipotentials and field lines in a gun of perveance $\simeq 2 \times 10^{-6}\ \mathrm{A\ V}^{-\frac{3}{2}}$

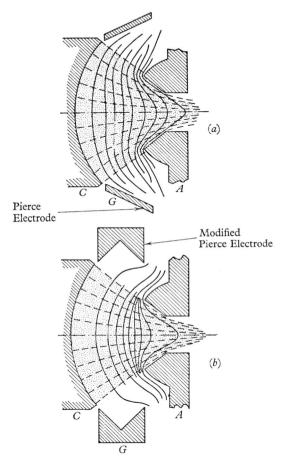

Fig. 9.21. Effect of the anode aperture on the field in a Pierce gun.

which was designed according to Pierce's principles. The distortion of the equipotential lines results in transverse electric fields which cause electrons to be intercepted by the anode, and also a serious loss of homocentricity. An examination of the profiles of beams emerging from such guns using methods explained in §7.7

reveals a characteristic 'hollow beam' in which the current density is quite low on the axis but shows sharp maxima at the beam edge.

Fig. 9.21 (b) shows a design by Brewer (1957b) which gives a more uniform field at the cathode and a more laminar flow through the anode aperture. This improvement is effected by modifying the shape of the grid electrode, so that the outer ends of the equipotential lines near the cathode are bent round towards the anode and are thus more nearly concentric with the cathode in the flow region. The problem of the exact design of the modified grid electrode is discussed by Müller (1955, 1956) and by Brewer (1957b). The approach is semi-empirical, and employs the electrolytic trough in the manner explained in §8.8.

A still better compensation of the distortion of the field at the anode aperture has been obtained by Frost, Purl and Johnson (1962). They modified the potential distribution along the beam edge experimentally by adjusting the potentials on a set of ring electrodes between cathode and anode until the beam was of uniform current density, and transverse velocities at the beam waist were minimized. In this way, they designed a gun of large area of convergence 300:1 and perveance $2 \cdot 2 \times 10^{-6}$. These authors also showed that 'hollow beam' formation in the type of gun shown in fig. 9.21 could be entirely eliminated by increasing the curvature of the cathode as explained in §8.8. It is obvious that the potential specification along the beam edge in all these methods of compensation deviates from the Langmuir–Blodgett distribution in the anode region.

A further disturbing effect of the anode aperture is the diverging lens action already discussed in §8.4 for plane parallel geometry. For low perveance guns, this can be treated simply by applying the Davisson–Calbick formula (4.8) for the focal length of an aperture lens (cf. §4.4). Space-charge forces tend to increase the diverging power of the anode lens, and in high perveance applications, the Davisson–Calbick formula needs to be corrected following the method given in §8.5.

In guns of high compression (i.e. high ratio of cathode area to beam area) the effects of transverse thermal emission energies are also important. The loss of laminarity due to thermal velocities has already been mentioned in §8.4. Thermal velocities also tend to limit the amount of beam compression obtainable in practice, to alter the position of the beam waist and to modify the current density profiles of the beam emerging from the gun. Cutler and Hines (1955) have analysed these effects for an ideal Pierce gun, and their results are in reasonable agreement with the experimental measurements of Climer (1962) who examined the current density profiles at the anode of a Pierce gun of small anode aperture. The more complex case of large anode aperture has been analysed by Danielson, Rosenfeld and Saloom (1956), and studied experimentally by Gregory and Beck (1966) who examined the effect of thermal velocities in a gun of low anode voltage (275 V) and perveance 8×10^{-7}.

According to Brewer (1967), the practically obtainable upper limit on the perveance of convergent solid beam guns of good optical quality (i.e. guns which produce a nearly homocentric beam) is about 3×10^{-6} A $V^{-\frac{3}{2}}$.

Electrostatic image converters. A technically important type of emission system with concave cathode is found in the electrostatic diode image converters.

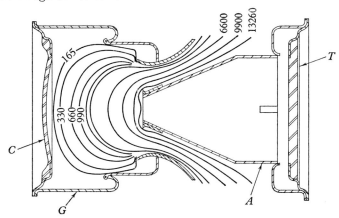

Fig. 9.22. Image converter.

These tubes are used for converting an infra-red or ultra-violet or X-ray image into a visible light image. As shown in fig. 9.22 for the example of an infra-red image converter, the infra-red image is projected onto the transparent photo-electric cathode C. The emitted photo-electrons are accelerated by the field between cathode and anode A and the electronic image is projected onto the fluorescent screen T, where it is converted into a visible image. The cathode is curved and the equipotentials drawn in the figure have an increasing curvature towards the anode in order to avoid, as much as possible, the field curvature (§9.7) of the system. The equipotentials shown correspond to an anode potential of 13·5 kV (Wreathall, 1966). The electrode G is at cathode potential and two factors dictate its shape, namely the need to have approximately spherical equipotentials in the space between the cathode and the tip of the anode, and the need to reduce the field at the cathode to about 500 V/cm. If the field is greater than this, spurious field emission can occur from the very low work function photo-cathode surface. In distinction to the cathode in the Pierce gun which is emitting under space-charge conditions, the emission of the cathode in the image converter is saturated (cf. §9.1). Referring to (9.18) of §9.7, it is clear that the low field at the cathode severely limits the maximum resolution ($\sim$ 20 line pairs mm^{-1}) obtainable in electrostatic diode image tubes.

Image converters with photo-electric cathodes, highly sensitive to visible light, yield a lumen gain of about 100 which can be further multiplied if such 'image intensifiers' are used in cascade, with the fluorescent screen of the first converter adjoining the cathode of the second converter and so on. More information about converter tubes is found in McGee *et al.* (1958, 1966 and 1969), also in Schagen and Woodhead (1967).

§9.9. Hairpin and point cathodes in emission systems

Hairpin cathodes. The guns used in electron microscopes, microanalysers and diffraction cameras are normally triode systems employing thermionic emission from a hairpin cathode (§9.1). These systems are characterized by high anode voltage and low beam current and are thus of low perveance (e.g. 10^{-10} A V$^{-\frac{3}{2}}$); the emitting area of the cathode is, however, very small and high current densities of the order of 10^5 A m^{-2} are drawn.

Fig. 9.23 shows two versions of the emission system as it is

employed in the electron microscope. Both versions have a hairpin cathode (C) and a flat anode (A). The grid electrode (G) of the
version in fig. 9.23(a) has a flat surface facing the anode, and the
tip of the cathode is arranged at a distance c_i from this surface,
inside the grid. On the other hand, the version of fig. 9.23(b) has
a re-entrant grid electrode and the tip of the cathode generally
protrudes into the grid aperture being at a distance $-c_i$ from the
inner flat surface of the grid.

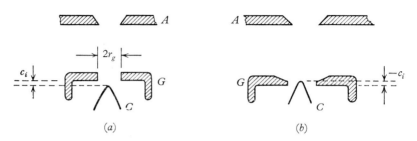

Fig. 9.23. Emission systems in electron microscopes.

The important parameters for gun performance are the cathode-
grid spacing c_i, the (negative) grid bias V_g and the filament temperature, which is governed by the filament current (normally 2–
3 amp). Haine and Einstein (1952) established that for a given
cathode-grid spacing c_i and temperature, there is an optimum
value of grid bias $(V_g)_{opt}$ which gives maximum brightness of
electron emission and maximum directional intensity of the beam
(cf. §1.6). As the grid bias becomes less negative, starting from the
'cut-off' value $^{bl}V_g$ (§9.5), the brightness increases rapidly from
zero, until a maximum β_{max} is reached at $(V_g)_{opt}$, and thereafter
falls off quite slowly. Table 9.1 refers to a flat grid (fig. 9.23(a))
and shows the values of $^{bl}V_g$ and $(V_g)_{opt}$ for various values of c_i,
for a filament of 0·125 mm diameter at a temperature of 2650° K,
a grid aperture diameter $2r_g = 1\cdot75$ mm and an anode voltage
$V_A = 50$ kV. The last two lines of the table show the values of total
beam current $(I_B)_{opt}$ and semi-angle $(\Theta)_{opt}$ of beam divergence obtained at the optimum grid bias voltage. As c_i varies, the maximum
directional intensity b_{max} does not change significantly, being

practically equal to the theoretical maximum given by (1.50) in §1.6. This shows that, under optimum conditions, the emission is not significantly reduced by space charge.

TABLE 9.1. *Grid spacing and grid bias for optimum performance of an electron microscope gun*

c_i	1·5	1·0	0·65	0·5	0·4 mm
$^{bi}V_g$	−150	−290	−460	−670	−860 volts
$(V_g)_{opt}$	−115	−245	−400	−600	−790 volts
$(I_B)_{opt}$	300	160	100	80	50 μA
$(\Theta)_{opt}$	20	7·5	5·5	5·0	4·5 milli-radians

The influence of filament temperature on gun operation is summarized in fig. 9.24, which applies to a wide range of gun geometries. The continuous curve shows the ratio of the maximum

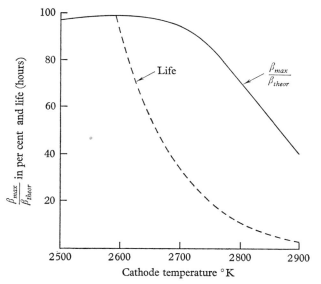

Fig. 9.24. Maximum brightness and filament life for
a hairpin cathode in an electron gun.

brightness obtained to the theoretical brightness ($\beta_{max}/\beta_{theor}$) as a function of cathode temperature. The fall-off of this ratio at higher cathode temperatures corresponds to the effect of space-charge repulsion on the size of the cross-over, formed in front of the cathode, which acts as a virtual source for the accelerated electrons. Table 9.2 gives β_{theor} as a function of filament temperature. The broken curve of fig. 9.24 shows filament life (as limited by evaporation) as a function of temperature. This indicates that if a useful lifetime (> 30 hours) is required, then it is inadvisable to operate above 2700° K. Thus the useful range of operation of the gun lies outside the region where space-charge is important.

TABLE 9.2. *Temperature and brightness of the cathode in a microscope gun*

T	2615	2680	2760	2840	° K
β_{theor}	$5\cdot8 \times 10^8$	$9\cdot6 \times 10^8$	$1\cdot8 \times 10^9$	$3\cdot1 \times 10^9$	amp m^{-2} ster^{-1}

Field plots and electron trajectory tracing (Haine and Einstein, 1952) have given some insight into the mechanism of beam formation. As in other triode systems (cf. §§9.5–6), the value of V_g controls the size of the emitting area, which, for satisfactory operation must be confined to a very small region, in the shape of a spherical cap, at the extreme tip of the hairpin. As V_g becomes less negative, the zero equipotential, which intersects the tip of the cathode along a circular line, moves backward from the vertex of the hairpin, causing the emitting area to increase, since emission occurs only from the cathode surface in front of the zero equipotential. The focusing effect of the field between grid and cathode weakens as V_g becomes less negative so that, with increasing beam current, the cross-over which acts as the effective electron source moves forward towards the anode. The effect of increasing c_i is also to move the cross-over towards the anode; this will be described below (cf. fig. 9.25) for the rather similar case of a point cathode. The diameter of the cross-over lies in the range 3–5 × 10^{-2} mm for most practical gun adjustments.

The current density profiles of the beam emerging from an electron microscope gun have been studied by André (1967). As V_g increases from $^{bl}V_g$, this profile corresponds to a roughly Gaussian curve whose amplitude increases until V_g reaches $(V_g)_{opt}$. With further increase in V_g, the current density curve acquires a flat top in the axial region, and the current density on axis falls until at small negative values of bias a hollow beam is formed, in which the maxima of current density are found at the beam edge. The Gaussian profiles are in accordance with (9.22) of §9.7, while the 'hollow beam' is a result of emission of electrons from areas behind the extreme tip of the point. Electrons emitted from these areas do not encounter a uniform radial field (cf. §9.8) and thus homocentricity is lost.

The behaviour of a triode gun with re-entrant grid geometry has been described by Borries (1948) and by Haine, Einstein and Borcherds (1958), it is very similar to that of the flat grid gun described above. The significant differences concern the filament position and the working range of grid potentials. It is usual to operate a re-entrant gun with the filament protruding slightly (e.g. $\frac{1}{4}$ mm) into the grid anode space, as indicated in fig. 9.23(b). Furthermore it is found that in general, smaller values of grid bias voltage are required than in the flat grid case, which can be a considerable advantage in practice.

Point Cathode. The effective source diameter of the hairpin cathode may be too large in applications where high coherence is required, and for this reason, considerable interest has been shown in 'point cathodes' in which thermal emission is drawn from a sharply pointed tungsten wire described in §9.1. The behaviour of this type of cathode when used as a thermal emitter in a conventional triode gun has been studied by Swift and Nixon (1962); it is quite similar to that of a hairpin cathode. These authors were, for instance, able to deduce information about the mechanism of beam formation by examining the emerging beam profiles and from the geometry of the projected image of a mesh placed in front of the gun anode. If, for any given grid cathode spacing c, the bias is reduced, the effective source position moves towards the anode. The effect of varying c at constant grid bias is illustrated in fig. 9.25. When the cathode is well forward, emergent electrons are converged slightly by the field but continue to travel away from the axis and give rise to a virtual source slightly behind the actual cathode. Moving the cathode further back increases the convergence, causing the virtual source to move rapidly backwards, until eventually the electrons are converged sufficiently to cross the axis. A real cross-over exists within the gun for all higher values of c.

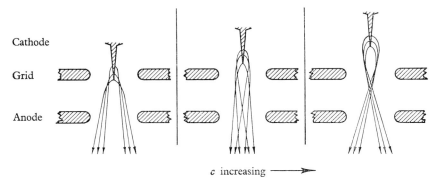

Cathode

Grid

Anode

c increasing ⟶

Fig. 9.25. Variation of effective source with increasing cathode to grid distance (c) in a point cathode triode gun.

According to some early investigations point cathodes were believed to yield greater brightness than hairpin cathodes. Temperature field emission (Schottky effect) was assumed to be the explanation for it. This has been disproved by an accurate investigation by Hanszen and Lauer (1967) who measured the directional intensity (b) of beams emitted from hairpin cathodes and from point cathodes of the same radius of curvature $r = $ 0.12 mm for various cathode temperatures and grid electrode potentials. The above point cathodes were then replaced in successive steps by other point cathodes with smaller and smaller radius of curvature down to $r = $ 0.7 μm. The measurements were taken with two fine apertures one behind the other, these were moved across the beam. The first aperture defined the current density i, the second aperture and its distance from the first defined the solid angle Ω. The directional intensity $b = i/\Omega$ (cf. (1.46) in §1.6) was found to increase rapidly with increasing grid drive $V_d = V_g - {}^0V_g$ (cf. §9.5) where ${}^0V_g$ is the cut-off grid voltage. b reached its theoretical value b_{max} (cf. (1.50) in §1.6) at relatively large V_g, i.e. at relatively high emission currents I. The smallest grid drive V_d for just obtaining the theoretical directional intensity in the paraxial beam, and the corresponding emission current, are given for points of various radii of curvature in table 9.3. The values given refer to an anode voltage of $V_A = 10$ kV, to a cut-off bias $V_g = -75$ V and to a cathode temperature $T_c \approx 2750°$ K, where $b_{max} \approx 10^9$ A m^{-2} sterad. At a given grid bias V_g, greater brightness is obtained with the finer points. This follows from the more rapid increase of b with increasing V_d for these finer points. Apparently, the smaller r, the more rapidly the space-charge in front of the point disappears with increasing V_d.

TABLE 9.3. *Geometry and performance of point cathodes*

Radius of curvature of cathode r (μm)	Grid drive V_d (volts) when b reaches b_{max}	Emission current I (μA) when b reaches b_{max}
125	> 30	—
65	~ 30	370
21	~ 20	160
7	~ 75	25
2·7	~ 6	17·5
0·7	~ 4·5	10

Much effort has been directed towards developing a practicable field emission source using point cathodes. Although field emission requires an ultra high vacuum (10^{-9} torr or less) for reliable operation, it has the advantage that the emitter brightness is no longer limited by the theoretical value β_{theor} set by the cathode temperature. Small tip radii lead to high fields at the cathode surface. In the case of a thermionic emitter, fields greater than about 10^7 V m^{-1} tend to depress the work-function of the cathode leading to what is termed 'field-enhanced' or Schottky emission. With cathode fields greater than 10^9 V m^{-1}, the cathode need not even be heated, but a cold cathode field emission source can be realized.

Some indication of the advantages to be expected from using a point cathode field emission can be obtained from the calculations of Everhart (1967). It appears that for a point cathode of radius of curvature 1μm, the beam brightness using cold cathode field emission should be approximately 10^4 times higher than the brightness using Schottky emission, which is itself 10^2 higher than the brightness obtainable from a conventional hairpin cathode. The effective source diameter is a function of the cathode field; it can be smaller than 1 000 Å for Schottky emission and approaches 100 Å for cold field emission.

Field emission cathodes have so far not found much application in electron optics since they are very sensitive to the slightest surface contamination and to any deterioration of the very high vacuum which is required for constancy of emission currents. An electron gun with field emission cathode by Crewe (1966) which was incorporated in a scanning microscope (Crewe *et al.* 1967) was perhaps the first successful application of such a system.

Electron microprobes. Beams converged through a focus of the order of a μm or less are required for instance in scanning microscopes (cf. Smith and Oatley, 1955) in micro-analysers, (cf. Cosslett and Nixon, 1961) and in electron interferometers (cf. Möllenstedt and Düker, 1956; Keller, 1961). For a description of these instruments, we have to refer to the above references; the electron microprobe, however, which is used in all of them will be discussed here. The microprobe is a highly demagnified image of the emitting area of a thermionic cathode usually of a hairpin cathode. A triode gun focuses a beam of say 10–40 keV on a diaphragm with an aperture of a few hundredth of a mm. A short focal length condenser lens, arranged with a large object to image distance, produces a strongly demagnified image, hence the aperture of the beam is considerably increased (cf. Abbe's sine law §1.5). The central part of this beam is selected by means of a diaphragm and a final, again highly demagnified image which is the electron probe, is projected by a second short focal length condenser lens. The total current in the probe varies with the square of the angular aperture α; it reaches a maximum when

$$\alpha = (D/4\,C_s)^{\frac{1}{3}}, \qquad (9.27)$$

where D is the diameter of the probe and C_s is the spherical aberration of the final lens, but the directional intensity of the beam of the successive images of the cathode remains constant.

The best probes reach a current of about a μamp at a diameter slightly above a μm and with a directional beam intensity of

5×10^8 amp m^{-2} sterad^{-1}. If α is decreased, then a spot diameter of 100 Å can be obtained with a current of 10^{-12} amp. This spot size and current level is suitable for a high resolution scanning microscope, while a probe carrying a μamp is suitable in an X-ray microanalyser.

Telefocus guns. Steigerwald (1949) showed that from a short emission system, consisting of a hairpin cathode, a specially shaped grid electrode and a flat anode diaphragm, a fine focus could be

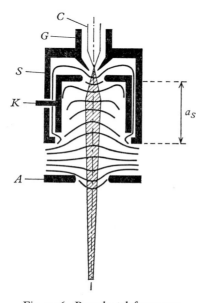

Fig. 9.26. Braucks telefocus gun.

obtained at a great distance without a previous cross-over. With -0.3 kV on the grid and 50 kV anode voltage he obtained a focus of 0.5 mm diameter at 90 cm distance from the cathode. The properties of the Steigerwald system have been described in great detail by Braucks (1958), who also improved the versatility of the system by introducing a control electrode (Braucks, 1959) which allows an easy adjustment of the position of the focused spot and also leads to an increased directional beam intensity. As shown in fig. 9.26, the Braucks gun contains a hairpin cathode C, a conical grid electrode G, connected to a cylindrical screen S, and a plane anode diaphragm A. Inside the screen, there is arranged the focal control electrode K, which controls the large distance L of the focal spot from the cathode. This distance L as a function of the control electrode voltage V_K and of the anode voltage V_A with respect to the filament tip $V_c = 0$ is plotted in fig. 9.27. Typical

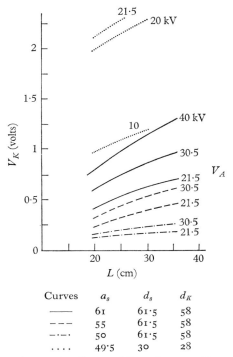

Curves	a_s	d_s	d_K
——	61	61·5	58
– – –	55	61·5	58
–·–·	50	61·5	58
· · · ·	49·5	30	28

Fig. 9.27. Dimensions and voltages of Braucks guns. Beam voltage V_A and focal spot distance L for various dimensions of screen aperture diameter d_s, control electrode depth a_s and control electrode large aperture diameter d_K in mm.

values for the resulting beam are for instance: a current $I = 5\ \mu$A in a spot of 0·07 mm diameter with 4 milliradians divergence of rays at $L = 23$ cm distance from the cathode at $V_A = 30$ kV, with $V_K = 0·7$ kV and with $V_G = -250$ V at the grid electrode.

Quantitative investigations on electron probes from telefocus tetrode guns have been made by Kanaya et al. (1968). They measured aberrations in the probes by a shadow projection of a fine mesh (cf. §6.4) and current density distributions by a Faraday cage. At 15 kV they obtained aberration coefficients of the order $C_s \simeq 20$ cm at directional beam intensities $b \simeq 2 \times 10^{-5}$ A cm^{-2} sterad^{-1}. 10^5 watt mm^{-2} could be concentrated into a spot of 1 μm.

A telefocus gun which can concentrate relatively large currents in the focus has been constructed by Bas and Gaydou (1959). As shown in fig. 9.28, it contains a bolt cathode C (cf. §9.1) with a

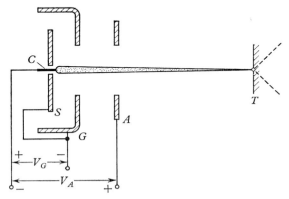

Fig. 9.28. Telefocus gun of Bas and Gaydou.

plane emitting surface of 0·3 mm diameter which protrudes through a flat diaphragm S which is connected to the grid electrode G at $V_G = -70$ V with respect to the cathode. A flat diaphragm A forms the anode at $V_A = 50$ kV. This gun projected 2 mA into a spot of 0·15 mm diameter at a distance of 17 cm.

After a detailed study and a further development of this gun by Bas et al. (Bas, Cremosnik and Wiedmer, 1967; Bas and Gaug, 1967), still greater currents (150 mA at 20 kV) could be projected into a focus of a few tenths mm at distances of 15–25 cm.

§9.10. Guns developed for special purposes

Slow electron guns. In the production of good intensity slow electron beams a great difficulty is experienced owing to the strong mutual repulsion of the electrons which grows with increasing perveance $I/V^{\frac{3}{2}}$ (cf. §8.1). This difficulty is overcome to some extent by the application of a relatively high field at the cathode, so that reasonably large currents can be drawn. This results in an initial acceleration of the beam which consequently has to be decelerated before leaving the gun. This principle has been applied, for instance, in a gun described by Simpson and Kuyatt (1963). This gun contains a plane oxide cathode and six plane diaphragms with circular apertures, these are: a grid electrode of variable bias, an anode at 300 volt followed by a small and a wide aperture both at 300 V. The beam is decelerated to 30 volts by the last two apertures, and a focus of 1 mm with 8 μamp is obtained at a distance of 4 cm. A gun with a tungsten spiral cathode incorporating initially a deflecting field and acceleration to 40 volts has been described by Lee (1968); it projects electrons, the energy of which can be controlled between 4 and 15 eV, into a focus of 1·5 and 1 mm respectively, but the current is of the order of 1 μamp only. Larger currents can be drawn if longitudinal magnetic fields are superimposed. For instance Hart and Weber (1961) have constructed a gun containing a plane oxide cathode, a grid at small positive bias and five plane diaphragms with apertures of various diameters. The voltages applied to these diaphragms were for example 15, 100, 30, 10 and 6 V. The space charge repulsion was counteracted by the magnetic field and parallel beams of 1 mm diameter up to 40 μA have been obtained.

The only arrangement known for producing beams of the order of 1 eV or less seems to be a positively charged diaphragm with an aperture directly in front of an emitter which is preferably an indirectly heated equipotential cathode. Marmet and Kerwin (1960), for instance, used a filament facing a 0·5 × 1 mm slit. At 0·5 mA total emission they could collimate a 0·5 eV electron current of 10^{-9} amp. From this current, they selected a beam of very small energy spread (0·05 eV half width) with a monochromator which will be described in §10.4. The shape of beams projected from slow electron guns can be investigated with the sliding target technique described in §2.1. However, a post accelerating target has to be employed. This consists of a fine wire mesh at the potential of the last gun electrode arranged closely in front of a fluorescent target at 1–2 kV, so that the electrons acquire sufficient energy to excite the fluorescent materials.

High velocity electron guns. The guns of high energy electron accelerators are required to inject currents of up to a few amp at about 100 kV into a synchrotron or a loaded line wave guide structure. The beam must be roughly parallel on injection and have a radius of a few mm only. Oxide cathodes are not favoured since they are poisoned easily in a demountable system. There is no standard design for the electrode structure; however, it is usual for the cathode to be slightly concave to give some degree of initial focusing, and for there to be mesh in the anode plane which serves both to shield the gun field from the accelerator field and to remove the defocusing effect of the anode aperture. In the Stanford Mk III linear accelerator (Chodorow *et al.* 1955) the cathode is a directly heated plane tungsten spiral, the initial concavity of the field being assured by a flared conical electrode surrounding the cathode, making an angle of $41°$ with the beam axis. An alternative cathode is described by Austin and Fultz (1959). This consists of a concave tantalum sheet which is heated by electron bombardment.

The electron optics of a high velocity electron gun constructed by Heinz (1966, 1967) has been studied in detail by Hoffmann and Heinz (1967, 1968). The electron beam of 200 kV and 10 amp is originally projected from a Pierce system (§9.8) in pulses of 50 Hz repetition rate. The tantalum cathode is indirectly heated by electron bombardment. A modification of the optimum dimensioning of the system is necessary owing to the relativisitic speed of the electrons, and is a function of the current density at the cathode and of the radii of curvature of equipotentials r_c at the cathode and r_A at the anode. E.g. $r_c/r_A = 2 \cdot 27$ was required instead of $2 \cdot 15$ for the non-relativistic case. Also a reduction of the perveance P is expected depending on r_c/r_A and on the current density at the cathode. Actually obtained was $P = 1 \cdot 12 \times 10^{-7}$ amp volts$^{-\frac{3}{2}}$, with radii $r_c = 10$ mm of the cathode and $r_A = 9 \cdot 6$ mm for the anode.

Production of high power electron beams. These are of technical importance for machining, welding and melting of metals by electron bombardment (see Refs: *Processes* 1960, 1961, 1962, 1963, 1965). The use of electron beams for machining of metals was originally demonstrated by Steigerwald (1953); it has now become a standard technique in which beams of 100–150 kV can be focused with a power of 10–50 kW into an area of a few mm². Electron beam machines are commercially produced by a number of firms. Schwarz (1962, 1964) describes the performance of a gun which, for welding or cutting purposes, can concentrate 10^{13} watts m^{-2} into a spot of 250 μm, he also explains the mechanism of the penetration into metal of such high power density beams. The depth of penetration is 1000 times greater than expected for the high density of the solid metal. The beam vaporizes the metal instantaneously along a narrow channel, a fraction of a mm wide, and by gas focusing due to an ionization of the metal vapour, it remains very narrow, though it enters the surface with an aperture of several degrees.

The telefocus gun of Bas *et al.* (Bas, Cremosnik and Lerch, 1962; Bas, Cremosnik and Wiedmer, 1967; Bas and Gaug, 1967) which we described at the end of §9.9 and also many other e.g. magnetically focused high power guns are suitable for machining etc. Descriptions can be found in the references (*Processes*).

The best way to obtain a fine, intense cathode ray was, up to about 1930, a gas discharge, e.g., in hydrogen of 10^{-3} torr. The electrons generated in the discharge were transmitted through a narrow aperture in the anode into a space of lower pressure (cf. Wierl, 1931). This way of producing electron beams was gradually superseded as the emission system with thermionic cathode was developed. We only mention here an investigation of the properties of a gas discharge emission system by Möllenstedt and Düker (1953) which has found an application in an electron microscope. More recently, however, gas discharges have become of interest for the production of extremely intense electron beams from the 'duoplasmatron' sources. These consist of a glow discharge or an arc discharge between a thermionic cathode and a watercooled anode, the electron beam from a fine aperture in the anode passes through an aperture in an accelerating 'extraction electrode'. A strong magnetic field is superimposed over the discharge and extends in the direction of the extracted electron beam into a drift space where a vacuum of 10^{-5} torr can be maintained. From the wide literature on duoplasmatron emission systems we mention here the source of Roberts, Cox and Bennett

(1966) which, from a hollow arc discharge of 1–12 amp produces a narrow beam of 1·6 mm diameter 0·5 amp and 30 kV which can be decelerated through a focusing electrode to 3 kV. Engelhaaf (1968) measured the directional intensity of beams from duo-plasmatron discharges between various metal electrodes. The best value obtained with magnesium electrodes was measured to be $5·7 \times 10^6$ A m^{-2} sterad^{-1} as compared with the directional intensities from systems with oxide cathodes (10^7) or with the best values from hairpin cathodes ($\sim 10^9$ A m^{-2} sterad^{-1}) (cf. §9.8 and §9.9).

Line-focus emission systems. Few of these systems have become of practical importance. It is, however, of some basic interest to compare their performance with that of the corresponding emission systems of circular symmetry. Take for instance the strip-cathode systems corresponding to the simple triodes of circular symmetry in §9.2. There figs. 9.1 and 9.2 could represent cross-sections of strip-cathode systems. Let the plane of two-dimensional symmetry be the xz-plane, coinciding with the plane of the drawing. The strip cathode now extends in the y-direction and the apertures of the grid and anode diaphragms are now rectangular slots. Comparing the properties of the two classes, it is obvious that the strip systems always yield appreciably larger currents than the circular emission systems of §9.2, even if the strip cathode is much restricted in length. The reason for this is found not only in the relatively larger area of the strip cathode but also in the relatively larger penetration of fields from the anode to the cathode surface. Two quantities are again of practical interest for the design of strip-cathode systems of a desired electrical performance: (i) the penetration factor D defined by (9.10) and (ii) the geometry factor G of Child's equation, applying to an equivalent-diode voltage as defined by (9.14). These significant parameters (D and G) are tabulated below for some strip-cathode systems characterized by their spacings: c/y_{gr} = grid top to cathode spacing measured in terms of slot semi-apertures, and a/y_{gr} = grid top to anode spacing also measured in terms of slot semi-apertures.

The rectangular slots all have the ratio of semi-apertures (x_{gr}/y_{gr}) = 20/1, i.e. their length is 20 times greater than their width. The penetration factors D are found to vary over a large range and, for strip systems, they are quite generally larger than for the corresponding circular systems. D increases with the slot length x_{gr} up to about $x_{gr}/y_{gr} \approx 5$, but for longer slots, D is very nearly independent of x_{gr}.

On the other hand, G is to some extent proportional to x_{gr}. For a given (c/y_{gr}) spacing, G does not depend very much upon the geometry of the anode. Table 9.5 contains only G values for the symmetrical two-slot system with equal grid-cathode and grid-anode spacings ($a = c$). All systems of table 9.4, however, have within 15 per cent the same G values as those given in table 9.5.

TABLE 9.4. *Penetration factors D for strip-cathode systems*

Grid top to cathode spacing (c/y_{gr})	1	2	3	5	8
($D \times 100$) for symmetrical two-slot systems (cf. fig. 9.1)					
$a = y_{gr}$	38	15	8·6	4·3	2·3
$a = 2y_{gr}$	21	9·3	5·2	2·8	1·5
$a = 6y_{gr}$	7	3·3	1·9	0·8	—
($D \times 100$) for open-anode system (cf. fig. 9·2)	2·2	0·88	0·55	0·30	0·16

TABLE 9.5. *Factors G for symmetrical two-slot systems*

Grid top to cathode spacing (c/y_{gr})	1	2	3	5
G	52	38	27	17

Since for any strip-cathode system G is much larger than for the corresponding circular system, the anode voltage V_A required for drawing a given emission current is relatively small. For the same reason the black-out voltage $^{bl}V_g$ of a strip system of a given D value is relatively small but the correction term (cf. §9.5) for the measured black-out voltages is quite substantial.

Of practical interest is the wide range of obtainable perveances $P = I_{em}/V_A^{\frac{3}{2}} \approx GD^{\frac{3}{2}}$ for different strip-cathode systems of a given G. For example, for the fixed grid spacing $c = y_{gr}$ one obtains the very different perveances of $10\ \mu\text{amp}\ V^{-\frac{3}{2}}$ per metre slot length for a system with wide, open anode and of $500\ \mu\text{amp}$ $V^{-\frac{3}{2}}$ per metre slot length for an unsymmetrical two-slot system with $y_a = 0\cdot5y_{gr}$ and $a = y_{gr}$. These data should be compared with those of the corresponding circular symmetrical structures namely the open-anode system and the symmetrical two-diaphragm system: these circular systems reach only the relatively small perveances of $0\cdot006$ and $0\cdot2\ \mu\text{A}\ V^{-\frac{3}{2}}$ respectively.

Angular current distribution from line-focus emission systems. Electrons emitted from the strip cathode form a beam waist or a cross-over line in x-direction and then fan out in shape of a wedge. Now, let a slot stop with adjustable y-aperture cut off a given external part of the emitted beam. The fraction of the transmitted beam current to the total emission is the current efficiency of the system for a given wedge angle 2Θ.

Experimental values for this current efficiency referring to the fixed angle $\Theta = \pm 0\cdot1$ radian and to full space-charge conditions are plotted in fig. 9.29 as a function of electrode spacings for a few characteristic representative systems (Klemperer, 1947a). The efficiency is high ($0\cdot8$) for an open, wide anode and low ($0\cdot24$) for a symmetrical two-slot system with anode spacing equal to one semi-aperture of the slot. Within the range of the figure, the efficiency of both these systems is not markedly dependent on the cathode-grid spacing c which is plotted as abscissa.

On the other hand, the efficiency of an unsymmetrical two-slot system with an anode slot half as wide as the grid slot ($y_a = \frac{1}{2}y_{gr}$) shows a very pronounced maximum for a critical grid-to-cathode spacing. The broken curves in fig. 9.29 belong to this unsymmetrical slot system and correspond to three different total emission currents of 1, 5 and 20 mA. Apparently, the maxima of these curves correspond to the critical conditions represented by fig. 8.3, where the beam waist is smallest and reaches its greatest elongation. Interpreted in this way, the spacings at which the efficiency maxima occur are 'preference' spacings of the kind explained in detail in our discussion of the influence of space-charge on the circular systems as given in §9.6. Again, the angular current distribution from strip-cathode systems depends a great

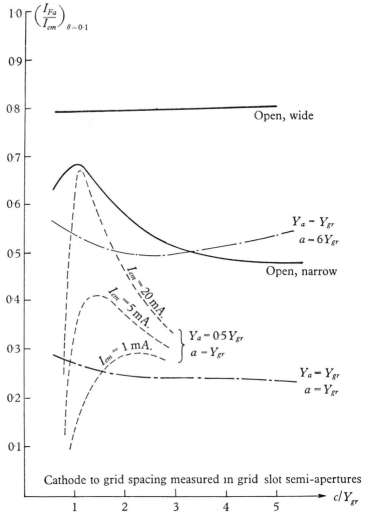

Fig. 9.29. Efficiency of line-focus emission systems; fraction of total emission projected into an angle $\Theta = \pm 0.1$ radian.

deal upon space-charge and often changes entirely when the emission is reduced from full space-charge conditions to tempera-ture-limited saturation.

Space-charge pressure in line-focus systems can be balanced by external fields in the way we described for circular structures in

§8.7. Electrodes have to be found which produce a potential distribution proportional to the 4/3 power of the distance from the cathode. According to Pierce (1940), fields of this kind are determined as above (§8.8) in the electrolytic trough with the electron beam being replaced by a piece of insulator, for the two-dimensional case the electrolyte having parallel bottom and top faces. The practical interest of the optics of line-focus systems lies in its application to valves and X-ray tubes and to the production of annular beams. Knoll (1934) has investigated the bunching of the electrons by the control grid in valves and his work has led the way to a valuable improvement of multi-electrode valves. For instance the screen current in a tetrode valve could be reduced by a factor 10 and the slope of its grid characteristic improved by a factor 2 by proper alignment of the screen with respect to the grid. More about electron optics in various kinds of valves is found in the papers by Schade (1938) by Bull (1945), and by Jonker (1949).

The emission system in X-ray tubes (cf. Clark, 1955) frequently consists of a tungsten helix serving as the thermionic cathode which is surrounded by a semi-circular cylinder at negative

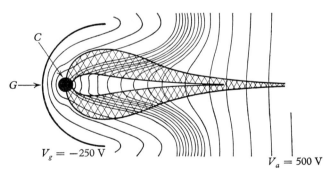

Fig. 9.30. Line-focus emission system with focusing cylinder.

potential, acting as a focusing electrode and projecting the emitted electrons into a line focus on the anode. The optics of such a system have been studied by Hellwig-Zwiessler, Stierstadt and Hellwig-Zwiessler (1962) with a model, the cross-section of which is shown in fig. 9.30. The helix is here replaced by a filament C, which is surrounded by the cylinder G at $V_g = -250$ volts. The line focus is projected onto the anode at $+500$ volts. From the deformation of the equipotentials shown in the figure, it can be concluded that space-charge is formed mainly at that part of the cathode which faces the anode and also at that part which faces the cylinder G. Thus, the emission is projected from the two sides into two different bands which are united in the line focus, as seen in the figure. In actual X-ray tubes, the voltages are of course very much scaled up from the above model experiment. These range from a few kV to a few hundred kV. The problem of dissipating the very high power concentration, e.g. of several kwatt in a small focal line is solved in some tubes by liquid cooling of the anode or by rotating the anode at high speed in order to avoid surface damage.

As an example of a design in accordance with electron optical principles, we quote here the fine-focus X-ray tube by Hosemann (1955). There, the emission from a helix of 1·2 mm diameter is projected on to an anode at 50 kV into a line focus of 16 mA with 0·7 mm width (measured where the intensity has dropped to 1/10). The helix cathode is surrounded by the usual negatively charged focusing cylinder. In addition, this gun contains a positive accelerating electrode at 6 kV.

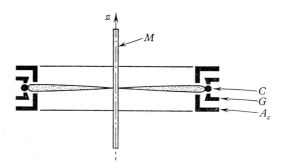

Fig. 9.31. Annular gun for 'floating-zone' melting.

Of technical importance are some ring shaped line-focus guns (cf. Bakish, 1962). These guns are quite common in the 'floating-zone' melting of metals and semi-conductors. Fig. 9.31 shows a schematic drawing of the cross-section through such a gun. There,

z represents the axis of rotational symmetry. Electrons are emitted from a circular tungsten wire cathode C. G and A_c are grid electrode and accelerator, both ring shaped and with annular slots focusing a disc-shaped electron beam at the centre of the circle onto the metal rod specimen M. A magnetic field, produced by the heating current of the cathode, would upset the proper electron optics, unless it is compensated by the field of a current in the opposite direction, situated near to the heating current of the cathode (Bas and Gaug, 1967). The focused beam of, say, 20 kV and 6 kW has sufficient power to melt the rod M in a relatively narrow zone. M is rotating about z and is slowly moved along the z-direction so that it is gradually melting and solidifying again, thus purifying the specimen by recrystallization.

Annular emission systems have also been developed with toroidal cathodes to produce hollow beams (Schwartz, 1958; Harris, 1959). There, however, the annular slit in the anode seriously affects the uniformity of cathode emission. These guns will not find widespread use until the properties of the annular lens are better understood. They are, however, attractive in principle since they offer the possibility of obtaining the high perveances associated with strip beams (see above), while retaining the convenience of cylindrical symmetry.

Of some technical importance again are emission systems which have been developed in order to inject electrons into crossed field devices. In these systems, the drift velocity of a stream of electrons, an electrostatic field E and a magnetic field B are mutually perpendicular. The electron drift velocity is E/B. As shown in fig. 9.32 electrons are emitted from a cathode C and attracted by the positive plate P. Currents of the order of 100 mA are deflected by the magnetic flux density B into the space between the positive anode A and the negative 'sole' S. The trajectory of a single electron is a cycloid which, however, is to some extent smoothed out when space-charge produces a laminar flow, in particular when the anode is shaped to have a tapered gap between anode and an electron emitting sole.

The crossed field gun has found wide application in microwave oscillators and amplifiers. In these, the anode is replaced by a slow wave structure, containing a series of cavities, in order to relate the velocity of the electromagnetic wave to the drift velocity of the electrons. The arrangement can be that of a plane diode as in fig. 9.32. Alternatively, both beam and an emitting sole are closed

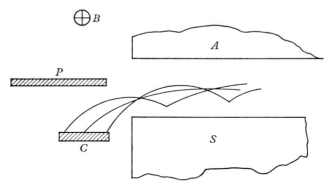

Fig. 9.32. Crossed field electron gun.

in on themselves forming a co-axial cylindrical magnetron structure. Electrons may then be extracted from the hollow beam along the magnetic field direction. For details we have to refer to the literature (cf. Kino, 1961).

CHAPTER 10

DEFLECTING FIELDS

§10.1. Homogeneous weak deflecting fields

Electrostatic deflexion. Let an electron *el* enter into the homogeneous electrostatic field between two plane parallel condenser plates of the voltages $\pm V_p$ as shown in fig. 10.1. Let the electron

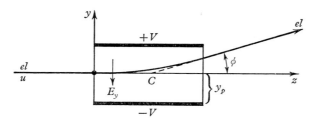

Fig. 10.1. Electrostatic deflexion.

move along the z-axis and let the two plates be arranged symmetrically to the z-axis, parallel to the xz-plane, the mutual distance of the plates being equal to $2y_p$. The electrostatic field vector is now in the $-y$-direction and is given by

$$E_y = -\frac{V_p}{y_p}. \tag{10.1}$$

The equations of motion (mass × acceleration = force) for the electron are:

$$\left. \begin{aligned} m\,\frac{\mathrm{d}^2 y}{\mathrm{d}t^2} &= (-e)E_y, \\[2mm] m\,\frac{\mathrm{d}^2 z}{\mathrm{d}t^2} &= 0, \end{aligned} \right\} \tag{10.2}$$

while the initial conditions are given by

$$(\mathrm{d}y/\mathrm{d}t)_0 = 0 \qquad\qquad (\mathrm{d}z/\mathrm{d}t)_0 = u_z$$

$$y_0 = 0 \qquad\qquad z_0 = 0.$$

By twice integrating the equations (10.2) one obtains

$$\left.\begin{aligned} y &= \frac{-e}{2m} E_y t^2, \\ z &= u_z t. \end{aligned}\right\} \qquad (10.3)$$

Elimination of the time t from equations (10.3) yields the equation of the electron orbit:

$$y = \frac{-e}{2m} E_y \left(\frac{z}{u_z}\right)^2. \qquad (10.4)$$

This is the equation of a parabola. We can now still substitute the initial electron energy V_{el} for u_z in (10.4) since, owing to the energy equation we have

$$u_z = \left(\frac{2e}{m} V_{el}\right)^{\frac{1}{2}} \qquad (10.5)$$

and with (10.1) and (10.5), (10.4) can be written

$$y = \frac{z^2}{V_{el}} \frac{V_p}{4y_p}. \qquad (10.6)$$

The angle ϕ of deflexion is obtained by differentiating (10.6) which yields

$$\tan \phi = \frac{dy}{dz} = \frac{z}{V_{el}} \frac{V_p}{2y_p}. \qquad (10.7)$$

This equation (10.7) can easily be applied to the determination of the electron energy V_{el} of a beam since the deflexion angle ϕ, the length (z) of the field, the voltage $(2V_p)$ between the condenser plates and their spacing $(2y_p)$ are all easily accessible to measurement. Elimination of $V_p/(y_p V_{el})$ between (10.6) and (10.7) yields

$$\tan \phi = \frac{y}{z/2}, \qquad (10.8)$$

which shows that the tangent to the beam leaving a condenser of length z can be produced back to the centre C $(y_c = 0, z_c = z/2)$ of the condenser.

Magnetic deflexion. The deflexion of an electron beam in a homogeneous transverse magnetic field of flux density B_y in the $+y$-direction can be derived from the equations of motion with only the Lorentz force (5.1) acting on the electron. These equations are

$$
\left.\begin{aligned}
m \frac{d^2x}{dt^2} &= eB_y \frac{dz}{dt}, \\
m \frac{d^2y}{dt^2} &= 0, \\
m \frac{d^2z}{dt^2} &= -eB_y \frac{dx}{dt}.
\end{aligned}\right\} \tag{10.9}
$$

By integration of (10.9) with the initial conditions $x_0 = 0$, $z_0 = 0$, $(dx/dt)_0 = 0$, $(dz/dt)_0 = u_0$, we obtain

$$
\left.\begin{aligned}
\frac{dx}{dt} &= \frac{e}{m} B_y z, \\
\frac{dz}{dt} &= -\frac{e}{m} B_y x + u_0.
\end{aligned}\right\} \tag{10.10}
$$

Now, the electron speed in a purely magnetic field, namely

$$
u_0 = [(dx/dt)^2 + (dz/dt)^2]^{\frac{1}{2}}
$$

is constant. Thus squaring and adding (10.10) yields

$$
x^2 - 2 \frac{m}{e} \frac{u_0}{B_y} x + z^2 = 0. \tag{10.11}
$$

This is the equation of a circle of radius

$$
r_e = mu_0/(eB_y) \tag{10.12}
$$

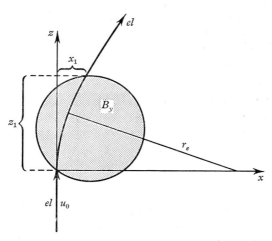

Fig. 10.2. Magnetic deflexion.

with the z-axis as tangent (see fig. 10.2). If a relatively weak field extends only over a relatively short, limited distance z_1, along the axis, the deflexion x_1 at the end of the field will be small in comparison with z_1 so that x^2 in (10.11) can be neglected against z^2 and we obtain

$$x_1 \approx \frac{eB_y}{2mu_0} z_1^2 = \frac{z_1^2}{2r_e} \qquad (10.13)$$

where again r_e is the radius of curvature of the electron path in the homogeneous field as given by (10.12); with (10.5), (10.13) becomes

$$x_1 = \sqrt{\frac{e}{8m}\frac{B_x}{\sqrt{V_{el}}}} z^2 = 1 \cdot 48 \times 10^5 \frac{B_x}{\sqrt{V_{el}}} z^2. \qquad (10.14)$$

The angle of deflexion ϕ at the point (y_1, z_1) is given by

$$\tan \phi = \frac{dx}{dz} = \frac{z_1}{r_e} = \frac{2x_1}{z_1}. \qquad (10.15)$$

§10.2. Practical electrostatic deflexion systems

The treatment of deflecting fields in §10.1 is an idealization of the actual case. Fields can never break off suddenly, but the actual boundary of a homogeneous field is a complicated, gradually decaying 'fringing field' which extends far into the field-free region. These fringing fields have been the object of extensive calculations and experiments in connexion with early precision measurements of the specific charge (e/m) of the electron by Bucherer (1909) and by Neumann (1914) respectively. According to their results the fringing field acts as though the deflecting condenser of length z_p were extended at the exit side by a relatively small amount Δz_p. Formulae have been derived giving Δz_p as a function of the separation $2y_p$ of the plates and of their shape. For the plane-parallel plate condenser with straight edge, Recknagel (1938) has derived

$$\Delta z_p = \frac{2y_p}{\pi}\left(1 - \log_e \frac{y_p}{\pi z_p}\right). \qquad (10.16)$$

For small plate distances, for example $y_p = 0 \cdot 05 z_p$ to $0 \cdot 1 z_p$, Δz_p is found to be of the order of $3y_p$.

The most important practical application of the weak electrostatic deflecting field is found in cathode-ray tubes for television and for measuring purposes. There it is desirable to obtain a maximum deflexion y_T at the target for any given deflecting voltage difference $2V_p$ applied to the plates. In other words, the field should be shaped in such a way that its 'deflecting power' becomes as large as possible. The deflecting power is usually defined (cf. Knoll, 1939) by

$$G = \tan \phi_p \frac{V_{el}}{V_p} = \frac{z_{eff}}{2y_{eff}}, \qquad (10.17)$$

where by comparison with (10.7) z_{eff} and $2y_{eff}$ are seen to be an 'effective' length and an 'effective' separation respectively of an ideal condenser producing a strictly homogeneous field of limited length, $2V_p$ being the potential difference between the plates.

Obviously, the deflecting power increases with decreasing plate separation. If, however, for a required deflexion the separation is reduced too much, the deflected beam will foul one of the deflecting plates. Hence, the minimum separation $2(y_p)_{min}$ and maximum deflexion $(y_T)_{max}$ are connected by the relation

$$\tan \phi = \frac{(y_T)_{max}}{z_T} = \frac{2}{z_p} [(y_p)_{min} - y_{el}] \qquad (10.18)$$

where y_{el} is the radius of the circular cross-section of the electron beam at the point of leaving the deflecting condenser. z_p and z_T are the length of the plates and the distance of the target from the centre of the condenser respectively.

In practical tube design, a desired deflexion should be produced by the smallest possible voltage $2V_p$ applied to the plates. For this purpose an optimum plate length $(z_p)_{opt}$ has to be determined for a given total beam length $(z_p + z_T) = c$. A relationship between the variables V_p and z_p with the constants y_{el}, y_p and c can be obtained by combining (10.7) with (10.18). Moreover, by minimizing V_p with respect to z_p, the optimum plate length can be calculated. According to Maloff and Epstein (1938) this is found to be

$$(z_p)_{opt} = \frac{3}{2} c \frac{y_{el}}{y_T - y_{el}} \left(\sqrt{\left[1 + \frac{8}{9} \frac{y_r - y_{el}}{y_{el}} \right]} - 1 \right). \qquad (10.19)$$

The corresponding optimum separation $2(y_p)_{opt}$ may then be calculated from (10.18) by substitution for z_p from (10.19).

The use of (10.19) may be illustrated by the following example: In a television tube with screen distance $c = 30$ cm, the size of the picture requires a maximum deflexion at the target $y_T = 10$ cm. The beam radius at the end of the deflecting condenser is $y_{el} = 2\cdot5$ mm. According to (10.19), the least deflecting voltage will be required if the length of the deflecting plates is $z_p = 5\cdot8$ cm and if their separation is $2y_p = 2\cdot7$ cm.

The plane-parallel plate condenser of dimensions as given by (10.19) does not yet represent the best electrode arrangement for maximum deflecting power. The separation of plane plates can be further reduced without danger of fouling if the plates are tilted towards the axis of the undeflected beam. At the entrance of the beam, the necessary plate distance has to be only slightly larger than the diameter of the beam. At the exit side, however, the plate distance has to be large enough to account for the deflexion of the beam after its passage through the field of the condenser.

We refrain from discussing the theory of the tilted plate condenser, but we present in the nomogram (fig. 10.3) some results about it after Knoll (1939). Though these results have been obtained neglecting the fringing fields they were found to be satisfactory for the purpose of practical design. The nomogram of fig. 10.3 contains the necessary information on the optimum geometry of the tilted plane plates for any given maximum angle of deflexion. The abscissae in fig. 10.3 represent the ratios of the condenser length (z_p) to the separation of the plates ($2y_1$) at the entrance of the beam. The ordinates represent the deflecting power

$$G = \tan \phi_{max} V_{el}/(V_p)_{max},$$

where ϕ_{max} is the maximum angle through which the beam can be deflected without fouling the plates, and $2(V_p)_{max}$ is the corresponding maximum possible deflecting voltage difference between the plates.

For instance, for parallel plates ($y_1/y_2 = 1$) and for a required $\tan \phi_{max} = 0\cdot2$, there can be found with the help of fig. 10.3:

   (i) the ratio of plate length to separation $z_p/2y_1 = 4\cdot8$;

   (ii) the deflecting power $G = 4\cdot8$.

Thus for an electron energy of 5 keV the plate voltages are

$$\pm V_p = \frac{0\cdot2 \times 5\,000}{4\cdot8} = 208 \text{ V}.$$

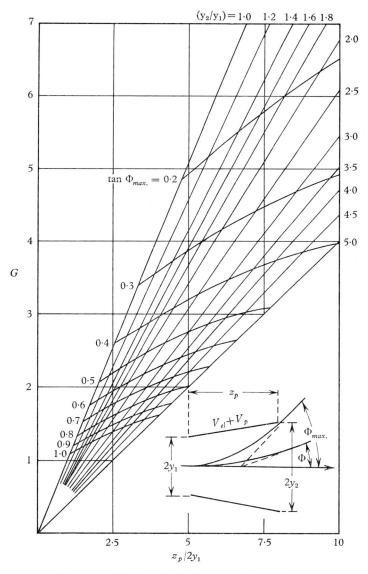

Fig. 10.3. Deflexion in the electrostatic field between tilted, plane condenser plates.

If the plates are tilted in the ratio $y_2/y_1 = 2$, the ratio $z_p/2y_1 = 9$ is obtained from the above ϕ_{max}. The corresponding deflecting power is now 6.4, and the maximum deflecting voltage is reduced to $V_p = \pm 156$ V. The increase in $z_p/2y_1$ implies that either the length z_p of the plates has to be increased or that their initial separation $2y_1$ has to be reduced. Since it is always desirable to reduce the overall dimensions of the condenser, the latter alternative will be chosen as far as possible, i.e. the initial separation $2y_1$ of tilted plates will be reduced until it reaches its natural limit, namely, the width of the beam diameter.

Further increase in deflecting power has been obtained by the introduction of 'kinked' deflecting plates, each consisting of two plane parts joined at an obtuse angle. In this way the beam passes in succession through two adjoining condensers, the plane plates of the second condenser having a greater tilt angle than the plane plates of the first one. The relative lengths of the two part-condensers are chosen to produce approximately equal beam deflexions (i.e. half the total deflexion each), and under these conditions their lengths and tilt angles can be determined from the nomogram in fig. 10.3.

The maximum theoretically possible deflecting power, however, is obtained when the condenser plates are shaped flush with the outer boundary of the deflected beam throughout the path. Maloff and Epstein (1938) have calculated the shape of deflexion plates under the assumption that the field everywhere along the z-axis is given by $E_y(z) = V_p/y_p(z)$, where y_p the ordinate of the plate is a function of z. They derived an optimum curve for the cross-sectional shape of the optimum deflecting plates which is similar to the general space-charge curve given in fig. 8.2. Though such curved deflecting plates would attain the maximum possible deflecting power, they are not frequently met in practice because the effort of shaping the plates is not sufficiently repaid by the relatively small increase in deflecting power that can be gained in comparison with the above kinked plates or even with plane deflecting plates tilted at the correct angle.

The relatively small advantage gained by curving the plates may be seen in the following example quoted from a paper by Borries (1941). A condenser with the plate length $z_p = 600$ mm and with the distance $z_T = 300$ mm between plates and target is used for deflecting a 1000 eV beam of width $2y_{el} = 4$ mm through an angle given by $\tan \phi = 0\cdot3$. The deflecting power (cf. (10.17)) of the condenser in terms of the theoretical optimum ($= 100$ per cent) may be stated as:

> 100 per cent for curved plates,
> 93 per cent for kinked plates,
> 75 per cent for tilted, plane plates,
> 41 per cent for plane, parallel plates.

The shorter the condenser (smaller z_p), the smaller becomes the difference in deflecting power for the different types of plates.

Great deflecting powers of the order of $G = 20$ have been observed with deflecting electron mirrors by Klemperer (1937). In these deflectors the electron beam is slowed down until it meets a reflecting equipotential surface at a sufficiently small angle of incidence, say $\alpha < 45°$, so that the remaining transverse velocity component of the electron is small. An auxiliary electrode is provided which changes the direction of the reflecting equi-potential by an angle β, say, if a potential $2V_p$ is applied to it. The angle of deflexion ϕ is equal to the change in angle of reflexion, which, however, is always twice as large as the angle by which the reflecting mirror is turned. Various types of deflexion mirrors have been proposed; however, no design with a satisfactory range of proportionality between ϕ and V_p has been obtained.

Post-deflexion acceleration. The great deflecting power of the deflexion mirror is mainly due to the fact that the electrons are exposed to the deflecting field when their velocity is very small, the deflexion being (cf. (10.4)) inversely proportional to the square of the velocity. In this way the deflexion mirror utilizes a principle which can be applied to every deflexion condenser or a deflecting magnetic field, namely, electrons may be deflected while they are of very small velocity, and they may be 'post-accelerated' later on to whatever speed is required, say, for producing sufficient light intensity on a fluorescent target. There arises, however, the problem that the spot size should not be increased by the post-accelerating field. In distinction to the deflecting power G defined by (10.17), the so-called 'deflexion sensitivity' is here characteristic of the quality of a cathode-ray tube. If D_{sp} is the diameter of the focused spot on the fluorescent screen and y_T is the deflexion of the spot, then the minimum observable ratio of the deflexion y_T at the target as measured in spot diameters D_{sp} will give the smallest deflexion which can be registered by the tube. If again $2V_p$ is the voltage applied to the deflecting condenser, the deflexion sensitivity is defined by

$$S_D = y_T/(D_{sp} \times 2V_p). \tag{10.20}$$

On the other hand, the deflexion sensitivity S_D can be estimated in advance from the geometry of the arrangement and from the voltage V_{el} of the electron beam (i.e. the anode voltage which is available for the electron gun), namely

$$S_D = Gz_T/(2V_{el}D_{sp}), \tag{10.21}$$

where z_T is the distance of the target from the deflexion plates. The sensitivity of a good commercial monitor tube is of the order of one spot diameter per volt.

According to (10.21), the sensitivity of a cathode-ray oscillograph cannot be increased simply by increasing the distance of the fluorescent screen from the deflecting field, i.e. by increasing the 'length of the pointer'. The spot is an electron-optical image. Image distance and thus magnification and consequently spot width would be increased in the same proportion as the target distance. Post-deflexion acceleration, however, appears to offer the means to increase the sensitivity.

Early tubes were built in which a metal mesh was placed parallel to and at a small distance in front of the target; thus post-accelerating fields could be maintained between mesh and target. These first attempts were not successful, since appreciable enlargement of the spot, due to the field penetration through the apertures of the mesh, largely upset the gain in sensitivity. Later on, Schwartz (1938) and Pierce (1941) showed theoretically that the scheme of post-acceleration was sound and that an ideal double layer should be efficient in increasing the sensitivity of a tube, but the gain would depend not only upon the voltage difference but also upon the shape and position of this double layer.

Lampert and Feld (1946) published details about a tube with post-acceleration voltage ratio 10:1 which was applied to several conducting bands deposited on the inside of the cylindrical glass tube. They found not only greatly increased light output but also somewhat increased deflexion sensitivity, since the decrease of spot size due to post-acceleration overbalanced the decrease of the deflexion.

On superficial examination, post-acceleration seems to recommend itself wherever an increase in deflexion sensitivity is desirable. In table 10.1 we give a survey of various types of acceleration arrangements with the approximate ratio of sizes of the scanned pattern with and without the acceleration. A distortion

TABLE 10.1. *Post-deflexion acceleration* (= *PDA*):

$$\frac{\text{Scan size with } PDA}{\text{Scan size without } PDA} \quad \text{for a } PDA\text{-ratio } (4:1)$$

PDA electrodes	Scan size
Single band	0·6
Spiral	0·7
Mesh in front of and parallel to fluorescent screen	1·0
Variable pitch spiral producing spherical equipotentials	1·3
Fine grid on cylinder behind deflexion plates	1·5

of the pattern and of the scanned spot can be largely avoided by suitable arrangement of the deflecting and the accelerating fields, for details, we have to refer to the literature. We should point out, however, that the best post-acceleration is effected by means of fine meshes. The problem of losing contrast in post-acceleration tubes due to secondary electron emission from these meshes has been met either by coating the meshes with an insulating layer (Schlesinger, 1956) or by applying a bias voltage which tends to draw these electrons back to the gun (Law, Davne and Ramberg, 1961).

For television tubes, economy in the scanning power that is supplied to the deflexion field is of technical importance. The question arises whether power could be saved by the use of a post-accelerating field. For a given beam velocity at the fluorescent screen, introduction of acceleration after the deflexion would allow us a decrease of beam velocity in the deflexion field and hence a decrease in field intensity. But for best economy in power the distance of the deflecting condenser plates (or the diameter of the deflecting coils) would have to be as small as feasible with respect to the maximum angle of deflexion and with respect to the given diameter of the beam. Decrease of beam velocity, on the other hand, implies, according to Lagrange's law (cf. (2.7)), a corresponding increase in beam diameter. Thus the separation of the deflecting plates (or the coil separation) would have to be increased correspondingly. The volume of the deflecting field would be increased and so would be the supply of power which is proportional to the volume of the field. Thus the advantage of a decrease in scanning power which may have been expected owing to decrease in deflexion field intensity on account of reduced beam velocity is cancelled out by the increased volume of the field.

High speed oscillograph. For registration of very rapidly changing voltages, cathode-ray oscillographs of high 'temporal resolution' are required. This latter property can be defined as the ratio of the spot diameter D_{sp} (or of the width of the written line) to the writing speed u_{sp}, or the ratio of necessary charge density q_{min} for exciting the fluorescent screen to the current density i_{sp} in the spot at the electron energy V_{el}, namely

$$t_0 = D_{sp}/u_{sp} \rightarrow \frac{q_{min}}{i_{sp}} . \qquad (10.22)$$

Thus t_0 is the time which the 'flying spot' needs to pass along a distance equal to its own width while moving with a maximum velocity that is still registered.

Numerous constructions of oscillographs have been described in the literature. We mention here only a paper by Haine and Jervis (1957) who have given a detailed analysis of the limitations of the high speed oscillograph in terms of gun performance, unavoidable aberrations and available recording media. These authors have also constructed an oscillograph which comes close to the theoretical limit of performance. They used a demountable tube, a gun with hairpin cathode (cf. §9.9) producing a beam of 10^9 amp m^{-2} sterad^{-1} directional intensity which was deflected by a condenser with straight tilted plates. A flying spot of 40 μ diameter was recorded by a fluorescent screen. A sensitivity of 0·5 spot widths per volt was obtained at $V_{el} = 30$ kV. At a current density $i \simeq 10^5$ amp m^{-2} a temporal resolution $t_0 \simeq 10^{-13}$ s was obtained at the oscillographic record, corresponding to a required charge density $q_{min} \simeq 10^{-3}$ coulomb m^{-2}. Thus the writing speed was of the order of the light velocity.

The best temporal resolution obtainable by external photography of the trace on a fluorescent screen is comparable to that obtainable by direct impact of the electron beam on a photographic emulsion. There should, however, still be room for improvement of the temporal resolution, if required. E.g. post deflexion acceleration could be incorporated, or the trace on the fluorescent screen could be registered by an image intensifier or a storage tube.

We mention finally an electrostatic system in which, by a special arrangement of plates the deflecting field simultaneously focuses the electron spot. This is found in the electrostatic 'Vidicon'

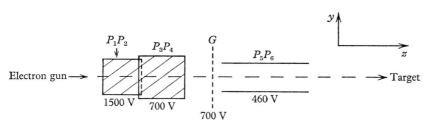

Fig. 10.4. Deflectors in the all-electrostatic Vidicon.

television pick-up tube of Lubszynski et al. (1969), and is shown in fig. 10.4. The x-deflexion is provided by two pairs of plates P_1P_2, P_3P_4 and x-focusing is assured by the decelerating line-focus lens due to the field between these plates (corresponding to a planar lens as in fig. 2.6). The scanning by P_5P_6 has little influence on the focusing action of the lens, since this is virtually complete before the transverse scanning field has had an appreciable effect. Focusing for the y-direction is due to the lens formed between the mesh G and the y-deflexion plates $P_5 P_6$; this is a line-focus analogue of a gauze-tube lens (cf. §4.8) and has low field-curvature (cf. §6.7).

§10.3. Practical magnetic deflexion systems

The design of these systems varies considerably according to their particular use. For instance, the scanning coils of a television tube should consume a minimum of power for building up the field, and their overall frequency response should be sufficiently high; moreover, the deflexion of the beam should be proportional to the coil current and it should not depend upon the initial coordinates of the beam. On the other hand, very different problems turn up in the design of deflexion magnets for electron spectrometers. For example, great flux density, or great field homogeneity, or a certain law of decrease of field strength with one of the coordinates, etc., may be required.

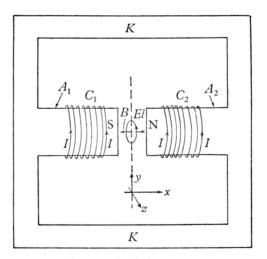

Fig. 10.5. Deflexion magnet.

The general scheme of a simple deflexion magnet is shown in fig. 10.5. This magnet is able to produce strong homogeneous fields over a relatively large region inside the gap between its pole-pieces A_1 and A_2. These pole-pieces are of cylindrical shape with flat end-faces, and they are made from soft, perfectly homogeneous iron, energized by the two current coils C_1 and C_2.. An 'iron-yoke' K is used to reduce the volume of the useless flux outside the magnetic gap, thus saving a substantial amount in energizing ampere-turns. The cylindrical pole-pieces and the yoke have approximately the same cross-sectional area. The cross-section of the cylindrical pole-pieces can be made to any shape required for the particular purpose. The distance between the flat ends should be as small as the aperture of the deflected electron beam allows, so that the stray field at the edges is as small as possible in comparison with the proper field inside the gap.

If a deflexion magnet of the type of fig. 10.5 has to be designed for producing a certain density of flux B in the gap, the number of ampere-turns, nI, necessary to produce B can be estimated from the relation

$$V_H = nI = \oint H \, ds = \frac{Bg}{\mu_0} + \int \frac{B}{\mu} \, ds, \qquad (10.23)$$

where V_H, the magnetomotive force is given by the line integral of the magnetic field H about a closed path s. V_H also equals the current (nI) threading a surface which is bounded by this path. The closed line integral in (10.23) may be replaced by two terms, the magnetic potential difference across the air gap of width g, and a line integral along the path of the field lines inside the iron. Since the permeability μ in the iron is many thousand times larger than the permeability μ_0 in air, the second term can be neglected and gives

$$Bg = \mu_0 In, \qquad (10.24)$$

where B is measured in Wb m^{-2}, g in metres and I in amperes.

Various designs for electromagnets, based on the scheme of fig. 10.5, have been described in the literature. For instance, constructional details for a large model are found in a paper by Hudson (1949). His magnet dissipated 100 kW in two water-cooled layer-wound coils, and it produced up to 2 Wb m^{-2} in a gap of 3–5 cm between circular pole-pieces of 10 cm diameter. Overall dimensions of the whole magnet were $60 \times 30 \times 30$ cm. Details of a very powerful permanent magnet for the deflexion of electron momenta up to 20×10^{-3} Wb m^{-1} ($= 20$ kgauss $\times$ cm) have been published by Cockcroft, Ellis and Kershaw (1932). The magnet

has plane-parallel pole-faces 17×29 cm, spaced by a gap of
5·5 cm. Its permanent part consisted of two rectangular prisms
made up from straight steel laminations weighing 500 lb. It was
closed magnetically by a mild steel yoke of 1 500 lb weight. The
field could easily be set to any desired value by passing suitable
current through field coils which surrounded the permanent part
of the magnet. This design has proved very useful for the photographic exposure of a β-spectrum in a deflexion spectrometer
where great constancy of field was required over long times.
A permanent magnet of similar dimensions (0·6 Wb m^{-2} over
150 cm^2 in a 5 cm gap) has been described by Shutt and Whittemore (1951). This magnet is composed of straight, relatively short
permanent bar magnets made from Alnico, and from mild steel
respectively, and it has found useful application for deflexion of
fast electrons in a cloud chamber.

Relatively small deflexion magnets are required for cathode-ray
scanning in television tubes. In the early stages of development,
magnets of the type shown in fig. 10.5 were in use. However, since
relatively large gaps were required, the effects of the inhomogeneous
stray fields could not be avoided, and the deflexion errors, which
will be discussed in §10.7, became troublesome.

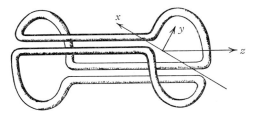

Fig. 10.6. Saddle coils with parallel conductors.

More satisfactory results have been obtained by the application
of 'saddle-coils' with parallel conductors (Knoll, 1939), as shown
in fig. 10.6. These coils can be placed over the glass neck of the
cathode-ray tube. In order to reduce the volume of the useless
flux, soft iron wire can be wound round the saddle coils, the
windings of the iron wire being co-axial with the undeflected
electron beam.

Equally good results are obtained by the use of elliptical saddle
coils (Marschall and Schröder, 1942)
shown in fig. 10.7. These coils are
particularly useful for slightly correcting the direction of an electron
beam in an experimental tube in the
laboratory. They are easily wound on
a cylinder of circular cross-section

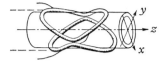

Fig. 10.7. Elliptical saddle coils.

and then pulled into shape over the neck of the experimental tube.

For the deflexion of the electron beam in high definition television tubes, toroidal coils are used with a non-uniform distribution
of turns, wound on a ferrite core. Such a coil is illustrated in

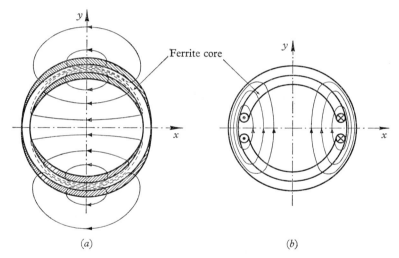

(a)   (b)

Fig. 10.8. Deflexion coils for television display tubes. (a) Toroidal coil with
distributed turns (for vertical deflexion); (b) cross-section perpendicular to axis
of saddle coil (for horizontal deflexion).

fig. 10.8(a); the shaded region corresponds to the coil windings,
which are most dense in the plane of deflexion. (The axis of the
undeflected beam is the z-axis which is perpendicular to the plane
of the paper in fig. 10.8.) The distribution of winding density is
arranged so as to introduce a slight pin-cushion inhomogeneity in
the resultant field, the flux lines of which, together with their
return paths through the core, are shown. The strength of the

deflexion field increases with increasing distance from the z-axis, a desirable property for the reduction of distortion (cf. Schlesinger, 1949).

In an advantageous arrangement, toroidal coils are used for vertical deflexion and saddle coils, an early example of which we have already shown in fig. 10.6, are used for horizontal (line) deflexion. The ferrite core of the toroidal coil encloses the parallel conductors of the saddle coils, thereby reducing the volume of useless flux associated with the saddle coils. Fig. 10.8(b) indicates the positioning of the saddle coil relative to the ferrite core, the toroidal windings being omitted for clarity. By use of distributed windings in the saddle coil, in which the highest turns density is found near to the x-axis, the desired degree of pin-cushion inhomogeneity is again introduced into the field. The whole deflexion system can thus be assembled as a compact unit with a single core. It is of course necessary to insulate the two coils from one another; this can be done using thin moulded plastic formers.

Of some technical interest is the power consumption of cathode-ray scanning fields in television; it is relatively much larger for a magnetic field than for the corresponding electrostatic field. A comparison of the forces exerted on the electron in magnetic and in electrostatic fields of equal energy density

$$\left(\frac{B^2}{2\mu} = \frac{\epsilon E^2}{2} \right)$$

yields
$$\frac{F_B}{F_E} = \frac{euB}{eE} = \frac{u}{c}, \tag{10.25}$$

where the ratio (u/c) of electron velocity to light velocity in television tubes is of the order of $\frac{1}{8}$. Hence a more powerful scanning oscillator will be needed to produce a given deflexion by magnetic fields than by electrostatic fields. This difference is increased by the fact that deflecting plates are put inside the vacuum envelope, while magnetic deflecting coils have to be arranged outside it and hence tend to produce stray fields and stored energy over a much greater volume that that occupied by the beam.

In spite of the greater power requirements, however, it is normal to use magnetic deflexion in television display tubes. This is because the required maximum deflexion voltage would be of the same order as the final anode voltage (e.g. 15 kV), and sawtooth voltage wave-forms of this amplitude are not easy to produce. Magnetic deflexion usually requires sawtooth currents with maximum values of about one ampere and these can be obtained quite easily. Magnetic deflexion is also superior to electrostatic deflexion from the point of view of spot distortion (cf. § 10.7). Electrostatic deflexion is preferred for oscilloscope tubes since virtually no current is drawn by the deflecting plates, and the input impedance of the device is therefore high. Moreover, the spot displacement on the screen is practically independent of signal frequency, providing this is not too high. In contrast, magnetic deflexion is influenced by the self-inductance of the coils, and is highly dependent on current and frequency, a fact that must be allowed for in the design of scanning circuits.

§ 10.4. Focusing deflexion in electrostatic fields

It can be shown that every deflecting field acts as a combination of a prism with a line-focus lens. For the following obvious reason every electrostatic field should have focusing properties; different rays of an electron beam, travelling at some angle to the field lines, will be at a given time at different potentials. Hence these rays will have different velocities and, according to (10.6), they will be subject to different deflexion. Thus rays of a homogeneous electron beam of finite aperture which enter a field E_y parallel to the z-axis will ultimately form a line-focus parallel to the x-axis. According to Recknagel (1938), the focusing power of a short field is given by

$$\frac{1}{f} = \left(\frac{1}{2V_{el}^2}\right) \int_{-\infty}^{+\infty} E_y^2 \, dz, \qquad (10.26)$$

where V_{el} is the initial energy of the electrons. For the 'chopped-off' homogeneous deflecting field of § 10.1, (10.26) yields

$$\frac{1}{f} = \left(\frac{V_p}{V_{el}}\right)^2 \frac{z_p}{4y_p^2} = \frac{\tan^2 \phi_p}{z_p}, \qquad (10.27)$$

where all the designations are those introduced in § 10.1.

According to Recknagel (1938) the focusing effect of the stray fields at the end of the parallel plate condenser may be taken into account by substituting the 'equivalent length' $(z_p + \Delta_1 z_p)$ for the length z_p of the homogeneous deflecting field, where

$$\Delta_1 z_p = 2y_p/\pi.$$

This correction $(\Delta_1 z_p)$ for focusing is usually smaller than the corresponding stray field correction (Δz_p) for deflexion introduced in § 10.2 (cf. (10.16)).

Focal lengths of weak electrostatic deflecting fields, as used in cathode-ray tubes, are always large in comparison with the dimensions of the deflecting plates. For example, for plate length $z_p = 2$ cm and for a deflexion angle $\phi = 0 \cdot 2$ radian the focal length is found to be as large as $f = 25$ cm.

On the other hand, strong fields resulting in large deflexions have very much shorter focal lengths. Yarnold and Bolton (1949), for instance, have used the homogeneous electrostatic field as a focusing velocity separator, deflecting the rays through a right angle. This separator is essentially a plane-parallel plate condenser, having two parallel slots representing entrance and exit aperture respectively, both cut in the positive plate at some given mutual distance z_f. The electron beam crosses this plate through the entrance slot at a slope angle α, it is then retarded by the deflecting field and bent back to the plate describing a parabolic trajectory just like the trajectory of a projectile in the earth's gravitational field. The distance at which the beam is returned to the plate is given by

$$z_f = 2V_{el}\frac{y_p}{V_p} \sin 2\alpha, \qquad (10.28)$$

where, as above, $2y_p$ and V_p are separation and potential difference respectively of the condenser plates, and V_{el} is the initial electron energy. Differentiation of (10.28) shows that slight changes $\pm \Delta \alpha$ in slope angle α have a relatively small influence upon the range z_f, if the deflexion angle $\phi = \frac{1}{2}\pi$. In this case, the slope angle is $\alpha = \frac{1}{4}\pi$, and a bundle of rays leaving the entrance slot with a given small divergence $\mp \Theta$ will be focused through an exit slot of sufficient width. The homogeneous field can be used for sorting electron velocities if certain electron energies V_{el} are selected for transmission by proper adjustment of the plate voltage V_p. An electron beam spectrometer of this kind has been described by Lassettre *et al.* (1964) and used for measuring energy spectra of 200–600 eV electrons. The theoretical resolution of the system, considering the principal rays only, is, according to Harrower (1955), given by the widths of entrance and exit slits divided by their mutual distance which, for the above spectrometer was 1/2500. The method, however, suffers from two disadvantages. First, the defining slots have to be arranged at the boundary of a strong field; thus unwanted electron lenses are set up with relatively large aberrations (cf. §6.10). Secondly, the electrons are slowed down appreciably while they are approaching the vertex of their parabolic path. This would involve the danger of space-charge disturbances for all beams of high intensity (cf. §8.2).

Both these disadvantages can be avoided to a large extent if the electrons are deflected and focused while they are moving approximately along equipotential lines. In this way, a frequently applied deflexion method uses the radial field of a cylinder condenser confined within a quadrant and deflecting the rays by an angle $\phi = \frac{1}{2}\pi$. For example, Löhner (1930) has given constructional details of a small energy selector for slow electrons. Aperture slots were arranged at the entrance and at the exit between the condenser plates, and any desired voltage of electrons was selected with an accuracy of $\pm$ 10 per cent from an initially continuous energy distribution.

Another model for a cylindrical electrostatic 90° deflector, described by Allison, Frankel *et al.* (1949), has been used for the sorting of very fast electrons. A voltage difference of 50 kV was applied to cylinder plates of 15 cm mean radius spaced by a mutual distance of only 0·5 cm. The width of the selected energy range depends on the ratio of the selecting slot width to the mean curvature of the cylinder condenser. In the described model, electrons of 0·75 MeV have been selected within a range of a fraction of 1 per cent.

The full advantages of the cylinder condenser, however, are realized only if the electrons are deflected through an arc of $\pi/\sqrt{2} = 127\cdot28°$, since only after such a deflexion is a line object projected into a line image. This has been shown analytically by

Hughes and Rojanski (1929). The discussion of the focusing properties of the cylinder condenser which is presented here is due to Brüche and Scherzer (1934).

Let the electric field half-way between the condenser plates be adjusted to a value E_0 to keep the principal ray on a circle of radius r_0. By balancing in this way the centrifugal force of the electrons on their circular orbit one obtains

$$mr_0\omega_0^2 = eE_0, \qquad (10.29)$$

where ω is the angular velocity of the electron, and the subscripts o refer to the circular orbit. The equation of motion for the radial component of an electron which does not move exactly on the circular orbit is

$$m\frac{\mathrm{d}^2r}{\mathrm{d}t^2} = mr\omega^2 - eE. \qquad (10.30)$$

For this orbit which deviates but little from the circular shape, the local field is given by

$$E = E_0\frac{r_0}{r} = E_0\frac{r_0}{r_0 + \Delta r} = E_0\left(1 - \frac{\Delta r}{r_0}\right). \qquad (10.31)$$

This value of E may be substituted into (10.30). Moreover, the law of conservation of angular momentum, $mr^2\omega = \text{const.} \approx mr_0^2\omega_0$ may be used to eliminate ω. Hence (10.30) yields

$$\frac{\mathrm{d}^2r}{\mathrm{d}t^2} = \omega_0^2\frac{r_0^4}{r^3} - \frac{e}{m}E_0\left(1 - \frac{\Delta r}{r_0}\right). \qquad (10.32)$$

Substituting $r = r_0 + \Delta r$ and neglecting higher terms, (10.32) becomes

$$\frac{\mathrm{d}^2(\Delta r)}{\mathrm{d}t^2} = r_0\omega_0^2 - 3\omega_0^2\Delta r - \frac{e}{m}E_0 + \frac{e}{m}E_0\frac{\Delta r}{r_0},$$

which reduces by substitution of (10.29) to

$$\frac{\mathrm{d}^2}{\mathrm{d}t^2}(\Delta r) + 2\omega_0^2\Delta r = 0. \qquad (10.33)$$

This is the differential equation of a simple harmonic motion with the solution

$$\Delta r = C\sin[\omega_0\sqrt{2}(t - t_0)]. \qquad (10.34)$$

The angular frequency of the simple harmonic motion is $\omega_0\sqrt{2}$. In a time interval of 2π sec, in which a principal electron ray completes ω_0 circles, the 'marginal' electrons oscillate about the circular orbit $\omega_0\sqrt{2}$ times, intersecting the circle twice in each period of oscillation. Thus the points of intersection are spaced by

$$\phi = \pi/\sqrt{2} = 127° \, 17'. \qquad (10.35)$$

Fig. 10.9 shows an application of this electrostatic focusing method in a velocity separator which has been used by Hughes and MacMillan (1929) for verifying experimentally the theoretical

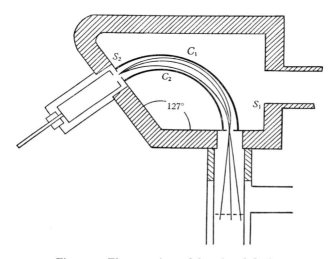

Fig. 10.9. Electrostatic 127° focusing deflexion.

value of (10.35) for the focusing deflexion angle. Electron rays diverging up to 12° enter the velocity analyser through a narrow slot S_1 and are focused through a narrow exit slot S_2. The voltage difference $2V_p$ which was applied to the plates C_1 and C_2 of the cylinder condenser in order to pass electrons of the velocity V_{el} through the device is given by

$$V_p = V_{el} \log_e \frac{r_1}{r_2}, \qquad (10.36)$$

where r_1 and r_2 are the radii of curvature of the outer and of the inner cylinder plates C_1 and C_2 respectively. For instance, radii $r_1 = 60$ mm, $r_2 = 50$ mm, slot widths $S_1 = 0\cdot3$ mm, $S_2 = 1$ mm, and slot lengths of 10 mm were used by the above authors. The

deflexion voltage was $2V_p = 0.3650V_{el}$ and rays of 99·1 and
100·7 eV could still be separated. The radius of the principal
trajectory of electrons travelling on the potential V_{el} is given by
$r_e = \sqrt{r_1 r_2}$.

127°-condensers have been very widely used both as mono-
chromators and as spectrometers for electron beams. We mention
here two examples only which show features of special interest.
A spectrometer due to Kunz (1962), shown here in fig. 10.10, was

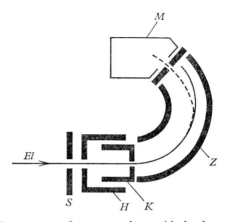

Fig. 10.10. 127°-energy analyser with decelerator.

constructed for analysing the energies of a very fine beam El of
about 50 keV entering through the slit S. In the retarding field
between the electrodes H and K, the beam was decelerated to
about 300 eV, and then deflected by the focusing field between the
cylindrical condenser plates Z. The deflecting voltage at Z was
adjusted to a fixed value so as to pass electrons of a selected
energy eV_0 into the electron multiplier M. The spectrum was
obtained by varying the retarding field, i.e. by adjusting the voltage
of the electrode K. The very high resolution of $\Delta V \leqslant 0.5$ eV of
the 50 keV beam was here obtained by the frequently used artifice
of deflexion of the decelerated beam. The resolving power of the
relatively small condenser itself was of the order of $V_0/\Delta V \simeq 10^3$
only.

The other 127°-condenser to be mentioned here was constructed
by Kerwin and Marmet (1960) in order to select beams of the

order of 0·5 eV only (cf. §9.10). It consisted of two solid cylindrical electrodes with internal shields of 90 per cent transparent wire gauze. The relatively dense space-charge of the very slow electrons in this monochromator was found to block the way for the beam unless the gauze shields were introduced through which stray electrons could leave the deflexion space.

One of the best analysers for non-relativistic electron energies is the field between two concentric spheres. In this spherical condenser, the electron orbits are Kepler ellipses, since the field varies according to an inverse square law. Focusing takes place after a deflexion of 180°. The spherical condenser has a very high collecting efficiency as focusing occurs in two mutually perpendicular directions. The voltage difference V_p applied to the two spherical plates of radii r_1 and r_2 to focus electrons of the energy eV_{el} is

$$V_p = \left(\frac{r_2}{r_1} - \frac{r_1}{r_2}\right) V_{el}. \qquad (10.37)$$

The image of a circular entrance aperture is projected onto a circular exit aperture. At electron energies well below 1 MeV, a spherical condenser of, say, $r_e = (r_1 + r_2)/2 = 7$ cm can have a resolution of about 1 per cent with a collecting efficiency of about 5 per cent. However, in such a condenser with 0·1 mm apertures, a resolution of 0·05 eV can, according to Lohff (1963), be obtained for 40 keV electrons if these are decelerated to 65 eV before entering the condenser. If, e.g. in β-ray spectroscopy, deceleration is not feasible, high resolving power is obtained by using condensers of very large radius. In this way, Browne, Craig and Williamson (1951) obtained with an orbital radius of 46 cm a resolution $p/\Delta p = 650$ with a collecting efficiency $\Omega = 3 \times 10^{-3}$ for 1 mm diameter apertures. It is, however, essential that the shape of the condenser plates is exactly spherical and concentric, a construction which requires refined workshop facilities. Purcell (1938) who was first to use the spherical condenser as an energy analyser has also investigated theoretically its focusing properties. He could estimate the relativistic effects by regarding the inverse square field of the condenser as a scaled up atom in which relativistic effects produce the well calculated Sommerfeld precession of the orbits around the appropriate foci. Since these foci for all, but for the central ray are not located at the centre of the orbit, the shift caused by the precession becomes rather large, and the resolving power is much reduced even though the orbits remain very nearly circular. The accurate analysis of the relativistic effect, however, is highly complicated and we have to refer to the extensive treatment by Rogers (1946, 1951) and Ashby (1958).

To obtain focusing deflexion, there is no need for the electrons to travel entirely in the deflecting field. Source and focus may lie outside given field boundaries. For instance, a focusing field sector (cylindrical or spherical condenser) may be bounded by two radial planes enclosing a given central angle ϕ, thus forming the so-called electrostatic prism. Electrons coming from an outside source travel along straight lines until they meet the sector boundary at right angles to the principal ray. Then they travel along curved orbits through the field. After leaving the field they again travel along straight lines until eventually they reach the focus. Herzog (1934, 1935) was the first to point out that lens properties can be ascribed to a field sector, and he derived relatively simple expressions for the focal length of the sector field between co-axial cylinder plates.

Besides the spherical condenser fields, there is only one electrostatic deflector which is able to project a faithful image. This is the cylindrical mirror analyser, shown in fig. 10.11. It consists of two co-axial cylinders T_1 and T_2 of radii r_1 and r_2 and at voltages V_1 and V_2 respectively. The electron trajectory passing through the axis at J is reflected by the electrostatic field and passes through the axis again at J'. Zashkavara, Korsunski and Kosmachev (1966) calculated that an electron bundle projected onto the axis at an angle of $\alpha = 42 \cdot 3°$ is reprojected onto the axis forming a faithful image in the radial plane at J' from an electron-optical object at J.

There, the voltage ratio at the cylinders for an incident electron energy eV_1 is connected with the ratio of the cylinder radii by

$$V_1/(V_2-V_1) = 1 \cdot 3 \ln r_2/r_1, \qquad (10.38)$$

the distance at which refocusing occurs is $JJ' = 6 \cdot 1r_1$, and the maximum elongation of the trajectory from the axis is $r_{max} = 1 \cdot 8r_1$. A fairly large angular spread, say $2\gamma \simeq 10°$ can be accepted without

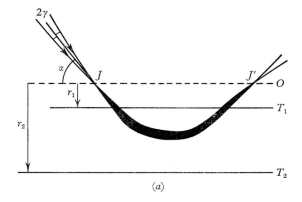

(a)

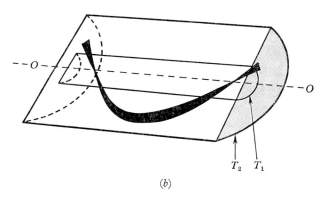

(b)

Fig. 10.11. Cylindrical mirror analyser. (a) Schematic;
(b) perspective drawing.

degrading the resolution beyond 1 per cent. Compared with a 180°
spherical condenser of the same object to image distance, the
dispersion of the mirror analyser is only very slightly smaller, but
it can accept a larger beam aperture without disturbance by aberrations. Hafner, Simpson and Kuyatt (1968) compared the performance of the cylindrical mirror analyser with that of the 127°
spherical condenser with the result that the ratio of dispersion to
aberration constant of the spherical condenser is slightly larger.
Kuyatt and Simpson (1967) used the mirror analyser as a monochromator for obtaining a narrow energy band at very low voltages,

they obtained at 1 eV a band of 0·007 eV half-width with 2×10^{-9} amp output.

Very suitable for the treatment of focusing deflexion problems is the introduction of curvilinear coordinates. The principal ray of the deflected bundle is chosen as the curvilinear axis (ζ). η- and ξ-ordinates are plotted along the normal and the binormal respectively of the principal ray. For example, in the simple case of focusing deflexion in the cylinder condenser, the ζ-axis is a circle half-way between the plates, the ξ-coordinates are parallel to the cylinder axis and the η-coordinates extend in radial direction. By transforming the curvilinear (ξ, η, ζ)-coordinates into a Cartesian system ($\bar{\xi}$, $\bar{\eta}$, $\bar{\zeta}$), the principal orbit of the deflected electron bundle appears to be stretched out into a straight ray serving as the $\bar{\zeta}$-axis, and the marginal orbits appear to be focused towards the $\bar{\zeta}$-axis just like the rays in an ordinary electron lens. For example, after transformation into the Cartesian system, the one-directionally focusing cylinder condenser would appear like an ordinary line-focus lens. On the other hand, the two-directionally focusing spherical condenser would appear like an ordinary lens of circular symmetry converging a homocentric beam into a point focus. In every case, the $\bar{\zeta}$-axis connects the centre of an object space with the centre of an image space. In the 'stretched-out' ($\bar{\xi}$, $\bar{\eta}$, $\bar{\zeta}$) system the laws of Gaussian dioptrics can be applied to determine the image point for any given paraxial object point. Hence in the 'stretched-out' system focal points and principal points ($\bar{F}$, $\bar{F}'$, $\bar{P}$, $\bar{P}'$) can be located in the customary way (cf. §2.2). If the transformation is reversed back into the curvilinear (ξ, η, ζ) system, the location of the cardinal points (F, P, etc.) is considered to determine the characteristic optical properties of the original deflecting field.

Wendt (1943) found that the electron orbits in a focusing deflecting field can be expressed in the above curvilinear coordinate system by relatively simple equations if the principal ray representing the curvilinear axis coincides with an equipotential line. In this case, the potential field about the ζ-axis which can be derived from Laplace's equation can be developed as a power series, and the equations of motions are obtained from Fermat's principle. Marschall (1944) and Hachenberg (1948) presented these equations in terms of the two radii of curvature r and ρ of the equipotential surface half-way between the condenser plates. They obtained the following pair of equations for an electron orbit:

$$\left.\begin{array}{l} \dfrac{d^2\xi}{d\zeta^2} + \dfrac{1}{r\rho}\,\xi = 0, \\[3mm] \dfrac{d^2\eta}{d\zeta^2} + \left(\dfrac{2}{r^2} - \dfrac{1}{r\rho}\right)\eta = 0. \end{array}\right\} \tag{10.39}$$

For example, for the cylinder condenser one radius is infinite ($\rho = \infty$), hence (10.39) are reduced to the following simple form

$$\left.\begin{array}{l} \dfrac{d^2\xi}{d\zeta^2} = 0 \\[3mm] \dfrac{d^2\eta}{d\zeta^2} + \dfrac{2\eta}{r^2} = 0, \end{array}\right\} \tag{10.40}$$

and

which by integration yield

$$\xi = A\zeta + B,$$

$$\eta = C \sin\left(\frac{\sqrt{2}}{r}\,\zeta\right) + D \cos\left(\frac{\sqrt{2}}{r}\,\zeta\right). \tag{10.41}$$

This result confirms that an electron beam in the cylinder condenser is not influenced in the ξ-direction but only in the η-direction, i.e. in the direction of its deflexion (one-directional focusing). The solution for η is periodical, and the rays starting from a point on the ζ-axis return to the ζ-axis after the argument

$$\left(\frac{\sqrt{2}}{r}\,\zeta\right) = \pi \tag{10.42}$$

of the sine or the cosine function respectively. Hence, the points of intersection are spaced by a distance ζ given by

$$\phi_0 = \zeta/r = \pi/\sqrt{2}$$

in agreement with (10.35), where ϕ_0 is the angle of deflexion. Moreover, if (10.41) are applied to a cylinder condenser of finite length subtending the arbitrary centre angle ϕ, the following expression can be derived for the focal length of this sector:

$$f = \frac{r}{\sqrt{2}}\,\frac{1}{\sin(\phi\sqrt{2})}. \tag{10.43}$$

Hence f is found to be a function of the geometry of the condenser.

The electron-optical properties of the spherical condenser can be derived from (10.39) by introducing $\rho = r$. There periodical solutions result both for ξ and η, i.e. two-directional focusing is obtained.

Quite generally, the focusing properties of all double-curved condensers can be derived from (10.39). Positive coefficients of both ξ and η in (10.39) which are obtained, either with $r > 0$ and $(\frac{1}{2}r) < \rho < \infty$, or with $r < 0$ and $(\frac{1}{2}r) > \rho > -\infty$, lead to periodical solutions composed of the sum of two trigonometric functions. Hence, the corresponding field not only deflects but also converges the bundle of rays. On the other hand, negative coefficients of ξ or of η in (10.39) lead to solutions with hyperbolic functions; consequently the distance of a ray from the ζ-axis should increase all the time. Hence the corresponding field diverges the deflected rays in one direction. In particular, systems with a relatively small positive ρ within the boundaries $0 < \rho < \frac{1}{2}r$ still converge the rays in the ξ-direction, but they diverge them in the η-direction. Examples of this class are given by some toroidal ring condensers; there ρ is small in comparison with the radius r of the electron orbit. On the other hand, all systems with negative ρ have diverging action in the ξ-direction but converging action in the η-direction.

If, no geometrical similarity is required, point-by-point projection can still be obtained by any two-directional focusing deflexion. Such a point projection is achieved sometimes after a relatively long orbit subtending a large centre angle since the rays generally first pass through an intermediate line-shaped cross-over before they are focused eventually into a point.

A great variety of deflecting fields (amongst them some with diverging action) can be employed in combination with electron lenses of circular symmetry. Again, some of these combinations lead to a final point-by-point projection via an astigmatic intermediate image. For example, rays diverging from an object point may be collected by an electron lens in order to enter the deflecting field as a convergent bundle. An astigmatic intermediate image is formed in the deflecting field, while a second lens arranged in the field-free space after the deflexion is used for projecting the image point on to a distant target.

§10.5. Focusing deflexion in magnetic fields

The most frequently used method for electron spectrometry has been the well-known method of semicircular focusing deflexion in the homogeneous magnetic field. Magnetic deflexion in a semi-circle has been utilized for a long time. We only mention the experimental investigations by Meyer and Schweidler (1899), by Classen (1908) and by Danysz (1911). However, Rutherford, Robinson and Rawlinson (1914) first fully realized the focusing effects of the semicircular deflexion. In fig. 10.12, J represents

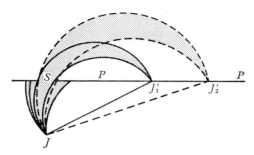

Fig. 10.12. Focusing deflexion in a homogeneous magnetic field.

a linear source of electrons which extends at right angles to the plane of the drawing. S is the intensity stop. The foci J'_1, J'_2, ... of electron beams of various velocities all lie on a straight line along which may be placed a photographic plate or a fluorescent screen PP. Let the distances of the images J' from the centre of the slot S be $SJ' = s$. If the plate PP is arranged at right angles to $JS = c$, the hypotenuses JJ'_1, JJ'_2, etc., are the diameters of the semicircular orbits given by

$$(2r_e)^2 = s^2 + c^2, \qquad (10.44)$$

from which the electron momentum $(Br)_e$ of the rays focused into a line at J' can be determined if the flux density B of the homogeneous magnetic deflecting field is known.

The need for focusing deflexion of particles with relatively large momenta by magnetic fields of limited flux density which originally arose in atomic mass spectroscopy, led to the investigation of the properties of bounded fields. Barber (1933) and Stephens (1934) recognized the focusing properties of a homogeneous magnetic field confined within a triangular cross-section in a plane at right angles to the field lines. Such a field, which may be called a 'magnetic prism', is produced—to some approximation—between closely spaced pole-pieces of triangular cross-section. The focusing deflexion by such a prism is shown in fig. 10.13 (*a*). The prism is enclosed between the two planes S and S' intersecting at the angle ϕ and extending perpendicularly to the plane of the drawing.

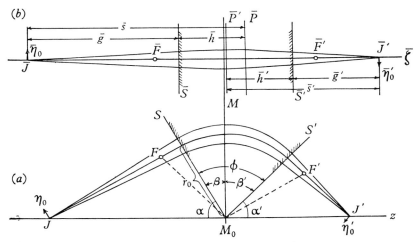

Fig. 10.13. (*a*), (*b*) Image formation by a homogeneous magnetic sector field.

A narrow wedge-shaped electron beam is emitted from a line-shaped source or from a slot J which extends parallel to the field lines but is situated in the field-free space. The chief ray of this beam enters and leaves the field at right angles to the boundary planes. All rays travel through the field along relatively short circular arcs. The focusing effect of the prism is apparent from the drawing which shows that the outer rays stay longer in the field than the inner rays. Barber calculated that rays of equal velocity are focused into a line image at J', so that J, J' and the apex M_0 of the prism lie on a straight line (the z-axis in fig. 10.13 (*a*)). The angle ϕ enclosed between the field boundaries is also the angle of deflexion of the beam, and this angle ϕ may have any value up to

$180°$. For $\phi = \pi$ the focusing deflexion is again semicircular. Another interesting case arises if $\phi = \frac{1}{2}\pi$. There, electrons starting from an x-line focus situated at one of the field boundaries will leave the other boundary as a parallel bundle.

There are obvious advantages in deriving electron optical properties of the focusing deflexion in homogeneous magnetic fields with the aid of curvilinear coordinates. An elementary treatment due to Walcher (1949) may be followed up here. As

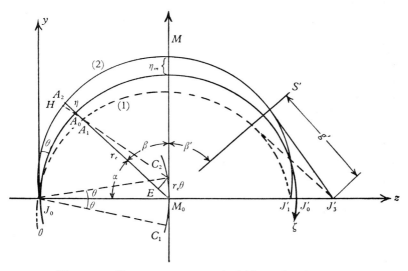

Fig. 10.14. Homogeneous magnetic field as a line-focus lens in a curvilinear coordinate system.

shown in fig. 10.14, a bundle of electrons with homogeneous velocity may be characterized by three circular orbits. The principal orbit crosses the z-axis at right angles at the object point J_0 and again at the image point J_0'. The two 'zonal' orbits (1) and (2) each make a small angle Θ with the principal orbit while passing through J_0. All orbits are strictly perpendicular to the magnetic flux which crosses the plane of the drawing at right angles.

The loci for the centres of the various circular orbits are on a circle of radius r_e about J_0. Now, draw a radius $r_e = J_0 C_2$ of this circle to make the angle Θ with the z-axis. The arc $C_2 M_0$ of this circle (M_0 is the centre of the principal orbit) has the length $r_e \Theta$.

If the perpendicular C_2E is dropped from C_2 on a radius vector M_0A_0 of the principal orbit, it will cut a length $M_0E \approx r_e\Theta \sin \alpha$ from the radius vector, where α is the angle between the radius vector M_0A_0 and the z-axis. On the other hand, the distance $\eta = A_0A_2$ between the principal orbit and the zonal orbit (2) (measured along the radius vector M_0A_0) equals very approximately M_0E, since the radius C_2H of orbit (2) equals the radius M_0A_0 of the normal orbit. Hence, the distance between the two orbits is

$$\eta = r_e\Theta \sin \alpha = r_e\Theta \sin \frac{\zeta}{r_e}, \qquad (10.45)$$

where ζ is the length of the arc of the principal orbit between J_0 and A_0. If the values of the curvilinear coordinates (η, ζ) of (10.45) are plotted in a Cartesian system $(\bar{\eta}, \bar{\zeta})$, the zonal rays would appear as simple sine curves, and the principal orbit would coincide with a straight optical axis $(\bar{\zeta})$ of the refracting system.

Returning to the curvilinear system it is clear that the distance η of the zonal orbit from the ζ-axis reaches a maximum η_m at a 'mid-plane' M_0M which intersects the z-axis at M_0. Since all orbits are parallel at the mid-plane, the arc length ζ between J_0 and M represents the focal length f of an optical system. This system is the homogeneous magnetic field enclosed within a rectangular sector between the mid-plane M_0M and the xz-plane.

It may be shown now that any sector field enclosed between two planes intersecting at the vertex M_0 and being perpendicular to the plane of the drawing of fig. 10.13 can act as a line-focus lens. In particular, it can be shown that a field sector is able to project an electron-optical image near the z-axis for any object near the z-axis. Let the homogeneous field in fig. 10.14 be terminated by any sector plane S'. After passing this boundary the rays will travel along straight lines through the field-free space, and they will cross over at J'_3 in the zx-plane. Now, the semi-aperture η' of a zonal ray at the sector plane S' may be represented as a function of the angle β' between mid-plane M and sector plane S'. For this purpose, $\alpha = (\beta' + \frac{1}{2}\pi)$ may be substituted into (10.45). Using η_m for the maximum η-value at $\alpha = \frac{1}{2}\pi$ (10.45) yields

$$\eta' = \eta_m \sin \frac{\zeta}{r_e} = \eta_m \cos \beta'. \qquad (10.46)$$

The slope of the zonal ray with respect to the principal ray when leaving the S'-plane is obtained by differentiating (10.46), namely,

$$-\tan \Theta' = \left(\frac{d\eta'}{d\zeta}\right)_{r_e, \beta'} = \left(\frac{\eta_m}{r_e}\right) \cos \left(\frac{\zeta}{r_e}\right) = \left(\frac{\eta_m}{r_e}\right) (-\sin \beta'). \tag{10.47}$$

On the other hand, it is evident from fig. 10.14 that the angular semi-aperture Θ' is also given by the relation

$$\tan \Theta' = \eta'/g', \tag{10.48}$$

where g' is the distance of the image J'_3 from the S'-plane. Substituting for η' from (10.46) and for $\tan \Theta'$ from (10.47), the following relation results from (10.48)

$$g' = r_e \cot \beta', \tag{10.49}$$

i.e. the distance of the focus J'_3 from the sector plane S' is independent of the beam aperture (η' or Θ'), and it depends only on the sector angle β' and the electron momentum given by r_e.

In Gaussian optics the focal length f' is defined (cf. §2.2) by

$$f' = \eta_m/\tan \Theta', \tag{10.50}$$

and by substituting for $\tan \Theta'$ from (10.47) one obtains

$$f' = r_e/\sin \beta' \tag{10.51}$$

The position of the principal plane P' is defined by this value of f'. Hence the distance of P' from the sector boundary S' is obtained by subtraction of (10.49) from (10.51), namely,

$$h' = f' - g' = r_e \tan \tfrac{1}{2}\beta'. \tag{10.52}$$

In a similar way focus and principal plane can be determined for parallel rays moving through the mid-plane in the opposite direction in a sector field terminated by a plane S making an angle β with the mid-plane M.

Now, formulae (10.49)–(10.52) which apply to a field sector with the vertical angle β' between the mid-plane M and the boundary S' will also apply to a sector with the vertical angle ϕ between the boundaries S and S'. In this sector the angular position of the mid-plane where all orbits are parallel and where $\eta = \eta_m$ is defined by the position of the object J, the straight line joining object and sector vertex being the z-axis, which has to be normal to the mid-plane. If $s = JP$ and $s' = J'P'$ are the distances of object J and image J' from their respective principal planes P and P', all measured along the principal ray representing the ζ-axis of the curvilinear system, then the well-known lens equation (cf. (2.4)) $(f/s) + (f'/s') = 1$ can be applied for finding the position of the image J'.

The analogy to ordinary lens optics, however, goes still further. If the object extends perpendicularly to the ζ-axis along the η-coordinate to a height η_0, the height of the image η_0' is determined by the usual relation for the magnification, viz. $\eta_0'/\eta_0 = s'/s$. All these relationships are well illustrated if the curvilinear system (ζ, η) is transformed into the Cartesian system $(\bar{\zeta}, \bar{\eta})$ as shown in fig. 10.13(b).

We have here confined our discussion to perpendicular incidence of the principal ray on the boundary of the sectorial field. For the more complicated case of an oblique entry and exit, we have to refer to a paper by Cartan (1937) where for this case a simple graphical construction for the location of the electron optical image is given. An application of the magnetic sector field can be seen in fig. 4.14 of §4.7. These fields have, however, scarcely been applied in electron spectrometry, but they are of much practical interest in the design of mass spectrometers.

Besides the magnetic field sectors, a great variety of bounded homogeneous fields is suitable for an electron-optical line-focus projection. For instance, homogeneous magnetic fields bounded by a circular cylindrical surface, the flux being parallel to the cylinder axis, have been used, for example, in Aston's mass spectrograph and in an electron spectrometer by Klemperer and Shepherd (1963). Such fields can be produced to some approximation, for instance, between closely spaced co-axial cylindrical pole-pieces of circular cross-section (cf. for instance the magnet shown in fig. 10.5). The focusing properties of these fields have been discussed by Siday (1947), by Jennings (1952) and by Ehrenberg and Jennings (1952). These authors found that a relatively sharp line focus parallel to the cylinder axis is formed by 'paraxial rays', i.e. by rays of an initial small (perpendicular) distance from the cylinder axis. The angle of deflexion ϕ and the focal length f in the yz-plane (perpendicular to the cylinder axis) are given by

$$\tan \tfrac{1}{2}\phi = R/r_e \tag{10.53}$$

and
$$f = \frac{R^2 + r_e^2}{2R}, \tag{10.54}$$

where R is the radius of the circular cross-section of the field boundary and r_e is the radius of curvature of the electron orbit in the field as obtained from the electron momentum $(Br)_e$ of the deflected electron ray. The principal points may be taken to coincide with the centre of the circular field boundary. The ordinary lens formulae of §2.2 are found to apply to object and image distance and to magnification.

The focal length of the homogeneous field with circular boundaries decreases, according to (10.53) and (10.54), inversely with the angle of deflexion. This is shown by table 10.2.

TABLE 10.2. *Angle of deflexion ϕ, cylinder radius R and focal length f of circular magnetic prism*

f/R	13·0	1·95	1·22	1·00	0·82	0·54
ϕ	23°	60°	80°	90°	102°	146°

It can be seen that initially parallel rays are focused outside the field boundary if $\phi < 90°$, but inside if $\phi > 90°$.

If the chief ray of an incident bundle is initially not directed towards the centre of the circular homogeneous field, focal length and deflexion will differ from the values given by (10.53) and (10.54). For instance, a bundle travelling at the convex side of, and near to the central bundle will suffer greater deflexion and it will be subject to shorter focal length than the bundle travelling at the concave side of it, since the focusing and deflecting power of a homogeneous field increases with increasing path length of the beam in the field. Also, the principal planes of the field for the eccentric bundles no longer coincide with the field centre; they are now located at the feet of the perpendiculars from the field centre to the straight continuation of the rectilinear part of the incident and of the emergent chief ray respectively.

The ideal case of the homogenous field with circular boundaries can of course only be approximated experimentally. If the gap between the pole-pieces is not kept very small in comparison with their radius, the field in the gap becomes inhomogeneous and the effect of fringing fields can no longer be neglected. Details about the orbits in the mid-plane between two plane-faced, circular pole-pieces spaced by one radius have been given by Siday (1947). Electron orbits between the end-faces of two long co-axial solenoids at various mutual distances have been studied by Jennings (1952). We shall come back to these particular cases in the discussion of focusing errors in §10.8.

Of much practical importance are circular-symmetrical fields which, in their equatorial plane, are characterized by the relation

$$\frac{B}{B_0} = \left(\frac{r_0}{r}\right)^n, \tag{10.55}$$

where B_0 is the flux density at a distance $r = r_0$ from the axis and where n is the 'fall-off index' which indicates how rapidly the field decays with increasing distance from the axis. With $n = 3$, for instance, (10.55) describes in the equatorial plane the rapidly decaying field of a magnetic dipole (Henneberg, 1934b), and the study of electron orbits in this dipole field has played an important part in the investigation of the aurora borealis (Störmer, 1933), of the van Allen belt (cf. §5.9), and in cosmic-ray research.

Slowly decaying fields of an index $n \approx \frac{1}{2}$ show important focusing properties which have been utilized for the stabilization of electron orbits in electron accelerators. In particular, the 'two-directional' focusing by such fields has found application in a beta-ray spectrometer, due to Siegbahn and Swartholm (1946a, b).

Two-directional deflexion-focusing may be explained here by considering the free oscillations of electron orbits about a line of constant field intensity. Such oscillations are well known from the study of orbital stability conditions established by Kerst and Serber (1941) for their betatron accelerator which will be described later in this section. Equilibrium between the centrifugal force and the constraint by the magnetic field (cf. §5.1) will lead to a stable electron orbit provided the flux density B_1 along this orbit and its radius r_1 are connected by the relation

$$er_1\omega_c B_1 = mr_1\omega_c^2, \tag{10.56}$$

where ω_c is the orbital frequency (cf. (5.2)). If an electron, moving in an inhomogeneous field $B(r)$ of circular symmetry, is to be restored to an equilibrium orbit, when displaced radially, the magnetic force $er\omega_c B$ should become larger than the centrifugal force for larger radii, and for smaller radii the converse should hold. Hence it is required that

$$\frac{dB}{dr} = -\left(\frac{B}{r}\right) n, \qquad (10.57)$$

where n is again the fall-off index. (10.57) is the differential equivalent to (10.55). According to (10.57) the fall-off index should be smaller than unity ($n < 1$) if the field is expected to decrease more slowly than with $1/r$.

For all positive values of n the magnetic field deflects the electron trajectories towards the equatorial mid-plane. This can be shown in the following way: For homogeneous fields ($n = 0$) the boundaries of flux tubes are given by straight lines in the meridional planes. If the field, with increasing distance from the axis, decreases in the mid-plane, the cross-section of the flux tubes will increase correspondingly due to the principle of conservation of flux (cf. §3.1). Hence, the flux lines in the meridional planes will become increasingly concave towards the axis as their distance from the axis increases. By this curvature of flux lines, however, all electrons with velocity components in equatorial planes are accelerated towards the mid-plane, since the Lorentz force always acts at right angles to the momentum of the electron and to the tangent of the flux line.

It follows that the restoring forces will set up oscillations about a principal orbit if

$$0 < n < 1. \qquad (10.58)$$

Kerst and Serber (1941) calculated the angular frequencies of these radial and axial oscillations of the orbits to be

$$
\left.
\begin{aligned}
\omega_r &= \omega_c (1-n)^{\frac{1}{2}} \\
\omega_a &= \omega_c n^{\frac{1}{2}},
\end{aligned}
\right\} \qquad (10.59)
$$

and

respectively, where ω_c is again, as in (10.56), the orbital (cyclotron) frequency. It is clear that if $n = \frac{1}{2}$, the radial and axial frequencies are equal. Thus, electrons diverging from a point source in such a magnetic field will take up their oscillations and will pass through a second point on the equilibrium orbit after half an oscillation period. Since both the axial and the radial frequencies are by a factor $1/\sqrt{2}$ smaller than the orbital frequency ($= 2\pi$), it is clear that this second or 'image' point will be at an angular distance of $\pi\sqrt{2} = 254\cdot56°$ from the first object point.

Equation (10.55) can be expanded to the following form:

$$\frac{B}{B_0} = 1 - C_1\left(\frac{r-r_0}{r_0}\right) + C_2\left(\frac{r-r_0}{r}\right)^2. \qquad (10.60)$$

For instance, when $n = \frac{1}{2}$, then $C_1 = \frac{1}{2}$, $C_2 = \frac{3}{8}$. Shull and Dennison (1947) pointed out that if for purposes of practical design C_1 and C_2 are considered as adjustable constants, for high-quality focusing C_1 is found to be necessarily $= \frac{1}{2}$, but C_2 is free to take a certain range of values. $C_2 = \frac{1}{4}$, for instance, gives an image of minimum size; however, with the view to keeping the gap between the pole-pieces short so that the deleterious effects of fringing fields are avoided as far as possible, a practical value of $C_2 \approx \frac{3}{8}$ might be more convenient.

Two-directional focusing fields as described by (10.60) can be produced between specially shaped pole-pieces. For example, the deflexion magnet of Siegbahn and Swartholm (1946b) is shown in fig. 10.15. The annular magnet M, which for sake of clarity is drawn cut open, shows circular symmetry about the x-axis. It is energized by the current coil C. The deflexion field is produced in the ring-shaped outer gap between the pole-pieces P_1 and P_2. The profiles of these pole-pieces were ground step by step to an approximately parabolic shape until the optimum focusing was obtained. In fig. 10.15 a small bundle of rays emitted from the point source J is shown to be focused in the image point J'.

Accurate two-dimensional focusing fields have also been produced by iron-free coils, which can be designed according to exact calculations. The first spectrometer of this kind is due to Siegbahn and Edwardson (1956). A very large air cored spectrometer which was two-directionally focusing up to 4 MeV, with orbital radius $r_e = 1$ m and an inherent precision of one part in 10^5 has been constructed by Graham, Ewan and Geiger (1960), (cf. §10.9, table 10.3).

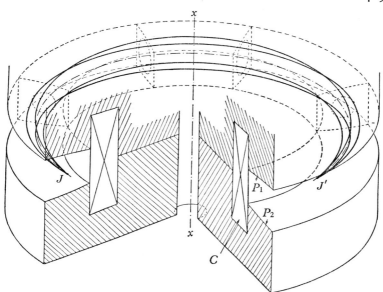

Fig. 10.15. Two-directional focusing deflexion.

A proper fall-off index is not the only requirement for the magnetic field of the 'Betatron' or 'Induction Accelerator', for which Kerst and Serber (1941) studied the orbital oscillations (cf. (10.56)). There are other electron optical conditions which must be fulfilled. In the betatron an electron beam moves, during all the time of the acceleration, in an orbit of constant radius r_e around an electro-magnet of circular cross-section. An alternating current produces a variable flux Φ in the magnet, linking the orbit and accelerating the electrons. In principle the betatron is a transformer in which the electron beam takes the place of the secondary winding. As in a transformer, the induced voltage per turn is $d\Phi/dt$, and the electric field (voltage per unit length) is given by

$$E = \frac{d\Phi}{dt} \frac{1}{2\pi r_e}. \tag{10.61}$$

The force on the electron is $-eE$ and the equation of motion can be written as

$$\frac{dp}{dt} = \frac{e}{2\pi r_e} \frac{d\Phi}{dt}. \tag{10.62}$$

On the other hand, by differentiation of $p/e = (rB)_e$ (cf. (5.1)) it follows that

$$\frac{\mathrm{d}p}{\mathrm{d}t} = er_e \frac{\mathrm{d}B}{\mathrm{d}t}.$$ (10.63)

The last two equations yield

$$2(\mathrm{d}B/\mathrm{d})\pi r_e^2 = \mathrm{d}\Phi/\mathrm{d}t.$$ (10.64)

(10.64) is often called the 'betatron $2:1$ rule', it shows that for accelerating the electron on an orbit of constant radius r_e the linking flux Φ must change at a rate twice that which would occur if the flux density inside the orbit were homogeneous and given by the value of B at the orbit.

It has been shown in the discussion of the focusing deflexion by a homogeneous magnetic field that a sector of the field is sufficient for the projection of a deflected electron image. The same can be shown to apply for axially symmetric fields of any fall-off index $n \geqslant 0$. In every case, object point, vertex of the field sector and image point lie on a straight line as in fig. 10.13(a). Moreover, (10.49), (10.51) and (10.52), which define the location of the cardinal points for a homogeneous field ($n = 0$), can be generalized to apply for all field indices $n \geqslant 0$. Generally the location of foci F, principal points P and images J' will be different for the focusing in radial η-direction and axial ξ-direction. General formulae derived by Judd (1950) giving the location of these points with respect to the field boundary (S) of a field sector with vertical angle ϕ (cf. fig. 10.13(a)) may be quoted as follows:

$$PF = f_\xi = \frac{r_e}{\sqrt{n}} \frac{1}{\sin(\phi\sqrt{n})},$$ (10.65)

$$SP = h_\xi = \frac{r_e}{\sqrt{n}} \tan\left(\frac{\phi}{2}\sqrt{n}\right),$$ (10.66)

$$SJ' = g_\xi = \frac{r_e}{\sqrt{n}} \tan\alpha.$$ (10.67)

These formulae apply for the focusing in the axial (ξ)-direction. For the focusing in the radial (η)-direction, the corresponding quantities, f_η, h_η, g_η are obtained by replacing $\sqrt{n}$ by $\sqrt{(1-n)}$. Radial and axial focus again coincide for $n = \frac{1}{2}$. As another special case, one-directional focusing occurs for $n = 0$, where the expressions for g_η, h_η and f_η coincide with (10.49), (10.51) and (10.52) respectively.

The technical problem of shaping the pole-pieces of a magnet to produce an axially symmetrical field sector of a given field index is not easy. However, if only approximate focusing is required, the proper shape of pole-pieces may be approximated by conical surfaces or even by plane end-faces, spaced at a given distance and tilted towards the mid-plane at a given angle. According to Swartholm (1950) a flux distribution

$$B = B_0 \frac{2r_0}{r+r_0} \qquad (10.68)$$

can be produced by high-permeability pole-pieces with conical surfaces given in cylindrical coordinates by the equation

$$2x = \frac{x_0}{r_0}(r+r_0), \qquad (10.69)$$

where $2x$ is the pole-piece distance as a function of the radial distance r from the vertex. A magnet producing a sector field of this kind has been designed by Mileikowsky (1952) and this has been applied in a high energy resolution spectrometer with two-directional focusing by Marton et al. (1958). Focusing errors due to the deviation of the actual shape of the pole faces from their theoretically exact form seem to be comparable in magnitude with the focusing errors produced by fringing fields (cf. §10.8).

§10.6. Superposition of electrostatic and magnetic fields

The path of an electron subjected to the combined action of both electrostatic and magnetic fields is generally quite complex. However, some relatively simple cases which have attained practical importance may be discussed in this section.

If an electron of velocity $u_z = u_0$ is injected into combined parallel electrostatic and magnetic fields which both act in the direction of the y-axis ($\check{E} \| \check{B} \| y$) it is simultaneously deflected by these two fields according to (10.4) and (10.13). The two deflexions are superimposed independently. We can eliminate e/m by taking the ratio of (10.13) and (10.4) and obtain

$$\frac{x}{y} = -\frac{B_y}{E_y} u_0 \qquad (10.70)$$

which is the ratio of the coordinates on a xy-plane intercepting the beam, say, on a photographic plate set up at right angles to its initial direction.

By elimination of u_0 from (10.13) and (10.4) a parabola

$$\frac{x^2}{y} = -\frac{e}{m} \frac{B_y}{2E_y} z^2 \qquad (10.71)$$

is traced on the xy-plane when this is set up at a given position z.

If an electron of given velocity u_0 is injected into combined homogeneous electrostatic and magnetic fields which are crossed at right angles and which are also perpendicular to the initial motion of the electron, a certain ratio of the two field strengths can be found for which the electron is not deflected in its path. In this case the electrostatic force $-e\breve{E}$ and magnetic force $e(\breve{u} \wedge \breve{B})$ acting upon the electron must be equal and opposite, hence

$$\frac{E_x}{B_y} = (u_0)_z. \qquad (10.72)$$

If an electron starts from rest in the homogeneous fields $\breve{E} = E_x$, $\breve{B} = B_y$ which are crossed at right angles, it is first accelerated by the electrostatic field but it is unaffected by the magnetic field since it has no velocity. As it gathers speed, it is deflected by the magnetic field, being gradually turned round until it moves against the electrostatic acceleration and eventually comes to rest at some distance z_L from its initial position. The net motion of the electron is a drift in the z-direction. The electron orbit can be obtained from the equations of motion:

$$m \frac{d^2x}{dt^2} = -eE_x - eB_y u_z, \qquad (10.73)$$

$$m \frac{d^2z}{dt^2} = -eB_y u_x \qquad (10.74)$$

with the initial conditions that the coordinates x, y, z of the electron and their derivatives u_x, u_y, u_z are zero for $t = 0$. Integrating (10.74) yields

$$\frac{dz}{dt} = \omega x, \tag{10.75}$$

where $\omega = \omega_c = (e/m)B$ is the cyclotron frequency (cf. §5.1) of the electron in the homogeneous field. Substituting from (10.75) into (10.73) leads to

$$\frac{d^2x}{dt^2} = -\left(\frac{e}{m}E_x + \omega^2 x\right). \tag{10.76}$$

A solution of this linear differential equation by standard methods yields

$$x = -\frac{e}{m}\frac{E_x}{\omega^2}(1 - \cos \omega t). \tag{10.77}$$

By integration of (10.75) and combination with (10.77) we obtain

$$z = -\frac{e}{m}\frac{E_x}{\omega^2}(\omega t - \sin \omega t). \tag{10.78}$$

(10.77) together with (10.78) represent a cycloid orbit in parametric form. The electron motion is seen to consist of a uniform translation in the z-direction with a superimposed circular motion. The radius of this circular motion is half the maximum displacement in the y-direction which, according to (10.77) is,

$$\tfrac{1}{2}x_{max} = r_b = \frac{e}{m}\frac{E_x}{\omega_c^2} = 5 \cdot 7 \times 10^{-12}\frac{E_x}{B_y^2}. \tag{10.79}$$

The displacement in the z-direction corresponding to one cycle of the motion is

$$z_L = 2\pi r_b. \tag{10.80}$$

It is produced by a point on the periphery of a disc of radius r_b rolling on the z-axis with a velocity

$$u_b = r_b \omega_c, \tag{10.81}$$

u_b being the velocity of the centre of the disc in its forward motion in the z-direction.

If the electron is injected into the field with an initial velocity u_0 in the $+z$ or $-z$-direction, u_0 can be considered to consist of two components, given by

$$\breve{u}_0 = \breve{u}_b + \breve{u}_e$$

where u_e is the velocity, the electron would have in a purely magnetic field of the above specified value B_y. There it would move on a circle of radius

$$r_e = \frac{mu_e}{eB_y}$$

given by (5.1) in §5.1. Hence, a point with radius r_e on the rolling disc will describe the cycloidal path of the electron. The motion of the electron in fig. 10.16(a) corresponds to zero initial velocity $u_0 = 0$ while fig. 10.16(b) and (c) represent the cases of an initial velocity in the $-z$-direction and in the $+z$-direction respectively.

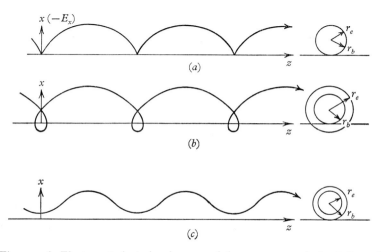

Fig. 10.16. Electron trajectories in crossed homogeneous electrostatic and magnetic fields: (a) for zero initial velocity; (b) for initial velocity in the $-z$-direction; (c) for initial velocity in the $+z$-direction.

The deflexion of electrons in crossed homogeneous magnetic and electric fields has found various applications in the past. It had been used very early by J. J. Thomson (1897) to measure the velocity of cathode rays preliminary to the determination of the specific charge of the electron. The straight motion of electrons of given velocity in crossed fields as in (10.72) was utilized for sorting electron velocities in the 'velocity filter' of Wien (1898) who realized already that beams of relatively large aperture could be monochromatized in this way. The theory of the Wien filter is contained in the papers by Herzog (1934) and by Henneberg (1934 a). A slightly divergent beam is focused in the direction of the electrostatic field. The focusing distance is half as long as that of the uniform, longitudinal field (cf. (5.5)) in §5.1 namely

$$z_f = \frac{\pi}{\omega_c} \frac{E}{B},\tag{10.82}$$

ω_c being again the cyclotron frequency. At this distance

$$z_f(\perp E \perp B)$$

an electron optical image is projected with unit magnification. The Wien filter, however, focuses in the direction of the electrostatic field only, so that a point source will be projected as a small line, the line being parallel to B.

A stigmatic projection can be effected by the additional use of a line focus lens. Boersch, Geiger and Stickel (1964) have used a Wien filter between two immersion lenses as a monochromator and another Wien filter in a similar arrangement as an energy analyser, both with a dispersion of 4 mm eV^{-1}. The monochromator transmitted a 50 keV beam of 0·05 eV half-width through a gas chamber. By the use of this analyser, energy losses of the electron beam in the gas atoms could be measured to within 0·02 eV. The high resolution was obtained by decelerating the beam in the immersion lens at the entrance of the filter to 20–300 eV and re-accelerating it again at the exit.

A stigmatic Wien filter has been constructed by Legler (1963). He superimposed the homogeneous magnetic field on to the electrostatic field of a cylinder condenser as shown in fig. 10.17. There, the y-axis coincides with the magnetic field B, the electron beam moves along the z-axis. The cylindrical equipotential surfaces of the electrostatic field are symmetrical to the xz-plane their radius r of curvature being perpendicular to the electron beam axis z.

It can be shown by integration of the equations of motion that the electron orbit performs harmonic oscillations in the xz-plane and in the yz-plane about the z-axis. In order to arrange for these oscillations to have the same period, the radius of curvature of the

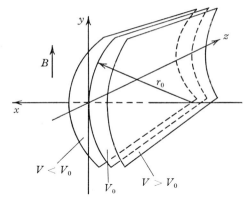

Fig. 10.17. Equipotential surfaces in the two-directional focusing Wien filter.

electrostatic equipotential surface at the z-axis must be given by

$$R_E = \frac{2}{\omega_c}\frac{E_0}{B}.$$ (10.83)

The common frequency of both oscillations is then

$$\omega = \omega_c/\sqrt{2}$$ (10.84)

and a beam focused on the z-axis will be refocused on this axis at a distance

$$z_f = \frac{\pi R_E}{\sqrt{2}} = \frac{\pi\sqrt{2}}{\omega_c}\frac{E_0}{B}.$$ (10.85)   .

This distance is a factor of $\sqrt{2}$ larger than that of the ordinary Wien filter (10.82).

If the energy of the beam is increased to $e(V_0+\Delta V)$, it can be shown that the orbit will oscillate about an axis which is displaced from the z-axis in the direction of the electrostatic field by the amount $(R_E/2)\Delta V/V_0$. Since the exit aperture is reached after half an oscillation, the electron there is at twice the distance from the z-axis. The dispersion of the filter is therefore given by

$$\frac{\Delta x}{\Delta V} = \frac{R_E}{V_0}.$$ (10.86)

The energy resolution of the beam by the filter is limited by an aperture error which causes every energy to be focused into a small disc of confusion. The shape of this disc can be controlled to some extent by the rate of change of the radius r of curvature of the equipotentials along the x-axis. For $dr/dx = 1$ one obtains a circular disc, for $dr/dx = -2$, the disc is an ellipse with its long axis in the y-direction. In the latter case, the aperture error causes less diffusion in the direction of the dispersion and therefore leads to an improved resolution which has been calculated by Legler (1963) to be

$$\Delta V/V_0 = 2\Theta^2, \tag{10.87}$$

where Θ is the angular semi-aperture of an incident beam of energy V_0.

Legler has constructed a practical filter with $E_0 = 4 \times 10^3 \text{ Vm}^{-1}$ produced by two condenser plates with radii of curvature of 88 and 112 mm respectively. The homogeneous field

$$B = 6.7 \times 10^{-4} \text{ Wbm}^{-2}$$

was produced by two current coils. With $r_0 = 100$ mm and $dr/dx = -2$ and an entrance slit of 0.05×2 mm a bandwidth $\Delta V = 0.06$ eV of the filtered beam of 5×10^{-11} amp was obtained. The adjustment and correction of the Wien filter has been discussed by Anderson (1967).

Of some technical importance for the design of the so-called Orthicon pick-up tube is the deflexion of electrons entering crossed homogeneous electrostatic and magnetic fields (E_y, B_z) with a velocity u_z in the direction of the magnetic flux only. It can be shown that an electron will travel through the flux in a slanting direction while performing a certain number n of loops, given by

$$2\pi n = \frac{\omega_c L}{u_z}, \tag{10.88}$$

where ω_c is again the cyclotron frequency in the flux density B_z, and L is the length of the electrostatic field as given, for instance, by the length of the deflexion condenser plates. Rose and Iams (1939) have shown that the deflexion will not eventually lead to a change in direction u_z but just to a shifting of the electron beam in the x-direction if the values of u_z and of ω_c are such that an integral number n of loops is performed during the traversal of the crossed fields. The displacement in the x-direction is proportional to the electrostatic field E_y, and it amounts to nx_L where x_L has the value given by (10.80) for z_L. By a reversal of the direction of the electrostatic field, the direction of the displacement will also be reversed.

Of great practical interest is a field combination called the 'magnetron field'. It consists of an electrostatic field between two co-axial cylinders, the smaller cylinder being the cathode, the larger, the anode; parallel to the cylinder axis extends a homogeneous magnetic flux. Electrons move from the cathode in cardioid-like orbits returning again to the cathode with zero velocity, provided that the orbits are not intercepted by the anode cylinder. The analysis of the orbits has been given by Hull (1921). For small cathode radii, the orbits are represented by

$$r = r_{max} \left(\sin \frac{2\Theta}{3} \right), \qquad (10.89)$$

where r and Θ are cylindrical coordinates, r_{max}, the maximum elongation from the axis, being given by

$$r_{max} = \frac{1}{B} \sqrt{\left(\frac{8e}{m} V \right)}, \qquad (10.90)$$

where V is the electric potential at the distance r_{max} from the axis. The curve representing (10.90) is called the 'cut-off parabola', since in a given flux density B the total electric current to an anode of radius r_{max} would be cut off as soon as the anode voltage dropped below the value of V.

The possibility of a focusing deflexion in superimposed electrostatic and magnetic fields was first observed experimentally by Bartky and Dempster (1929). They used the electrostatic field between co-axial cylinders in combination with a homogeneous magnetic field parallel to the axis of the cylinders. According to Henneberg (1934a) the focusing properties of this field combination can be easily derived from the equation of motion of the radial component in the purely electrostatic field (cf. (10.30)) while accounting for the magnetic field by adding the magnetic contribution of the total force upon the electron. Hence

$$\frac{m \, d^2 r}{dt^2} = mr\omega^2 - eE - eBr\omega. \qquad (10.91)$$

The law of conservation of angular momentum now leads to

$$mr^2\omega - \frac{eBr^2}{2} = \text{const.} \qquad (10.92)$$

Combination of (10.91) and (10.92) and application of the new equation to the circular orbit, where $d^2r/dt^2 = 0$, gives

$$mr_0\omega_0^2 = eE_0 + er_0\omega_0 B, \qquad (10.93)$$

which for $E_0 = 0$ or for $B = 0$ corresponds to (5.2) or to (10.29) respectively, the subscript (0) referring again to the circular orbit.

Now take
$$eE_0 = Kmr_0\omega_0^2, \qquad (10.94)$$

where K is the fraction of the centrifugal force which is compensated by the electrostatic field. Then, a calculation quite analogous to that given by (10.30)–(10.34) leads to the angle of deflexion ($=$ angle between the radius vectors through two foci)

$$\phi = \frac{\pi}{\sqrt{(1 + K^2)}}. \qquad (10.95)$$

For positive K, the electrostatic force, $-eE$, is directed towards the centre of the circular orbit, i.e. $-eE$ is acting against the centrifugal force. For negative K, $-eE$ is directed outwards, i.e. acting in direction of the centrifugal force. For $K = 0$, E vanishes and semicircular focusing in the purely magnetic field is obtained. For $K = +1$ purely electrostatic focusing with $\phi = \pi/\sqrt{2}$ follows in agreement with the earlier result (10.35).

Another particular case, given by $K = \pm\infty$, occurs according to (10.94) if the centrifugal force $mr_0\omega_0^2 = mu_0^2/r_0 = 0$, i.e. either if $u_0 \to 0$ or if $r \to \infty$. In the second alternative the orbit is straight, i.e. electrostatic and magnetic forces on the electron are compensated and the field intensities are given again by (10.72), the condition applying to Wien's velocity filter.

The focusing deflexion of very fast electrons in the cylinder condenser with superimposed magnetic field has been studied by Millet (1948), who extended Henneberg's calculations into the relativistic region. Equations (10.91)–(10.95) are found to apply as before if for m the relativistic mass $m_0/[1 - (u/c)^2]^{\frac{1}{2}}$ is substituted. Two-directional focusing can be obtained if the magnetic field

superimposed over the electrostatic field of the cylinder condenser
falls off with the radial coordinate. The case of a field with a fall-off
index (cf. 10.55)

$$n_1 = \frac{1 + K^2}{2(1 - K)},$$ (10.96)

where K is given by (10.94), has been studied by Fischer (1952).
There, two-directional focusing is obtained for every value of K.
The velocity dispersion and the resolving power is twice as large
as in the case of the superimposed homogeneous magnetic field,
i.e. for a fall-off index $n_0 = 0$. The azimuthal focusing deflexion
angle ϕ is larger by a factor $\sqrt{2}$ than that given by (10.95). In
fig. 10.18 these angles ϕ are plotted against K for the two fall-off
indices n_1 and n_0.

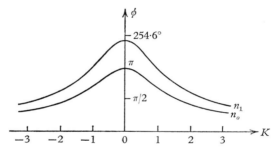

Fig. 10.18. Azimuthal focusing angles ϕ in superimposed fields as a function
of K (10.96) and of n (10.55).

Other two-directional focusing field combinations have been
discussed by Swartholm (1948). There, the decay $B(r)/B_0$ of the
magnetic field is given by the two parameters C_1 and C_2 of (10.60)
which determine the fall-off index n. The decay $E(r)/E_0$ of the
electrostatic field, however, is given by corresponding parameters
C_1' and C_2'. The focusing magnetic field superimposed over the
cylinder condenser is contained as a special solution in Swartholm's
results with the characteristic parameters assuming the values
$C_1 = C_2 = 0$ and $C_1' = C_2' = -1$.

§10.7. Deflexion errors

For small deflexions (cf. §10.1) the angle of deflexion is expected to be proportional to the deflecting field. This proportionality, however, gradually ceases to hold when the deflexion becomes large. Moreover, for a given field, the deflexion is only to a first approximation independent of the x- and y-coordinate at which the ray enters (parallel to z) into the deflecting field. Thus different rays of an electron beam of finite aperture are equally deflected only when the deflexion is small. However, peripheral rays of beams of finite aperture are subject to different deflexion as soon as the angle of deflexion is sufficiently increased. It has already been shown in §§10.4–10.6 that focusing effects are connected with large-angle deflexions. The focusing effect is generally not symmetrical with respect to the axis of the electron beam. If a beam is focused on a fluorescent screen it may produce (when undeflected) a small circular spot. This spot will be distorted as the deflexion of the beam is increased. All these deflexion errors can be demonstrated, for instance, in a television tube, when a picture is produced by the scanning of an intensity-modulated electron beam over a fluorescent target.

In a first-order 'Gaussian' theory (cf. §1.3) collinear projection is considered to apply independently of the image aperture. Hence, deformations of the spot due to large deflexion and the lack of proportionality between deflecting field and deflexion angle can be treated as aberrations (cf. §6.1), i.e. as deviations from the first-order theory. Theoretically and experimentally, there have been found deflexion errors corresponding to distortion, coma, astigmatism and field-curvature. There are no deflexion errors corresponding to spherical aberration or to anisotropic coma. The general theory of deflexion errors belongs to the most complicated problems in electron optics. We have to refrain from giving details but we refer to the papers by Picht and Hipman (1941), Hutter (1947) and by Glaser (1949) which contain a very comprehensive treatment.

Some special cases of practical interest, however, may now be discussed. Deserno (1935) and Wallraff (1935) investigated the errors of weak, substantially homogeneous deflecting fields. In the electrostatic condenser the deflexion must depend on the path of the ray, since the potential in the condenser is a function of the coordinates. Hence the velocity of the electrons is altered according to the position of the beam and according to the position of the electron within the beam. The alteration in electron velocities causes both a change in deflecting power and a defocusing of the spot off the Gaussian plane which increases as the deflected beam approaches one of the deflecting electrodes. In practice, however, it is found that the defocusing effect of the field inside the deflecting condenser is cancelled to some extent by the action of the stray field at the edge of it.

According to Recknagel (1938) the change in deflexion sensitivity can be represented by an additive correction term which increases with the third power of the deflecting voltage. On the other hand, the 'spurious' focusing depends according to (10.27) upon the square of the deflecting voltage.

Some more detailed theoretical papers on the electric deflexion condenser, given by Herzog (1934–40), can only be quoted. We mention here that least deflexion errors are experienced if for geometrically symmetrical condenser electrodes the deflecting voltages can be kept symmetrical about V_{el}, i.e. if the plate voltages are $(V_{el}+V_p)$ and $(V_{el}-V_p)$ respectively, where V_{el}, the energy of the electrons as they enter the deflecting field, is given by the anode voltage V_A of the electron gun $(V_{el}=V_A)$. If, however, unsymmetrical voltage distribution has to be applied, the shape of the deflecting electrodes too has to be chosen to be unsymmetrical for compensation. For instance, if one deflexion plate has to be connected with the anode of the gun so that its voltage is permanently $=V_{el}$, while the other plate is connected to the full deflecting voltage $(V_{el}+2V_p)$, very serious errors will be observed unless diaphragms are placed at the entrance and at the exit of the deflexion condenser, these diaphragms being connected to the voltages V_{el} and $(V_{el}+2V_p)$ respectively.

Attempts at correcting the deflexion condenser have had only limited success. However, according to a proposal by Hutter and Harrison (1950), the difficulty of field correction can be avoided by the introduction of a complementary 'predistortion' of the beam before it enters the condenser. In practical cathode-ray tubes line-focus lenses (cf. §2.5 and §10.2) have been introduced into the electron gun. For proper correction of both size and shape of the focused spot on the fluorescent screen, these line-focus lenses are operated by a certain fraction of the voltage applied to the deflexion plates.

Both electric deflexion and magnetic deflexion suffer from defocusing due to curvature of field and spot distortion due to astigmatism, these errors being in both cases proportional to the square of the tangent of the angle of deflexion. For the magnetic case, however, the constant of proportionality is several times smaller, a fact which facilitates the design of television display tubes, where large deflexion angles (e.g. $\phi = 55°$) are common.

In a short homogeneous magnetic field, the angle of deflexion ϕ equals the angle between the two radius vectors r_e of the circular electron orbit at the beginning and at the end of the field. If (as in §10.1) the first radius vector is perpendicular to the z-axis, the deflexion is geometrically given by

$$\sin \phi = \frac{z_p}{r_e}, \tag{10.97}$$

where z_p is the length of the deflecting field. The ratio z_p/r_e which, according to §10.1, is proportional to the deflexion y_T of the spot on the target, approaches the value $\tan \phi$ of (10.15) for small angles ϕ only, i.e. when $r_e \gg z_p$, otherwise it is distinguished from it by a factor $\cos \phi$. In practice, however, this cosine error which is characteristic of the homogeneous field is always small in comparison with the error produced by the inhomogeneous fringing fields.

The influence of the inhomogeneity of the magnetic field is schematically shown in fig. 10.19(a) and (b) after Zworykin and Morton (1940). In fig. 10.19(a) a beam of circular cross-section is seen passing through the barrel-shaped pattern of field lines between the pole-pieces N and S. An enlarged transverse section in fig. 10.19(b) shows that the rays 1 and 3 on the horizontal diameter will suffer different displacements as indicated by the force vectors. The vertical field is less spread out and therefore stronger at 1 than at 3. Hence, in being deflected to the right, the beam will also be contracted in horizontal direction. On the other hand, rays 2 and 4 on a vertical diameter through the beam experience the same horizontal force but vertical forces in opposite senses, since the horizontal component of the magnetic field is finite and oppositely directed at the two points. Hence the beam will suffer an elongation in the vertical plane and the resulting shape of the deflected spot will be elliptical. If the beam is not in

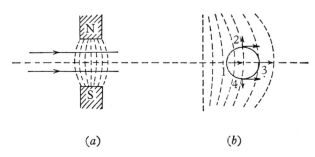

(a) (b)

Fig. 10.19. Distortion in a magnetic deflecting field: (a) longitudinal section; (b) enlarged transverse section of an initially circular beam.

the horizontal plane of symmetry, as assumed here, the ellipse will be rotated about the axis also.

Loss of proportionality between deflecting current and deflexion arises from variation of flux density with distance from the axis. In a barrel-shaped pattern of field lines as shown in fig. 10.19(a) the beam will enter a region of smaller flux density as it moves away from the axis, and thus the deflexion will grow less than proportional to the deflecting current. Increasing deflecting power with increasing current, however, will be experienced if the pattern of field lines is pin-cushion shaped.

The magnetic deflexion error has been treated quite generally by Glaser (1938), by Wendt (1942) and by Kaashoek (1968) and some experimental tests of the theory have been described by Marschall and Schröder (1942). It appears that for practical purposes, in particular for the scanning in television tubes, field curvature and astigmatism are by far the most serious errors, The effect of the latter error, due to which the undeflected, sharp circular spot on the fluorescent target is changed into a large elliptical spot, is shown in fig. 10.20. The tangential, sagittal and mean image surfaces differ here from the Gaussian plane in the same manner as was explained by fig. 6.19 for the corresponding lens errors. The principal plane of deflexion which is shown in the figure is obtained as locus of intersection of the undeflected rays with the backwards produced direction of the deflected rays.

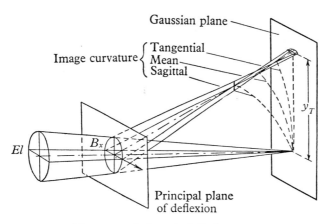

Fig. 10.20. Astigmatic deflexion error.

The effect of coma which, however, is much less important gives rise to elliptical spots which have been observed by Wendt (1953). The combined effect of coma and astigmatism has been studied in detail by Wang (1967). He computed the arrival points at the screen of a large number of trajectories, distributed uniformly over the entrance aperture. In this way he was able to plot out the current density distribution in spots suffering from astigmatism and coma in varying degrees. Very complicated aberration patterns can arise, which are only slightly modified if a Gaussian distribution at the entrance aperture is assumed.

Some technically important magnetic deflecting fields which show a substantial degree of correction, have already been discussed in §10.3. We add here, that if in these coils the length z_p of the deflecting field is defined as the distance between the points at which the flux density has decreased to one-tenth of its maximum value, the radius of the mean image field curvature is found to be of the order of

$$R \approx \tfrac{2}{3} z_p$$

under the best available conditions. According to Haantjes and Lubben (1957) R is of the order of z_p for all deflecting field distributions, thus it is not possible to correct field curvature by means of modifying the deflecting field distribution. Now while television screens are normally curved, their radii of curvature are usually considerably greater than R, so that some defocusing due to field curvature must be expected. In practice, owing to the small convergence angle of the beam, the depth of focus is generally sufficient for this to be tolerably small.

Astigmatism of the spot and pin-cushion distortion of the raster can both be corrected in principle by using a pin-cushion deflecting field such as indicated by the field lines in fig. 10.8. However, the small degree of pin-cushion distortion of the field lines which is required to correct astigmatism is not sufficient to completely correct raster distortion. The residual raster distortion is corrected by either two or four adjustable permanent magnets attached to the deflection unit.

§10.8. Deflexion-focusing errors, monochromatic line-width and field correction

It has been shown in §§10.4 and 10.5 that some deflecting fields act like electron lenses having definite foci and principal points for all paraxial rays moving within small apertures from a curvilinear axis. It seems quite obvious that lens errors, such as those treated in chapters 6 and 7 for ordinary electron lenses, should also occur for focusing deflexion fields. Spherical aberration, for instance, should be noticed as a difference in focus for paraxial and for marginal rays. Or, if the focus of a marginal pair of rays is observed at some distance from the principal ray, this could be taken as the coma of the deflecting field, and so on. So far, however, the optical treatment of focusing deflexion errors has only been developed systematically for applications involving the transport of high energy, heavy particle beams used in experiments in elementary particle physics. This important application lies largely outside the scope of this book, but the interested reader is referred to the specialized literature on this topic, e.g. to the book by Steffen (1965). In electron optics, as opposed to ion optics, errors of deflexion focusing are practically important in spectrometry only. Hence attention has been drawn rather to the general question of confusion of spectral line width in the plane of the photographic plate or of the collector slot. For a discussion of this question, however, a simple investigation by means of ballistic laws is generally sufficient.

Take, for example, the deflexion focusing in the homogeneous electrostatic field which has been discussed in the beginning of §10.4. The principal orbit may be compared with the ballistic parabola rising from the 'horizontal' xz-plane at a slope angle α. The least change Δz in 'range' z_f for a given small change in α is known to occur for $\alpha = 45°$. Now, the changes $\pm\Theta$ in α are equivalent to a semi-aperture Θ of a bundle of rays and the corresponding shortening Δz in range is equivalent to the departure from perfect focusing. From (10.28) with $\alpha = \frac{1}{4}\pi$, there follows immediately

$$\Delta z \approx 2z_f\Theta^2, \tag{10.98}$$

e.g. a semi-aperture $\Theta = 0·1$ of the beam gives rise to a width of confusion of the line focus of about 2 per cent of the ideal range z_f.

The trace width in the cylindrical mirror analyser (fig. 10.11) has been calculated by Hafner, Simpson and Kuyatt (1968) to be related to the radius r_1 of the inner cylinder by the expression

$$\frac{2\Delta L}{r_1} \simeq 5·6 \frac{\Delta V}{V} - 15(\Delta\alpha)^3. \tag{10.99}$$

Next may be considered the important case of semicircular focusing in the homogeneous magnetic field. As shown in fig. 10.14 a diverging bundle of rays of semi-aperture Θ may be emitted from the coordinate origin at J_0. All rays having exactly the same velocity describe circles of radius r_e in the strictly homogeneous field. The principal ray may travel initially along the y-axis, hence the centre M_0 of this circular orbit has the coordinates ($y_0 = 0, z_0 = r_e$). The centres C_1 and C_2 of the two limiting rays (1) and (2) of the bundle are given by the coordinates ($y_1 = -r_e \sin\Theta, z_1 = r_e \cos\Theta$) and ($y_2 = r_e \sin\Theta, z_2 = r_e \cos\Theta$) respectively.

The two marginal rays (1) and (2) cut the z-axis at the common point J_1' having from J_0 the distance $z = 2r_e \cos\Theta$, while the principal orbit intersects the z-axis at J_0' given by $J_0J_0' = 2r_e$.

Hence, the point object J_0 is projected into a confused image of width

$$J_0'J_1' = \Delta r = 2r_e (\mathrm{1} - \cos \Theta), \qquad (\mathrm{10.100})$$

which for small Θ is approximated by

$$\Delta r = r_e \Theta^2. \qquad (\mathrm{10.101})$$

As the confusion Δr extends to the inside of the principal circular orbit only, the image width obtained from an object of finite width b is found to be

$$b' = b + \Delta r = b + 2r_e(\mathrm{1} - \cos \Theta). \qquad (\mathrm{10.102})$$

This result is of interest in electron beam spectrometry. If the value of the electron momentum $(Br)_e$ of a monochromatic spectral line has to be determined from the position of an experimental line image, the radius r_e of the electron orbit which is required for the determination of the $(Br)_e$-value is given as half the distance from the outer edge of the line image to the inner edge of the source.

A detailed analysis of the effect of finite apertures of the electron beam in space and of the length and width of the electron source has been given by Wooster (1927) and by Li (1937). As an essential result it may be reported here that a filament object or an emitting straight slot is projected into a slightly curved line image. The inner edge of the object again corresponds to a sharp outer edge of the image, i.e. at the high-velocity side the image is limited sharply. The outer edge of the object, however, is represented by the maximum of an intensity distribution in the image. Towards the low-velocity side this intensity distribution falls off steeply, ending eventually in a long tail of low intensity which for most purposes in spectrometry can be disregarded. In the three-dimensional problem the electron orbits can, in general, not be considered as circles but rather as spirals, the pitch of which is limited by the length of the source and the aperture stops. If, however, the maximum slope angles γ of the spiral orbits with the equatorial plane ($\perp B$) are small as compared with the beam semi-aperture Θ or practically, if

$$\gamma \lesssim \mathrm{0 \cdot 1} \Theta, \qquad (\mathrm{10.103})$$

the intensity distribution in the image is found to be little affected by the extension of source and stops in the direction of the magnetic flux B. Thus, the image width calculated for the two-dimensional case (cf. (10.102)) gives a fair approximation to the actual width of the line image.

Owen (1949) has given a general analytic derivation of intensity distributions across the line image. He presented his results by a set of curves for various conditions of lengths and widths for source and stop slots respectively; all these curves are useful for the spectrometer designer. Detailed descriptions of wide-angle, short-source semicircular spectrometers to which condition (10.103) applies, have been given by Gurney (1925) and by Lawson and Tyler (1940). On the other hand, Geoffrion (1949) has pointed out that by increasing the source length and slightly decreasing its width the intensity of the line image can be greatly increased without affecting its width.

The confusion of the line width can always be reduced by cutting down the marginal rays, i.e. by reducing slot dimensions. On the other hand, (10.100) shows a possibility of bringing the focus of the marginal rays into coincidence with the paraxial focus by the correct increase of the radii curvature (r_e) of the marginal rays. For rays of given electron momentum $(Br)_e$ this can be done be reducing the deflecting field B at both sides of the principal ray. The field distribution $B(r)$ as a function of the distance r from the centre of the principal ray, as required for correct focusing, was calculated by Beiduck and Konopinski (1948), who arrived at the solution

$$\frac{B(r)}{B_0} = 1 - \frac{3}{4}\left(\frac{r-r_0}{r_0}\right)^2 + \frac{7}{8}\left(\frac{r-r_0}{r_0}\right)^3 - \frac{9}{16}\left(\frac{r-r_0}{r_0}\right)^4 + \dots \quad (10.104)$$

where r_0 is the radius of curvature of the principal orbit as determined by the flux density B_0.

The first experimental arrangement for correcting the focusing errors of the homogeneous magnetic field was described by Voges and Ruthemann (1939). These authors modified the field in the gap between two co-axial cylindrical air coils by the superposition of the field of two circular co-axial strip conductors, and in this way they succeeded in developing a slow-electron spectrometer of very high resolving power (cf. §10.9). In a different manner, the proper departure from uniformity of the magnetic deflecting field has been obtained by Korsunksy, Kelman and Petrov (1945). These authors gradually changed the profile of the pole faces of their magnet until they reached the desired sharpness of focus. The maximum aperture of rays still reaching this focus was $2\Theta = 40°$.

A correction of focusing by bounded homogeneous fields such as, for instance, Barber's magnetic prism (cf. §10.5), can be obtained by shaping of the field boundaries. Hintenberger (1948, 1949) and Kerwin (1949) investigated theoretically the great variety of boundary curves that give perfect focusing of an electron beam of wide divergence. In some cases the boundary correction simultaneously enables a reduction of the necessary size of the magnet. The discussion, however, deals with focusing in the equatorial plane ($\perp B$) only. Practically it would apply to short sources and short aperture slots only, and this restriction is not likely to lead to the optimum intensity of a line image. Moreover, the theoretical field boundary is not easily simulated by the shape of the magnetic pole-pieces. Coggeshall (1947) has shown how the fringing flux can be accounted for by using virtual field boundaries displaced outwards from the edges of the pole-pieces. However, even if the method of the virtual field boundaries is helpful for finding the boundary design leading to the desired orbits, it leaves the absolute energy determination in the electron spectrum as an unsolved difficulty.

The homogeneous field prism of circular symmetry (cf. §10.5) offers the interesting possibility of reducing focusing errors to a minimum by a proper choice of the angle of deflexion. Korsunsky (1945) realized that for smallest spacings of the pole-pieces focusing errors disappear completely at 90° deflexion. Siday (1947), who calculated the focusing errors for such vanishingly small spacing of pole-pieces over a large range of deflexion angles, found, in agreement with Korsunksy, minimum errors about the rectangular deflexion.

Electron orbits in the equatorial plane half-way between pole-pieces separated by one radius of their circular cross-section have been studied by Siday (1947) by means of numerical computation. There, the fringing flux was properly taken into account. The deflexion focusing aberrations were found—in distinction to lens aberrations—to be unsymmetrical, changing sign with the ray aperture. For example, the aberrations are positive or negative (cf. §6.3) for rays travelling outside or inside, respectively, of a principal ray which is initially directed towards the centre of the field. The aberrations depend greatly upon the angle of deflexion. For initially parallel rays the aberrations pass through a minimum at a deflexion angle of 76°.

According to Jennings (1952) the deflexion angle with minimum aberration gradually decreases with increasing pole spacing. However, by choosing bundles with eccentric incidence, conditions for minimum aberration can be found at any given pole spacing for a wide range of deflexion angles. For this focusing with least aberration, the eccentricity of the principal ray of the bundle is found to increase with increasing angle of deflexion.

The image confusion for two-directional focusing has been discussed by Swartholm (1946) and by Shull and Dennison (1947). In a field given by (10.55) with $n = \frac{1}{2}$ the intersection between principal and zonal ray in the equatorial plane ($\perp B$) of symmetry may be traced or calculated as a function of the semi-aperture Θ. The distance ζ between this intersection and the object point as measured along the principal ray is approximately given by

$$\zeta(\Theta) = r_0(\phi_0 - \tfrac{4}{3}\Theta), \tag{10.105}$$

where r_0 is the distance of the object from the field axis and where $\phi_0 = \pi\sqrt{2}$ is the angle of deflexion for paraxial focusing as given in §10.5. According to (10.105) the deflexion focusing suffers from positive spherical aberration, and the image in the 'Gaussian plane' will be confused towards the field axis by an amount (cf. transverse aberration §6.3) given by

$$\Delta r = \tfrac{4}{3}r_0\Theta^2. \tag{10.106}$$

(10.106) also gives a fair estimate of the monochromatic line width for a negligibly narrow object in the two-directional focusing deflexion fields described by (10.60). Only slight influence is exerted on the focused line width by a finite value of the coefficient C_2 ($< \frac{3}{8}$) of this equation (cf. §10.5). Other slight modifications of (10.106) are due to a finite beam aperture at right angles to the equatorial plane.

The focusing deflexion by a cylindrical condenser on to which a homogeneous magnetic field is superimposed (B parallel to the cylinder axis) can be adjusted by proper choice of the electric and magnetic field ratio. If this ratio is again (cf. §10.6) defined by the parameter K of (10.94) the monochromatic line width for a vanishingly small source is, according to Henneberg (1934a), to a first approximation given by

$$\Delta r = -r_0\Theta^2(1+\tfrac{1}{3}K+K^2+3K^3)/(1+K^2)^2, \quad (10.107)$$

where r_0 is the radius of the principal orbit and Θ the semi-aperture of the beam. It can be seen that Δr, as calculated from the above approximation formula, vanishes for $K = -0.76$. Actually, the value of the line width becomes very small for this particular field combination.

For the purely electrostatic case ($K = 1$), (10.107) yields

$$\Delta r = -\tfrac{4}{3}r_0\Theta^2. \quad (10.108)$$

For the purely magnetic case ($K = 0$) one obtains $\Delta r = r_0\Theta^2$, which is the line width given by (10.101). Of further interest is the line width in Wien's velocity filter (cf. §10.6). Taking $K \to \infty$, (10.107) yields

$$\Delta r \atop \left[{r\to\infty \atop K\to\infty}\right] = -3\Theta^2 \lim\left(\frac{r_0}{K}\right) = -3\frac{mu_0^2}{eE_0}\Theta^2, \quad (10.109)$$

the last transition in this equation having been obtained by introducing (10.94) for K and substituting $\omega_0 = u_0/r_0$, where u_0 is the linear velocity of the rays which travel at right angles to E and B.

§10.9. Dispersion, resolving power and collecting efficiency of focusing deflexion fields

The application of deflecting fields for analysing a distribution of electron velocities is based upon the different deflexion of beams of different velocities. Separate images are formed for separate velocities. In distinction to the chromatic aberration of electron lenses (cf. lens spectrometry, §7.2) the images are not located along the axis, i.e. along the principal rays of the deflected beam but along its radius of curvature.

Two images with the coordinates s and $(s+ds)$ may correspond to the velocities u and $u+du$. The displacement ds per velocity change du is the dispersion of velocities by the deflecting field; it is given by ds/du. Analogously, the dispersion of momenta and the dispersion of energies are defined by ds/dp and by ds/dW respectively.

As the dispersion for a given geometry of rays still depends on the velocity or on the momentum of the beam, let a 'specific dispersion' be defined as

$$D = |p(ds/dp)|. \qquad (10.110)$$

It is well known in spectrometry that dispersion is necessary but not sufficient for the resolution of a spectrum. It is also required that the line width, i.e. the width of the image of the source, be small enough, so that two images of sufficiently small displacement may be separated.

The smallest possible displacement Δr that still can be separated is of the order of the width of the image, the total image width being composed of that of a Gaussian (first order) image and of the confusion by the deflexion focusing errors (cf. §10.8). For images with a wide intensity distribution the 'half-maximum width' (i.e. the full width of the distribution curve at half the height of the maximum) is generally taken as a practical value of Δr. The expression

$$\frac{D}{\Delta r} = \frac{p}{\Delta p} \qquad (10.111)$$

is the resolving power, where Δp is the difference in momenta that corresponds to the smallest resolvable displacement Δr. For a vanishingly small object, Δp is a function of the focusing deflexion errors only.

Equation (10.111) gives a practical prescription for finding the resolving power for a given focusing deflexion field. Take, for instance, a homogeneous magnetic field spectrometer with the source in the plane of the photographic plate (cf. fig. 10.14) and with constant field B_0. The line width $\Delta r = r_0\Theta^2$ is given by (10.101). Moreover, $p = eB_0r$ and $s = 2r$, hence

$$D = p\frac{ds}{dp} = 2r$$

and, according to (10.111),

$$p/\Delta p = 2/\Theta^2. \qquad (10.112)$$

On the other hand, if the photographic plate is arranged in line with the aperture slot (cf. fig. 10.12) the radius r of an electron orbit is given by (10.44). Hence the momentum is

$$p = \tfrac{1}{2}eB_0(s^2+c^2)^{\frac{1}{2}}.$$

The dispersion has to be taken along the photographic plate in the direction of s, and one obtains

$$D = p \bigg/ \frac{dp}{ds} = \frac{s^2+c^2}{s} = \frac{(2r)^2}{s}.$$

With $\Delta s \approx \Delta r$ of (10.101) the resolving power becomes

$$\frac{p}{\Delta p} = \frac{4r^2}{s}\,\frac{1}{r\Theta^2}. \qquad (10.113)$$

Unless $s \approx 2r$, i.e. unless the perpendicular distance c between source and photographic plate is very small, the resolving power (10.113) is just slightly better than that of (10.112).

If the registration is carried out with a Faraday cage or Geiger counter, it is convenient to define the resolving power in another manner. The dimensions of source and slit may be negligible, but owing to the finite image-width electrons of homogeneous momentum p_0 will still reach the slit if the flux density is slightly different from the value for exact focusing given by

$$B_0 = \frac{p_0}{er_0}. \qquad (10.114)$$

Suppose the flux density is reduced to $B < B_0$. Then the principal ray is displaced outwards to $r > r_0$, and

$$B = \frac{p_0}{er}. \qquad (10.115)$$

If $p_0/e \,(= Br)$ is independent of r one obtains on differentiation

$$\frac{dB}{dr} = -\frac{p_0}{e}\frac{1}{r^2} = -\frac{Br}{r^2}, \qquad (10.116)$$

or considering magnitudes only

$$\frac{B}{dB} = \frac{r}{dr}. \tag{10.117}$$

After decreasing B_0 by dB, the new principal orbit is displaced outside both source and detector slit by a distance dr. However, the orbit which starts from the source and leads to the displaced image would require half this change in radius only, i.e. $\frac{1}{2}dr$. Hence for constant object position with an image displacement by one line width Δr, (10.117) may be used to give the following equation for the resolving power of the homogeneous field:

$$\frac{B_0}{\Delta B} = \frac{2r_0}{\Delta r}, \tag{10.118}$$

which on substitution of (10.101) for Δr yields again the result of (10.112). Thus the two results for the resolving power of the homogeneous field, namely, $B_0/\Delta B$ for fixed collector slot and variable field, and of $p/\Delta p$ for fixed flux density and variable image position, are identical.

The resolving power of a circular-symmetrical magnetic field decaying with a field index $0 < n < 1$ as given by (10.55), namely

$$B = B_0(r_0/r)^n,$$

could be derived analogously to the above calculation as follows. (10.57) would replace (10.116) and instead of (10.118) one would obtain

$$\frac{B_0}{\Delta B} = \frac{2r_0}{(1-n)\,\Delta r}. \tag{10.119}$$

The resolving power is seen to increase with increasing n. This seems to be quite plausible if one considers electrons of the momenta p and $p+dp$. In a field with $n > 0$, the radius of curvature of the ray of the larger momentum will be increased on two accounts, namely, the greater momentum and the weaker field, whereas in the homogeneous field with $n = 0$ only the first effect would be present. For example, for the two-directional focusing spectrometer by Siegbahn and Swartholm (cf. fig. 10.15) n is nearly enough $\frac{1}{2}$, and $\Delta r = \frac{4}{3}r_0\Theta^2$ is given by (10.106). By substitution of these values in (10.119) we obtain

$$B_0/\Delta B = 3/\Theta^2. \tag{10.120}$$

Values of resolving power for other arrangements will be given below in table 10.3.

For obtaining high resolving power the beam aperture should be kept as small as the requirements for intensity in the image will allow. This intensity $I(\Theta)$ may be written as the product of the total emission I_{em} from the source and of the collecting efficiency Ω of the field. In agreement with a definition given in §7.2 the collecting efficiency may be noted as

$$\Omega = I(\Theta)/I_{em}. \qquad (10.121)$$

The collecting efficiency is an important feature in practical spectrometer design. Approximate values of Ω for the important focusing deflexion fields are given in table 10.3 as a function of the geometry of the arrangement. For a point source, Ω simply equals $1/(4\pi)$ of the solid angle Ω_0 into which the focused electron beam is emitted and which is subtended by the aperture stops. For example, for two-directional focusing arrangements with circular aperture stops the solid angle of acceptance is $\Omega_0 \approx \pi\Theta^2$, where Θ is the semi-vertical angle of the collected bundles at the source.

Small sources with approximately isotropic emission can be treated like point sources. Thus the collecting efficiency is independent of the source area a as long as the source dimensions are small compared with the length of the electron orbit.

The collecting efficiency for one-directional focusing deflexion may be estimated from the product of the two maximum semi-apertures of the rays, namely, Θ in the equatorial plane and γ at right angles to this plane, since the solid angle is given simply by

$$\Omega_0 \approx 4\Theta\gamma. \qquad (10.122)$$

In one-directional focusing arrangements the meridional aperture 2γ is approximately defined by

$$2\gamma = \frac{h}{\phi_0 r_e}, \qquad (10.123)$$

where h is the length of the aperture slot (or of the collector slot) and $\phi_0 r_e$ is the length of the electron orbit from source to line image, ϕ_0 being the angle of deflexion and r_e being the radius of the electron orbit. Hence follow the values of Ω for one-directional focusing given in table 10.3; e.g. a value $\Omega = \Theta h/(2\pi r_e)$ is obtained for semicircular magnetic deflexion (cf. Gurney, 1925).

TABLE 10.3. *Focusing deflexion angle ϕ_0, resolving power $p/\Delta p$ and collecting efficiency Ω of some practically important focusing deflexion fields*

Field	ϕ_0	$p/\Delta p$	Ω
(1) $B = $ const.	π	$2/\Theta^2$	$\dfrac{\Theta h}{2\pi r_e}$
(2) $E = E_0\dfrac{r_0}{r}$ (cylindrical condenser)	$\dfrac{\pi}{\sqrt{2}}$	$3/(2\Theta^2)$	$\dfrac{\Theta h}{\pi r_e \sqrt{2}}$
(3) $E \perp B \begin{cases} B = \text{const.} \\ E = E_0\dfrac{r_0}{r} \end{cases}$	$\dfrac{\pi}{\sqrt{(1+K^2)}}$	$\dfrac{2}{\Theta^2}\left[\dfrac{1+K+K^2+K^3}{1+K/3+K^2+3K^3}\right]$	$\dfrac{\sqrt{(1+K^2)}\,\Theta h}{2\pi r_e}$
(4) $E(r) \perp B\left(\eta_1 = \dfrac{1+K^2}{2-2K}\right)$ (Fischer)	$\dfrac{\pi}{\sqrt{\frac{1}{2}(1+K^2)}}$	$\dfrac{4}{\Theta^2}\left[\dfrac{1+K+K^2+K^3}{1+K/3+K^2+3K^3}\right]$	$\pi\Theta^2$
(5) $E \perp B \begin{cases} B = \text{const.} \\ E = \text{const.} \end{cases}$ (Wien filter)	0	$2/(3\Theta^2)$	$\dfrac{\Theta h \pi m E}{eB^2}$
(6) $E(r) \perp B = $ const. $R_E = \dfrac{2m}{e}\dfrac{E}{B^2}$ (two-directional Wien filter)	0	$\dfrac{1}{\Theta^2}$	$\pi\Theta^2$
(7) $E = E_0(r_0/r)^2$ (spherical condenser)	π	$2/\Theta^2$	$\pi\Theta^2$
(8) $B = B_0(r_0/r)^{\frac{1}{2}}$ two-directional magnetic focusing deflexion	$\pi\sqrt{2}$	$3/\Theta^2$	$\pi\Theta^2$
(9) $E(r) = \dfrac{V_{el}}{1\cdot3\ln(r_2/r_1)}$ (cylindrical mirror analyser)	$95\cdot3°$ ($\alpha = 42\cdot3°$)	Exact focusing in radial plane, second-order focusing in axial plane of trajectory	$\pi\Theta^2$

$\Theta = $ angular semi-aperture in equatorial plane.
$r_e = $ radius of principal orbit.
$h = $ length of source and of collector slit.
$K = $ parameter defined by (10.94).
$\alpha = $ slope angle, see fig. 10.11.

The values for resolving power and collecting efficiency presented in table 10.3 are seen to be independent of the absolute size of the spectrometer, in particular, the resolving power of a spectrometer seems to be quite independent of the radius of the electron orbit. It should be pointed out, however, that all results in table 10.3 refer to negligibly small electron sources, and for sources of finite size the optical image size has always to be added to the line width that is due to confusion. From this point of view it is practically always worth while to make a spectrometer as large as economic considerations allow.

Since in spectrometer design the use of bounded fields implies certain economic advantages, such fields can be recommended for reaching high resolving power. For instance, with Barber's homogeneous magnetic field sector (cf. §10.5) a maximum possible resolving power $p/\Delta p = 8/3\Theta^2$ is obtained for $\phi = 120°$ deflexion with the object placed directly at the field boundary. As another example the spherical condenser may be mentioned which for a portion subtending $90°$ reaches a maximum resolving power $p/\Delta p = 8/3\Theta^2$ if the image distance is twice the object distance.

Both examples show that compared with the data of table 10.3 higher resolving power can be obtained if source and collector are placed outside the deflecting fields. The advantage of using bounded fields should become still greater if correction of focusing errors is achieved by proper shaping of the boundary. However, bounded deflecting fields have found little application in electron spectrometry because of the greatly upsetting influence of the fringing field which in practice is by no means under control.

Another important aspect not touched in this section concerns the increase in resolving power by proper modification of a field distribution, as is done for instance in the exact semicircular focusing as discussed in §10.8 (cf. (10.104)). However, the resolving power that can be achieved appears to be limited rather by the technical difficulties in the production of the prescribed field and by the economical maximum of the size of the spectrometer than by any theoretical reasons. Some practical data on resolving power and collecting efficiency which have been obtained in actual spectrometers are given in table 10.4. There, the figures presented in columns (2) and (3) refer to the principal electron

TABLE 10.4. *Comparison of deflexion spectrometers*

Spectrometer	Electron orbit r_e (cm)	Length ζ (cm)	Source (mm)	$\Omega \times 10^{-3}$	$p/\Delta p$	Reference
Semicircular magnetic	9	28	6 × 0·2	≈ 0·1	≈ 200	Ellis (1920–30)
	3	10	10 × 3	1	10	Gurney (1925)
	12	38	20 × 3	1·0	60	Lawson and Tyler (1940)
	12	38	10 × 1	0·1	220	Siegbahn (1944)
Semicircular magnetic (long source)	30	95	20 × 1·4	0·22	300	Geoffrion (1949)
	30	95	60 × 1·2	0·45	300	Geoffrion (1949)
Semicircular magnetic (α-ray)	40	126	10 × 0·1	0·03	3 200	Cockcroft (1933)
Corrected semicircular (air coils)	17	55	1·5 × 0·1	0·1	4 000	Ruthemann (1948)
Corrected semicircular (ground pole pieces)	40	126	25 × 4	1·0	200	Langer and Cook (1948)
Two-directional magnetic (iron pole pieces) {	12·5	54	8 × 0·2	1·4	150	Siegbahn and Swartholm (1946 a, b)
	50	222	20 × 3	10	100	Hedgran, Siegbahn and Swartholm (1950)
			20 × 1	2	600	
Two-directional magnetic (air coils)	100	445	0·34 × 37	0·8	10 000	Graham, Ewan and Geiger (1960)
			34 × 370	11	100	
Spherical condenser 90°-portion	46	150	1	0·4	1 000	Browne, Craig and Williamson (1951)
Circular cylindrical magnetic prism, 76° deflexion	7	160	0·1	2	$10^{3}\dfrac{V_1^{*}}{V_2}$	Klemperer and Shepherd (1963)

* Deceleration by lens before deflexion, e.g. $V_1/V_2 = 10^2$.

orbit giving its radius of curvature, r_e, and its length, ζ, between source and collector. Column (4) refers to the dimensions of the source, e.g. height × width of the emitter slot (or strip or wire) or, in case of a circular source, to the diameter. The next two columns refer to the collecting efficiency Ω and to the resolving power $p/\Delta p$, respectively, of the spectrometers. The figures for $p/\Delta p$ have either been taken directly from the references or they have been estimated here from other information (geometry of apparatus, spectral distribution curves, etc.) contained in the papers.

The product $\Omega \times p/\Delta p$ may be considered as a figure of merit for any given spectrometer design. By changing the aperture of a given spectrometer, one of the factors of this product is always increased at the expense of the other factor. By scaling up the dimensions of a given spectrometer type the product $(\Omega \times p/\Delta p)$ can generally be improved until the object (which for intensity reasons has to be of a given size) can be considered to be small in comparison with the dimensions of the electron orbit.

For the investigation of a substance of low emission density, however, sources of relatively large size have to be employed in order to obtain sufficient intensity at a given collecting efficiency. There the influence of the size of the source on the resolving power cannot be neglected, and the value of the collecting efficiency is no longer adequate for describing the merits of a spectrometer. Geoffrion (1949) introduced, for this reason, the product of source area and collecting efficiency as the 'luminosity' of the spectrometer. It is seen, for example, in table 10.3 that the luminosity of the semicircular magnetic spectrometer can be increased about six times by increasing the source length in the direction of the field by a factor three. The significance of the luminosity is appreciated, when it is realized that the specific activity of a source material is frequently the limiting factor in achieving good resolution.

All spectrometers listed in table 10.4 have been designed originally for electrons, excepting Cockcroft's apparatus, which has been used by Rutherford and his collaborators for their celebrated measurements on the fine-structure of α-ray spectra.

The data of table 10.4 on deflexion spectrometers can be compared with the data on lens spectrometers given in table 7.2 of §7.2. The two types of spectrometers have also been compared by means of the graph of fig. 7.5. It appears that better performance can be obtained with deflexion spectrometers. The widespread application of lens spectrometers is not due to a superior collecting efficiency or resolving power but to their technical simplicity and to the easy accessibility of the components (counter, source, etc.).

POTENTIAL DISTRIBUTION ALONG THE AXIS AND OFF THE AXIS IN FIELDS OF CIRCULAR SYMMETRY

A potential V at any point of an electron lens can be determined from the potential distribution along the axis, with the help of Laplace's equation. This equation can be written in cylindrical coordinates (r, z, ψ) as follows:

$$\nabla^2 V = \frac{1}{r}\frac{\partial}{\partial r}\left(r\frac{\partial V}{\partial r}\right) + \frac{1}{r^2}\frac{\partial^2 V}{\partial \psi^2} + \frac{\partial^2 V}{\partial z^2} = 0. \qquad (A. 1)$$

For circular symmetry, V becomes independent of ψ, hence the term with ψ vanishes and (A. 1) becomes

$$\frac{\partial}{\partial r}\left(r\frac{\partial V}{\partial r}\right) + r\frac{\partial^2 V}{\partial z^2} = 0. \qquad (A. 2)$$

Now, let the potential $V = V(r, z)$ in space be expressed by a power series in r of the potential $V(o, z) = \phi$ along the axis, namely

$$V(r, z) = \phi_0 + r\phi_1 + r^2\phi_2 + r^3\phi_3 + \ldots = \sum_{n=0}^{n} r^n\phi_n. \qquad (A. 3)$$

The coefficients ϕ_n can be determined by substitution of (A. 3) into (A. 2). For this purpose (A. 3) is differentiated with respect to r, then multiplied by r and we obtain:

$$r\frac{\partial V}{\partial r} = r\sum_{n=0}^{\infty} nr^{n-1}\phi_n = \sum_{n=0}^{\infty} nr^n\phi_n. \qquad (A. 4)$$

Equation (A. 4) differentiated with respect to r yields

$$\frac{\partial}{\partial r}\left(r\frac{\partial V}{\partial r}\right) = \sum_{n=0}^{\infty} n^2 r^{n-1}\phi_n. \qquad (A. 5)$$

On the other hand if we differentiate (A. 3) twice with respect to z we obtain after multiplication with r

$$r\frac{\partial^2 V}{\partial z^2} = \sum_{n=0}^{\infty} r^{n+1}\phi_n''. \qquad (A. 6)$$

Substitution of (A. 5) and (A. 6) respectively for the two terms of (A. 2) yields after a slight rearrangement

$$\left(0\times\phi_0+\phi_1+\sum_{n=2}^{\infty} n^2 r^{n-1}\phi_n\right)+\sum_{n=1}^{\infty} r^{n-1}\phi''_{n-2} = 0. \quad (A.\ 7)$$

It can now be shown first, that ϕ_1 must be zero. If (A. 3) is differentiated with respect to r, and if $\partial V/\partial r$ is taken on the axis where $r = 0$, all terms containing r will vanish and there remains

$$\phi_1 = \left(\frac{\partial V}{\partial r}\right)_{r=0} = -(E_r)_{r=0} = 0 \quad (A.\ 8)$$

where $(E_r)_{r=0}$ stands for the radial field component on the axis which must be zero owing to the postulated axial symmetry of the lens field. Equation (A. 7) with $\phi_1 = 0$ can then be written

$$\sum_{n=2}^{\infty} r^{n-1}(n^2\phi_n+\phi''_{n-2}) = 0 \quad (A.\ 9)$$

and since off the axis $r \neq 0$, it follows that

$$\phi_n = \frac{1}{n^2}\,\phi''_{n-2}. \quad (A.\ 10)$$

If this recursion formula (A. 10) is applied, starting with $\phi_1 = 0$ (see (A. 8)) the conclusion can be drawn that all ϕ with odd indices ($n = 1, 3, 5 \ldots$) must vanish. We find for instance

$$\phi_3 = -\frac{1}{3^2}\,\phi''_1$$

where

$$\phi''_1 = -\frac{\partial^2}{\partial z^2}\,(E_r)_{r=0}$$

vanishes since, owing to (A. 8), $(E_r)_{r=0}$ does not change along the z-axis. Hence all derivatives of $(E_r)_{r=0}$ with regard to z must be zero. Moreover, with the help of the recursion formula (A. 10), all terms of even index n in (A. 3) can be reduced to derivatives of ϕ_0 and we obtain

$$V(r, z) = \phi_0-\frac{r^2}{4}\,\phi''_0+\frac{r^4}{64}\,\phi_0^{iv}+\ldots \quad (A.\ 11)$$

i.e. the potential $V(r, z)$ in space is uniquely determined by the potential $V(0, z) = \phi_0$ on the axis. Equation (A. 11) applies to any scalar potential V, for instance to an electrostatic potential or to a scalar magnetic potential V_H.

It is easily seen, that (A. 11) must also apply to the axial component of the magnetic flux density B_z. For, we have

$$B_z = -\mu_0 \frac{\partial V_H}{\partial z}.$$

Equation (A. 11) as applied to the magnetic potential V_H, can be differentiated with respect to z and after multiplication with the permeability μ_0 the following equation is obtained

$$B_z(r, z) = B_z(0, z) - \frac{r^2}{4} B_z''(0, z) + \frac{r^4}{64} B_0^{\mathrm{iv}}(0, z) + \dots \quad \text{(A. 12)}$$

GEOMETRY OF AN EMISSION SYSTEM AND THE OBSERVED ENERGY DISTRIBUTION OF ELECTRONS

The energy distribution of electrons projected by an electron gun generally is approximately Maxwellian (cf. §7.1: (7.4) and (7.5)). Since it is of greatest importance for the design of high resolution electron beam spectrometers to employ guns which project beams of the smallest obtainable energy bandwidth $e\Delta V_b$ it is worthwhile to have some definition for ΔV_b which is accessible to experimental determination.

The distribution function which we expect to observe, for example with a deflection analyser, depends on the geometry of the emission system. If this is of spherical symmetry, as for example with electrons emitted from the tip of a hairpin cathode, the paraxial electron trajectories are perpendicular to the equipotentials. Hence, with some good approximation, we can expect a differential Maxwellian distribution of the form

$$f(x)\,dx = x\,e^{-x}\,dx \qquad\qquad \text{(B. 1)}$$

where
$$x = \frac{eV}{kT} \qquad\qquad \text{(B. 2)}$$

and
$$f(x) = n(eV)/n_e.$$

Here eV is the energy of an electron, k is the Boltzmann constant, T is the absolute temperature, n is the number of electrons with energies between eV and $eV + ed V$ and n_e is the total number of electrons.

It is easily seen that the maximum of the distribution has the ordinate $f_{max} = 1/e$, hence

$$\frac{f_{max}}{2} = x_1\,e^{-x_1} = \frac{1}{2e}. \qquad\qquad \text{(B. 3)}$$

Numerical evaluation of equation (B. 3) yields the two solutions

$$x_1 = 0\cdot231 \quad \text{and} \quad x_1' = 2\cdot68.$$

Substitution of $\Delta x = x_1' - x_1$, into equation (B. 2) gives us the half-width of the energy distribution (e.g. in electron volts) in terms of the cathode temperature, namely

$$\Delta V_b \simeq 0\cdot 21 \times 10^{-3} T. \tag{B. 4}$$

If the energy distribution of the electrons is determined by a retarding field analyser, we expect to obtain the integral distribution corresponding to equation (B. 1), namely

$$F(x)\,\mathrm{d}x = (1 + x)\,\mathrm{e}^{-x}\mathrm{d}x \tag{B. 5}$$

where $F(x) = N(eV)/n_e$, and $N(eV)$ is the number of all velocity components with energies greater than eV. The integral width ΔV of the energy distribution in a beam can be defined (cf. Legler, 1963) by

$$\Delta V \{f(V)\}_{max} = \int_0^\infty f(V)\,\mathrm{d}V \tag{B. 6}$$

where $\{f(V)\}_{max}$ corresponds to the most frequent energy of the differential distribution. ΔV can be determined from the intercepts of the steepest tangent to the integral energy distribution curve with the abscissa and with the line $F(V) = 100$ per cent respectively.

According to Boersch (1954), however, it is most convenient to characterize the energy bandwidth of the integral Maxwellian distribution by ΔV_{63} corresponding to the *smallest* energy interval which contains $(1 - \mathrm{e}^{-1})n_e \simeq 63$ per cent of the total number of electrons, and this definition has been adopted by various other authors. Numerical evaluation of $F(x_1) - F(x_2) = 1 - \mathrm{e}^{-1} = 0\cdot 632$ yields a minimum $x_2 - x_1 = \Delta x = 2\cdot 0$ between $x_1 = 0\cdot 3$ and $x_2 = 2\cdot 3$. Using (B. 2) we obtain

$$\Delta V_{63} = 2kT/e = 0\cdot 172 \times 10^{-3} T. \tag{B. 7}$$

The best resolution of a spectrometer, equipped with a tungsten cathode of $2\,500^\circ$ K and with a perfect energy analyser, would, according to (B. 4) and to (B. 7), be

$$\Delta V_{63} = 0\cdot 43 \text{ eV} \quad \text{and} \quad \Delta V_b = 0\cdot 51 \text{ eV}$$

respectively, i.e. no energy differences smaller than about half an electron volt could be resolved.

It is surprising, however, that the bandwidths of energy distributions in electron beams are usually even much greater than those calculated from the temperature of the cathode. The bandwidth actually corresponding to the cathode temperature is obtained only at very small current densities in the beam, say, of the order of $i \lesssim 10^{-9}$ A mm^{-2} (cf. Boersch, 1954 and Hartwig and Ulmer, 1963 a, b).

PLOTTING OF SPECTRAL DISTRIBUTION CURVES

This is quite elementary; however, as Kollath (1936) pointed out, mistakes about it are met so frequently in the literature that a short discussion of it seems worthwhile.

The momentum distribution curve of a given total number of electrons indicates the number of electrons with momenta between the values p and $p + dp$ and may be represented by

$$N(p)\,dp = dN. \tag{C. 1}$$

The distribution of energies W in the same assembly of electrons may be written as

$$n(W)\,dW = dn. \tag{C. 2}$$

Now since

$$W = \frac{p^2}{2m}, \tag{C. 3}$$

the bandwidths dp and dW in (C. 1) and (C. 2) are connected by

$$dW = \frac{p}{m}\,dp, \tag{C. 4}$$

i.e. the abscissa interval of the momentum band dp is changed into the abscissa interval of the energy band dW by multiplication with p/m. However, if the band area, i.e. the number of electrons belonging to the band, is supposed to be unchanged by the transformation, dN in (C. 1) should equal dn in (C. 2). Hence the ordinate values of the two bands are connected by

$$n(W) = \frac{m}{p}\,N(p). \tag{C. 5}$$

A general survey of the plotting of momentum and energy distribution curves respectively from the result of the four important experimental methods is given in table C. 1. There, column (2) refers to a magnetic deflexion spectrometer with constant field. The current indicator may be a photographic plate of known sensitivity or any movable electron collector of given aperture. Thus, B = const., dr = const. and r = variable. Since $p = (eB)\,r_e$ and $dp = (eB)\,dr$, the collector current $I = f(r)\,dr$ is, according to (C. 1), proportional to the ordinate of the momentum distribution curve, i.e. $N(p) \propto I$, its abscissa being proportional to the radius of the electron orbit, i.e.

$$p \propto r_e.$$

TABLE C. 1. *The plotting of spectral distribution curves*

	Magnetic spectrometry		Electric spectrometry	
	B = const.	B = variable	Variable deflecting field	Variable retarding field
Values measured directly:				
Abscissa	Orbital radius r_e	Flux density B	Condenser voltage V_p	Retarding voltage V_z
Ordinate	Collector current I	Collector current I	Collector current I	Collector current I
Plotted as momentum distribution curve:				
Abscissa	r_e	B	$\sqrt{V_p}$	$\sqrt{V_z}$
Ordinate	I	I/B	$I/\sqrt{V_p}$	$dI/d\sqrt{V_z}$
Plotted as energy distribution curve:				
Abscissa	r_e^2	B^2	V_p	V_z
Ordinate	I/r_e	I/B^2	I/V_p	dI/dV_z

On the other hand, for plotting the energy distribution curve from measurements with B = const., dr = const. and r = variable, (C. 3) and (C. 4) show that

$$W \propto r_e^2, \qquad (C.\ 6)$$

hence, $dW \propto r_e\,dr$, and since $dn = dN$ is postulated, it follows that

$$n(W) \propto \frac{1}{r_e}\,N(p) \propto \frac{1}{r_e}\,I, \qquad (C.\ 7)$$

i.e. the abscissae W of the energy distribution curve are proportional to the square of the measured radii r_e of the orbits, while the ordinates $n(W)$ are proportional to the collector currents divided by these radii.

Column (3) of table C. 1 refers to any magnetic spectrometer (deflexion field, lens, solenoidal field, etc.) with fixed collector position and constant collector aperture. There the electron momentum is proportional to the variable flux density B. The current which, as a function of B, enters the collector aperture is $I \propto f(p)\,dp$. However, since here $dp \propto dB \propto B$, the ordinate of the momentum distribution curve $f(p)$ should be proportional to the collector current I divided by the flux density B, as given in the table. Again, in the present case of variable B, abscissae ($\propto B^2$) and ordinates ($\propto I/B^2$) have to be plotted for the energy distribution curve, as can be shown by considerations quite analogous to those given above.

The functions to be plotted as coordinates of spectral distribution curves derived from electric deflecting field measurements are given in column (4) of table C. 1. In practice, one generally uses variable electric deflexion fields, hence the current I, picked up by a collector of given aperture ds, is measured as a function $f(V_p)$ of a condenser plate voltage V_p.

In any case, the coordinate s of the collector slot (s may stand, for example, for the range z_f of the parabolic path in (10.28) or it may stand for the radius r_0 of the circular path in (10.29)) is proportional to the kinetic energy W of the electron and inversely proportional to the condenser voltage V_p, namely,

$$s \propto W/V_p. \tag{C. 8}$$

Hence
$$ds \propto (W/V_p^2)\,dV_p. \tag{C. 9}$$

In particular for fixed collector position $s = $ const. there follows from (C. 8)
$$W \propto V_p. \tag{C. 10}$$

(C. 9) and (C. 10) yield
$$ds \propto dV_p/V_p, \tag{C. 11}$$

which shows that for fixed collector aperture ds one obtains $dV_p \propto V_p$. The ordinate of the energy distribution (C. 2) is consequently given by

$$n(W) = dn/dW \propto I/dV_p \propto I/V_p. \tag{C. 12}$$

(C. 10) and (C. 12) show that the measured function $I = f(V_p)$ leads to an energy distribution the abscissa of which is proportional to the potential difference at the condenser plate and the ordinate of which is proportional to the collector current divided by the corresponding condenser plate voltage.

Again, calculation of the momentum distribution (cf. (C. 1)–(C. 5)) from (C. 10) and (C. 12) leads to an abscissa $p \propto \sqrt{V_p}$ and to an ordinate $N(p) \propto I/\sqrt{V_p}$ as shown in the table.

The last column of table C. 1 deals with the determination of the coordinates of momentum and energy distribution curves from measurements with the electric retarding field method. The electron-optical problem connected with this important method is given by the requirement of keeping the retarded electrons exactly parallel with the electric field lines. In practice, the electron beam often passes through two grid electrodes between which is maintained a decelerating electric field, though such a field cannot be directed everywhere exactly parallel and opposite to the electron motion. An electron can overcome the field only if its kinetic energy W surpasses the voltage difference V_z between the two grid electrodes. All electrons which are able to pass the field enter the collector and contribute to the current.

$$I = \int_{W_1=eV_z}^{W_2=\infty} n(W)\,\mathrm{d}W. \tag{C. 13}$$

Hence, to find the ordinate $n(W)$ of the energy distribution curve the measured currents have to be differentiated, i.e.

$$n(W) = \mathrm{d}I/\mathrm{d}V_z, \tag{C. 14}$$

the abscissae $W \propto V_z$ being proportional to the retarding voltage. The calculation (cf. (C. 1)–(C. 5)) of the corresponding momentum distribution leads to values $p \propto V_z$ for the abscissa and

$$N(p) \propto \mathrm{d}I/\mathrm{d}\sqrt{V_z} \tag{C. 15}$$

for the ordinate as shown in table C. 1.

Table C. 1 applies to moderate electron velocities for which the relativistic change in mass can be neglected. The relatively fast electrons of β-spectra are generally investigated with magnetic spectrometers which immediately yield momentum distribution curves with coordinates as specified in table C. 1. Spectrometric investigations, however, frequently aim at the determination of atomic or nuclear energy levels. Hence energy distributions have to be calculated from the well-known relativistic formulae.

On the other hand, the electric deflexion or the retarding field method immediately yield energy distribution curves with co-ordinates as specified in table C. 1. However, electric methods are in practice not frequently applied to the spectrometry of fast electrons.

Another problem arising in the transition from momenta to energies and vice versa is concerned with the resolving power. From

$$p^2 = (mu)^2 = 2mW = 2meV, \qquad \text{(C. 16)}$$

there follows

$$p/dp = G(W/dW). \qquad \text{(C. 17)}$$

The value of the factor G for moderate electron energies, i.e. for $m = m_0$, is $G = 2$. There, the momentum resolving power has twice the value of the energy-resolving power. In the relativistic region, however, the total energy of the electron is given by

$$U = eV + m_0 c^2. \qquad \text{(C. 18)}$$

It is connected with the momentum by

$$U^2/c^2 = p^2 + m_0^2 c^2, \qquad \text{(C. 19)}$$

where c and m_0 are the velocity of light and the rest mass of the electron respectively. Differentiating (C. 19) and combining with (C. 17) and (C. 18) yields

$$G = \left(1 + \frac{2m_0 c^2}{eV}\right) \Big/ \left(1 + \frac{m_0 c^2}{eV}\right). \qquad \text{(C. 20)}$$

Hence one finds, for instance,

Energy in 10^6 eV	V	0·1	0·5	1	10
Factor G in (C. 17)	G	1·81	1·51	1·34	1·05

TRAJECTORY COMPUTATION IN ELECTROSTATIC FIELDS

As a supplement to the discussion of ray tracing in §3.6, we give here the mathematical details of the numerical integration of the trajectory equations using a procedure which is suitable for digital computation. The method is applicable to the general form of the equations of motion in two-dimensions and can thus be used for zonal ray tracing.

The Newtonian equations of motion (3.22) refer to the meridional (r, z) plane of systems of rotational symmetry, or, with a change of variables, to two dimensional (x, y) coordinates. They can be integrated using a method outlined by Hechtel (1962).

Assuming that, at time zero, the axial and radial components of an electron are z_0 and r_0 respectively, and its velocity components are $\left(\dfrac{dz}{dt}\right)_0$ and $\left(\dfrac{dr}{dt}\right)_0$, z and r at time t can be written as power series, viz.:

$$z = z_0 + \left(\frac{dz}{dt}\right)_0 t + \frac{1}{2}\left(\frac{d^2z}{dt^2}\right)_0 t^2 + \frac{1}{6}\left(\frac{d^3z}{dt^3}\right)_0 t^3, \qquad \text{(D. 1)}$$

$$r = r_0 + \left(\frac{dr}{dt}\right)_0 t + \frac{1}{2}\left(\frac{d^2r}{dt^2}\right)_0 t^2 + \frac{1}{6}\left(\frac{d^3z}{dt^3}\right)_0 t^3 \qquad \text{(D. 2)}$$

and hence the velocity components at time t are given by

$$\frac{dz}{dt} = \left(\frac{dz}{dt}\right)_0 + \left(\frac{d^2z}{dt^2}\right)_0 t + \frac{1}{2}\left(\frac{d^3z}{dt^3}\right)_0 t^3, \qquad \text{(D. 3)}$$

$$\frac{dr}{dt} = \left(\frac{dr}{dt}\right)_0 + \left(\frac{d^2r}{dt^2}\right)_0 t + \frac{1}{2}\left(\frac{d^3r}{dt^3}\right)_0 t^3. \qquad \text{(D. 4)}$$

The accelerations at time zero are given from (3.22) by

$$\left(\frac{d^2z}{dt^2}\right)_0 = \frac{e}{m}\left(\frac{\partial V}{\partial z}\right)_0, \qquad \text{(D. 5)}$$

$$\left(\frac{d^2r}{dt^2}\right)_0 = \frac{e}{m}\left(\frac{\partial V}{\partial r}\right)_0. \qquad \text{(D. 6)}$$

Differentiation of (D. 5) and (D. 6) yields

$$\left(\frac{\mathrm{d}^3 z}{\mathrm{d}t^3}\right)_0 = \frac{e}{m}\left(\frac{\partial^2 V}{\partial z^2}\right)_0\left(\frac{\mathrm{d}z}{\mathrm{d}t}\right)_0 + \frac{e}{m}\left(\frac{\partial^2 V}{\partial z\,\partial r}\right)_0\left(\frac{\mathrm{d}r}{\mathrm{d}t}\right)_0, \quad \text{(D. 7)}$$

$$\left(\frac{\mathrm{d}^3 r}{\mathrm{d}t^3}\right)_0 = \frac{e}{m}\left(\frac{\partial^2 V}{\partial z\,\partial r}\right)_0\left(\frac{\mathrm{d}z}{\mathrm{d}t}\right)_0 + \frac{e}{m}\left(\frac{\partial^2 V}{\partial r^2}\right)_0\left(\frac{\mathrm{d}r}{\mathrm{d}t}\right)_0. \quad \text{(D. 8)}$$

It is convenient to work in terms of a normalized time which is defined by

$$\tau = \left(\frac{2e}{m}\right)^{\frac{1}{2}} t. \quad \text{(D. 9)}$$

But the initial velocity of the electron u_0 is given in terms of the initial potential V_0 by

$$u_0 = \sqrt{\frac{2e}{m}}\, V_0 \quad \text{(D. 10)}$$

so that if the trajectory initially makes an angle Θ_0 with the z-axis, then

$$\left(\frac{\mathrm{d}z}{\mathrm{d}\tau}\right)_0 = (V_0)^{\frac{1}{2}} \cos \Theta_0, \quad \text{(D. 11)}$$

$$\left(\frac{\mathrm{d}r}{\mathrm{d}\tau}\right)_0 = (V_0)^{\frac{1}{2}} \sin \Theta_0. \quad \text{(D. 12)}$$

Substitution of (D. 5)–(D. 9) into (D. 1)–(D. 4) leads to the following expressions for the coordinates z_1, r_1 and the velocity components

$$\left(\frac{\mathrm{d}z}{\mathrm{d}\tau}\right)_1, \quad \left(\frac{\mathrm{d}r}{\mathrm{d}\tau}\right)_1,$$

of an electron after a short interval of normalized time:

$$z_1 = z_0 + \left(\frac{\mathrm{d}z}{\mathrm{d}\tau}\right)_0 \Delta\tau + \frac{1}{4}\left(\frac{\partial V}{\partial z}\right)_0 \Delta\tau^2$$
$$+ \frac{1}{12}\left[\left(\frac{\partial^2 V}{\partial z^2}\right)_0\left(\frac{\mathrm{d}z}{\mathrm{d}\tau}\right)_0 + \left(\frac{\partial^2 V}{\partial z\,\partial r}\right)_0\left(\frac{\mathrm{d}r}{\mathrm{d}\tau}\right)_0\right]\Delta\tau^3, \quad \text{(D. 13)}$$

$$r_1 = r_0 + \left(\frac{\mathrm{d}r}{\mathrm{d}\tau}\right)_0 \Delta\tau + \frac{1}{4}\left(\frac{\partial V}{\partial r}\right)_0 \Delta\tau^2$$
$$+ \frac{1}{12}\left[\left(\frac{\partial^2 V}{\partial r^2}\right)_0\left(\frac{\mathrm{d}r}{\mathrm{d}\tau}\right)_0 + \left(\frac{\partial^2 V}{\partial z\,\partial r}\right)_0\left(\frac{\mathrm{d}z}{\mathrm{d}\tau}\right)_0\right]\Delta\tau^3, \quad \text{(D. 14)}$$

$$\left(\frac{dz}{d\tau}\right)_1 = \left(\frac{dz}{d\tau}\right)_0 + \frac{1}{2}\left(\frac{\partial V}{\partial z}\right)_0 \Delta\tau$$

$$+ \frac{1}{4}\left[\left(\frac{\partial^2 V}{\partial z^2}\right)_0 \left(\frac{dz}{d\tau}\right)_0 + \left(\frac{\partial^2 V}{\partial z\,\partial r}\right)_0 \left(\frac{dr}{d\tau}\right)_0\right]\Delta\tau^2, \quad \text{(D. 15)}$$

$$\left(\frac{dr}{d\tau}\right)_1 = \left(\frac{dr}{d\tau}\right)_0 + \frac{1}{2}\left(\frac{\partial V}{\partial r}\right)_0 \Delta\tau$$

$$+ \frac{1}{4}\left[\left(\frac{\partial^2 V}{\partial r^2}\right)_0 \left(\frac{dr}{d\tau}\right)_0 + \left(\frac{\partial^2 V}{\partial z\,\partial r}\right)_0 \left(\frac{dr}{d\tau}\right)_0\right]\Delta\tau^2. \quad \text{(D. 16)}$$

The initial values, $\left(\frac{dz}{d\tau}\right)_0$ and $\left(\frac{dr}{d\tau}\right)_0$ are deduced from V_0 and Θ_0 using (D. 11) and (D. 12). The new values z_1, r_1, $\left(\frac{dr}{d\tau}\right)_1$, $\left(\frac{dz}{d\tau}\right)_1$ are taken as the initial conditions for the second step.

The accuracy of the method is dependent on the choice of time interval (cf. §3.6), and in this connexion it is useful to note that the length of the trajectory Δs during a given step is related to the normalized time interval and the local potential value V according to

$$\Delta\tau = \frac{\Delta s}{\sqrt{V}}. \quad \text{(D. 17)}$$

The potential and its various derivatives used in (D. 13)–(D. 16) must be calculated from the array of potential values (assumed to be already computed by one of the methods described in §3.1–§3.3) which are stored in the computer memory. We may suppose that the potential is known at each node of a network covering the region of interest. In calculating the derivatives of V at an arbitrary point (z, r) of the trajectory, the computer must first determine which node of the network is closest to the point (z, r). If the coordinates of this node are $z = m$, $r = n$ and its potential $V_{m,n}$, then the potential $V(z, r)$ is given approximately by the following second-order interpolation formula

$$V(z, r) = V_{m,n} + \left(\frac{\partial V}{\partial z}\right)_{m,n} (z-m) + \left(\frac{\partial V}{\partial r}\right)_{m,n} (r-n) + \frac{1}{2}\left(\frac{\partial^2 V}{\partial z^2}\right)_{m,n}$$

$$\times (z-m)^2 + \frac{1}{2}\left(\frac{\partial^2 V}{\partial r^2}\right)_{m,n} (r-n)^2 + \left(\frac{\partial^2 V}{\partial z\,\partial r}\right)_{m,n} (z-m)(r-n).$$

$$\text{(D. 18)}$$

The first derivatives of V at the same point are

$$\frac{\partial V}{\partial z}(z,r) = \left(\frac{\partial V}{\partial z}\right)_{m,n} + \left(\frac{\partial^2 V}{\partial z^2}\right)_{m,n}(z-m) + \left(\frac{\partial^2 V}{\partial z\,\partial r}\right)_{m,n}(r-n),$$
(D. 19)

$$\frac{\partial V}{\partial r}(z,r) = \left(\frac{\partial V}{\partial r}\right)_{m,n} + \left(\frac{\partial^2 V}{\partial r^2}\right)_{m,n}(r-n) + \left(\frac{\partial^2 V}{\partial z\,\partial r}\right)_{m,n}(z-m).$$
(D. 20)

The second derivatives of V at (z,r) are taken as being identical with the second derivatives at (m,n).

Finally, the derivatives of V at the node m,n which are required in (D. 19) and (D. 20) are obtained from the potential of the nine neighbouring nodes by means of the following difference equations, the unit of distance being one mesh length of the network,

$$\left(\frac{\partial V}{\partial z}\right)_{m,n} = \tfrac{1}{2}(V_{m+1,n} - V_{m-1,n}),$$
(D. 21)

$$\left(\frac{\partial V}{\partial r}\right)_{m,n} = \tfrac{1}{2}(V_{m,n+1} - V_{m,n-1}),$$
(D. 22)

$$\left(\frac{\partial^2 V}{\partial z^2}\right)_{m,n} = V_{m+1,n} - 2V_{m,n} + V_{m-1,n},$$
(D. 23)

$$\left(\frac{\partial^2 V}{\partial r^2}\right)_{m,n} = V_{m,n+1} - 2V_{m,n} + V_{m,n-1},$$
(D. 24)

$$\left(\frac{\partial^2 V}{\partial z\,\partial r}\right)_{m,n} = \tfrac{1}{4}(V_{m+1,n+1} - V_{m-1,n+1} + V_{m-1,n-1} - V_{m+1,n-1}).$$
(D. 25)

For a point on the electron trajectory very close to a boundary electrode, some of the neighbouring points may be missing. In this case the equations (D. 21)–(D. 25) are subject to modification. Appropriate alternative sets of difference equations have been listed by Hechtel (1962).

We now summarize the use of the sets of equations derived above in a trajectory computation. The electron trajectory is determined in a step-by-step manner by repeated application of the equations (D. 13)–(D. 16). These equations give the position and velocity of an electron at the end of a short time interval $\Delta\tau$ in terms of its position and velocity at the beginning of the interval.

The final coordinates of one step are used as the initial coordinates for the next.

In order to evaluate the right-hand sides of (D. 13)–(D. 16), it is necessary to compute the value of the derivatives of the electrostatic potential $\dfrac{\partial V}{\partial z}$, $\dfrac{\partial V}{\partial r}$ etc., at the initial point of each step, e.g. at (z_0, r_0). This is a two-stage process, since the potential is not normally known at every point in space, but only at a finite array of mesh points. Thus equations (D. 19) and (D. 20) are used to express the required partial derivatives in terms of partial derivatives $\left(\dfrac{\partial V}{\partial z}\right)_{m,n}$, $\left(\dfrac{\partial V}{\partial r}\right)_{m,n}$, $\left(\dfrac{\partial^2 V}{\partial z^2}\right)_{m,n}$ at the mesh point (m, n) nearest to the trajectory point, and these derivatives at the mesh point are themselves given in terms of the known potential values at surrounding mesh points by the equations (D. 21)–(D. 25). The array of potential values $V_{m,n}$ are stored in the computer memory.

At each step, therefore, the equations (D. 21)–(D. 25) are substituted into (D. 19) and (D. 20), and the resulting expressions for the partial derivatives of V at the trajectory point are substituted into (D. 13)–(D. 16).

BIBLIOGRAPHY AND AUTHOR INDEX

[Numbers in brackets following each reference indicate
the text pages where the reference has been cited.]

Agnew, H. M. and Anderson, H. L. (1949). *Rev. Sci. Instr.* **20**, 869.
[194, 243, 245]

Allison, S. K., Frankel, S. P., Hall, T. A., Montegue, J. H., Morrish,
A. H. and Warshaw, S. D. (1949). *Rev. Sci. Instr.* **20**, 735. [409]

Amboss, K. (1964). *I.E.E.E. Trans.* ED-11, 479. [289]

Anderson, H. L. *See under* Agnew, H. M.

Anderson, W. H. J. (1967). *Br. J. Appl. Phys.* **18**, 1573. [437]

Ando, K., Kamagaito, O., Kamiya, Y., Takashi, S. and Uyeda, R. (1959).
J. Phys. Soc. Japan **14**, 181. [322]

André, B. (1967). *C.R. Acad. Sci.* B **264**, 18. [370]

Archard, G. D. (1953). *J. Sci. Instr.* **30**, 352. [163]

Archard, G. D. (1954). *Proc. 3rd Intl. Conf. El. Microsc.* p. 97. [104, 157]

Archard, G. D. (1955). *Proc. Phys. Soc.* B **68**, 156. [217]

Archard, G. D. (1958). *Rev. Sci. Instr.* **29**, 1049. [173]

Archard, G. D. (1959). *Proc. Phys. Soc.* **74**, 177. [309]

Ardenne, M. v. (1939). *Z. Phys.* **112**, 744. [189]

Ash, E. A. (1955). *J. Appl. Phys.* **26**, 327. [295]

Ashby, N. (1958). *Nucl. Instr. Meth.* **3**, 90. [414]

Ashkin, A. (1957). *J. Appl. Phys.* **28**, 564. [360]

Austin, N. A. and Fultz, C. S. (1959). *Rev. Sci. Instr.* **30**, 284. [379]

Bachman, C. H. and Ramo, S. (1943). *J. Appl. Phys.* **14**, 8, 69, 155.
[91]

Baker, W. R. *See under* Panofsky, W. K. H.

Bakish, R. (ed.) (1962). *Introduction to Electron Beam Technology*, pp. 168,
184. New York. [386]

Barber, M. R. and Sander, K. F. (1958). *Proc. Inst. Elec. Engrs* **105**, 901.
[309]

Barber, N. F. (1933). *Proc. Leeds Phil. Lit. Soc.* **2**, 427. [419]

Barford, N. C. (1957). *J. El'onics and Control* **3**, 63. [299]

Barford, N. C. *See also under* Lubszynski, H. G.

Barham, P. M. *See under* Hopkins, H. H.

Barnes, R. L. and Openshaw, I. K. (1968). *J. Sci. Instr.* **1**, 628. [190]

Barnett, M. E., Bates, C. W. and England, L. (1969). *Adv. El'onics* **28** A,
545. [96]

Barnett, M. E. and Nixon, W. C. (1967). *Optik* **26**, 310. [267]

Barthere, J. *See under* Durandeau, P.

Bartky, W. and Dempster A. J. (1929). *Phys. Rev.* **33**, 1019. [438]

Bartz, G., Weissenberg, G. and Wiskott, D. (1954). *3rd Intl. Conf. El.
Micros.* p. 395. [96]

Bas, E. B., Cremosnik, G. and Lerch, H. (1962). *Trans. 8th Vac. Symp.*,
p. 817. [322, 380]

Bas, E. B., Cremosnik, R. N. and Wiedmer, H. B. (1967). *Z. angew. Math. Phys.* **18**, 379. [377, 380]

Bas, E. B. and Gaug, H. (1967). *Z. angew. Math. Phys.* **18**, 557; *cf. also* 605, 747. [377, 380, 387]

Bas, E. B. and Gaydou, F. (1959). *Z. angew. Phys.* **11**, 370. [377]

Bas, E. B. and Preuss, L. (1959). *Z. angew. Math. Phys.* **10**, 533. [70, 81]

Bas, E. B. and Preuss, L. (1963). *Z. angew. Math. Phys.* **14**, 84. [70]

Bas, E. B. and Preuss, L. (1964). *Optik* **21**, 261. [337]

Bas, E. B. *See also under* Preuss, L.

Bates, C. W. *See under* Barnett, M. E.

Bauer, H. D. (1966). *Optik* **23**, 596. [105, 221, 222]

Beck, A. H. (1966). *Handbook of Vacuum Physics* vol. 2, pt. 2: *Thermionic Emission*. Oxford. [322]

Beck, A. H. and Maloney, C. E. (1967). *Br. J. Appl. Phys.* **18**, 845.

Beck, A. H. *See also under* Gregory, B. C.

Becker, H. and Wallraff, A. (1938). *Arch. Elektrotech.* **32**, 664. [187, 188]

Becker, H. and Wallraff, A. (1939). *Arch. Elektrotech.* **33**, 491. [203]

Becker, H. and Wallraff, A. (1940*a*). *Arch. Elektrotech.* **34**, 43. [203, 210]

Becker, H. and Wallraff, A. (1940*b*). *Arch. Elektrotech.* **34**, 115. [119, 187, 213]

Becker, H. and Wallraff, A. (1940*c*). *Arch. Elektrotech.* **34**, 230. [203, 210]

Bedford, L. H. (1936). *J. Sci. Instr.* **13**, 177. [274]

Beiduck, F. M. and Konopinski, E. (1948). *Rev. Sci. Instr.* **19**, 594. [449]

Bennett, W. H. *See under* Roberts, A. S.

Berman, A. S. *See under* Lassettre, E. N.

Bernard, M. (1954). *Ann. Phys. Paris* **9**, 633. [30, 102]

Bernard, M. *See also under* Grivet, P.

Bertram, S. (1940). *Proc. Inst. Radio Engrs N.Y.* **28**, 418. [66]

Bevc, V., Palmer, J. L. and Susskind, C. (1958). *J. Br. Inst. Radio Engrs* **18**, 696. [304]

Birdsall, C. K. and Bridges, W. B. (1961). *J. Appl. Phys.* **32**, 2611. [311, 316]

Birdsall, C. K. and Bridges, W. B. (1963). *J. Appl. Phys.* **34**, 2946. [311, 316]

Bloch, J. (1936). *J. Sci. Instr.* **13**, 302. [55]

Blodgett, K. B. *See under* Langmuir, I.

Boersch, H. (1939). *Z. Tech. Phys.* **20**, 346. [178]

Boersch, H. (1940). *Naturwiss.* **28**, 709. [206]

Boersch, H. (1943). *Phys. Z.* **44**, 32. [260]

Boersch, H. (1949). *Optik* **5**, 436. [250]

Boersch, H. (1953). *Z. Phys.* **134**, 156. [250]

Boersch, H. (1954). *Z. Phys.* **139**, 118. [250, 466, 467]

Boersch, H., Bostanjoglo, O. and Lischke, B. (1966). *Optik* **24**, 460. [136]

Boersch, H., Geiger, J. and Stickel, W. (1964). *Z. Phys.* **180**, 415. [435]

Boersch, H. and Miessner, H. (1962). *Z. Phys.* **168**, 304. [248]

Bok, A., Kramer, J. and Le Poole, L. B. (1964). *Proc. European Regional Conference on Electron Microscopy, Prague*, appendix p. 9. [97]

Bol, K. *See under* Marton, L.

Bolton, H. C. *See under* Yarnold, G. D.

Borcherds, P. H. *See under* Haine, M. E.

Borgnis, F. (1943). *Ann. Phys. Lpz.* **43**, 616. [290]

Born, M. and Wolf, E. (1964). *Principles of Optics*. Oxford. [160]

Borries, B. v. (1941). *Tel. Fernsp. Funk.* **30**, 295. [396]

Borries, B. v. (1948). *Optik* **3**, 321, 389. [371]

Borries, B. v. and Dosse, J. (1938). *Arch. Elektrotech.* **32**, 221. [285]

Borries, B. v., Ruska, E., Krumm, J. and Müller, H. (1940). *Naturwiss*, **28**, 350. [134]

Bostanjoglo, O. *See under* Boersch, H.

Bothe, W. (1950a). *Naturwiss.* **37**, 41. [243]

Bothe, W. (1950b). *Heidelb. Akad.*, p. 191. [243]

Bouwers, A. (1935). *Physica* **2**, 145. [280]

Brack, K. (1962). *Z. Naturforsch.* **17**a, 1066. [248]

Bradley, D. E. (1961). *Nature, Lond.* **189**, 298. [322]

Braucks, F. W. (1958). *Optik* **15**, 242. [375]

Braucks, F. W. (1959). *Optik* **16**, 304. [375]

Brewer, G. R. (1955). *Hughes Aircraft Corp.* ETL 55–26. [319]

Brewer, G. R. (1957a). *Trans. Inst. Radio Engrs* ED–4, 132. [305]

Brewer, G. R. (1957b). *J. Appl. Phys.* **28**, 7. [360, 365]

Brewer, G. R. (1959). *J. Appl. Phys.* **30**, 1022. [289, 303]

Brewer, G. R. (1967). In *Focusing of Charged Particles* vol. 2 (ed. A. Septier). [317, 366]

Brewer, G. R. *See also under* Van Duzer, T.

Bridges, W. B. *See under* Birdsall, C. K.

Brillouin, L. (1945). *Phys. Rev.* **67**, 260. [301]

Broers, A. N. (1967). *J. Appl. Phys. (U.S.)* **38**, 1991. [322]

Broers, A. N. (1969). *J. Sci. Instr.* **2**, 273. [322]

Brown, W. F. and Sweer, J. H. (1945). *Rev. Sci. Instr.* **16**, 276. [138]

Browne, C. P., Craig, D. S. and Williamson, R. M. (1951). *Rev. Sci. Instr.* **22**, 952. [413, 459]

Brüche, E. and Johannson, H. (1932). *Naturwiss.* **20**, 353. [1, 325]

Brüche, E. and Scherzer, O. (1934). *Geometrische Elektronenoptik*. Berlin. [1, 83, 87]

Bruck, H. and Romani, L. (1944). *Cahiers de Physique* **24**, 15. [91]

Bucherer, A. H. (1909). *Ann. Phys. Lpz.* **28**, 513. [392]

Bull, C. S. (1945). *J. Inst. El. Engrs* pt. III **92**, 86. [385]

Bullock, M. L. (1955). *Am. J. Phys.* **23**, 264. [30]

Burfoot, J. C. (1952). *Br. J. Appl. Phys.* **3**, 22. [149]

Burfoot, J. C. (1954). *Proc. Phys. Soc.* B **67**, 523. [160, 214, 215]

Busch, H. (1926). *Ann. Phys. Lpz.* **81**, 974. [1, 115]

Calbick, C. J. *See under* Davisson, C. J.

Carré, B. A. (1961). *Computer J.* **4**, 73. [46]

Carré, B. A. and Wreathall, W. M. (1964). *Inst. El'onic and Radio Engrs.* **27**, 446. [45, 61, 149].

Cartan, L. (1937). *J. de Phys.* **8**, 453. [424]

Castaing, R. and Henry, L. (1962). *C.R. Acad. Paris* **255**B, 76. [97]

Castaing, R. and Henry, L. (1963). *J. Microscopie* **2**, 5. [97]

Castaing, R. and Henry, L. (1964). *J. Microscopie* **3**, 133. [97]

Chancon, P. (1947). *Ann. Phys. Paris* **2**, 333. [83, 186].

Child, D. C. (1911). *Phys. Rev.* **32**, 492. [312]

Chodorow, M., Ginzton, E. L., Hansen, W. W., Kyhl, R. L., Neal, R. B., and Panofsky, W. K. H. (1955). *Rev. Sci. Instr.* **26**, 134. [379]

Citron, A., Farley, F. J., Michaelis, E. and Øverås, H. (1959). *CERN Report* 59–8. [148]

Clark, G. L. (1955). *Applied X-rays.* New York. [385]

Classen, J. (1908). *Ver. Deu. Phys. Ges.* **10**, 700. [419]

Climer, B. J. (1962). *J. El'onics and Cont.* **13**, 385. [366]

Clogston, A. M. and Heffner, H. (1954). *J. Appl. Phys.* **25**, 436. [306]

Cockroft, J. D. (1933). *J. Sci. Instr.* **10**, 71. [459]

Cockroft, J. D., Ellis, C. D. and Kershaw, H. (1932). *Proc. Roy. Soc.* A **135**, 228. [403]

Coggeshall, N. D. (1947). *J. Appl. Phys.* **18**, 855. [450]

Cole, R. H. (1938). *Rev. Sci. Instr.* **9**, 215. [138].

Compton, K. T. *See under* Langmuir, I.

Conrady, A. E. (1929). *Applied Optical Design.* Oxford. [172]

Cook, C. S. *See under* Langer, L. M.

Cosslett, V. E. (1940). *Proc. Phys. Soc.* **52**, 511. [235]

Cosslett, V. E. (1946). *Introduction to Electron Optics.* Oxford. [76, 116, 266]

Cosslett, V. E. and Nixon, W. C. (1961). *X-ray Microscopy.* Cambridge. [374]

Cosslett, V. E. *See also under* Deltrap, J. H. M.; Haine, M. E.

Courant, E. D., Livingstone, M. S. and Snyder, H. S. (1952). *Phys. Rev.* **88**, 1190. [158]

Cox, J. L. *See under* Roberts, A. S.

Craig, D. S. *See under* Browne, C. P.

Craig, H. (1947). *Proc. Phys. Soc.* **59**, 804. [166]

Cremosnik, G. *See under* Bas, E. B.

Cremosnik, R. N. *See under* Bas, E. B.

Crewe, A. V. (1966). *Science* **154**, 729. [374]

Crewe, A. V., Eggenberger, D. W., Walter, L. M. and Wall, I. (1967). *J. Appl. Phys.* **38**, 4257; *Rev. Sci. Instr.* **39**, 576. [374]

Cutler, C. C. and Hines, M. E. (1955). *Proc. Inst. Radio Engrs* **43**, 307. [366]

Czarczynski, W., Ryley, J. E. and Gambling, W. A. (1966). *Int. J. Electr.* **21**, 51. [317]

Danielson, W. E., Rosenfeld, J. L. and Saloom, J. A. (1956). *Bell. Syst. Tech. J.* **35**, 375. [366]

Danysz, J. (1911). *C.R. Acad. Paris* **153**, 339. [419]

Davisson, C. J. and Calbick, C. J. (1931). *Phys. Rev.* **38**, 585. [1, 78]

Davne, L. *See under* Law, H. B.

Dawson, H. G. (1898). *Cf.* Jahnke–Emde: *Tables of Functions.* Leipzig (1938); New York (1945). [272]

De Beer, A. J., Groendijk, H. and Verster, J. L. (1961/2). *Philps Tech. Rev.* **23**, 352. [62]

Deltrap, J. H. M. (1964). Thesis, Cambridge. *Also cf.* Hawkes (1966, p. 78 ff.) [217]

Deltrap, J. H. M. and Cosslett, V. E. (1962). *Intl. Congr. El. Miscrosc.* **5**, KK-8. [217]

Dempster, A. J. *See under* Bartky, W.

Dennison, D. M. *See under* Shull, F. G.

Deserno, P. (1935). *Arch. Elektrotech.* **29**, 139. [441]

Deutsch, M., Elliot, L. G. and Evans, R. D. (1944). *Rev. Sci. Instr.* **15**, 178. [116, 233, 244]

Diels, K. and Wendt. G. (1937). *Z. Tech. Phys.* **18**, 65. [214]

Dietrich, W. (1958). *Z. Physik* **151**, 519. [243]

Dolder, K. T. and Klemperer, O. (1955). *J. Appl. Phys.* **26**, 1461. [286, 288]

Dolmatova, K. A. and Kelman, V. M. (1959). *Nucl. Instr. Meth.* **5**, 269. [195]

Dornfield, E. G. *See under* Reisner, J. H.

Dorrenstein, R. *See under* Francken, J. C.

Dosse, J. (1936). *Z. Tech. Phys.* **17**, 315. [109]

Dosse, J. (1940). *Z. Phys.* **115**, 530. [17, 332, 335, 353, 354, 355]

Dosse, J. (1941). *Z. Phys.* **117**, 316, 437, 722. [30, 137, 189, 226]

Dosse, J. *See also under* Borries, B. v.

Drude, P. (1925). *Theory of Optics* (trans. R. H. Millikan). New York. [10, 336]

Düker, H. *See under* Möllenstedt, G.

Dungey, J. W. and Hull, C. R. (1947). *Proc. Phys. Soc.* **59**, 828. [192]

Durandeau, P., Fagot, B., Barthere, J. and Laudet, M. (1959). *J. Phys. et Rad. Appl.* **20**, 80A. [143]

Durandeau, P. and Fert, C. (1957). *Rev. d'Optique* **36**, 205. [126, 129, 144, 189]

Dymikov, A. D., Fishkova, T. Ya., Ovsyanikova, L. P. and Yavor, S. (1966). *Nucl. Instr. Meth.* **42**, 293. [221]

Dymikov, A. D. and Yavor, S. (1964). *Zh. Tekh. Fiz.* **34**, 2008; *Sov. Phys. Tech. Phys.* **9**, 1544. [228]

Dymikov, A. D. *See also under* Yavor, S.

Edwardson, K. *See under* Siegbahn, K.

Eggenberger, D. W. *See under* Crewe, A. V.

Ehrenberg, W. and Jennings, J. C. E. (1952). *Proc. Phys. Soc.* B **65**, 265. [424]

Ehrenberg, W. and Siday, R. E. (1949). *Proc. Phys. Soc.* B **62**, 8. [8]

Einstein, P. A. (1951). *Br. J. Appl. Phys.* **2**, 49. [48]

Einstein, P. A. and Jacob, L. (1948). *Phil. Mag.* **39**, 20. [332]

Einstein, P. A. *See also under* Haine, M. E.

El-Kareh, A. (1961). *Rev. Sci. Instr.* **32**, 421. [105]

Elliot, L. G. *See under* Deutsch, M.

Ellis, C. D. (1930). In *Radiations from Radioactive Substances* (ed. E. Rutherford, J. Chadwick and C. D. Ellis). Cambridge. [459]

Ellis, C. D. *See also under* Cockcroft, J. D.

Engelhaaf, P. (1968). *Optik* **27**, 387. [381]

England, L. *See under* Barnett, M. E.

Epstein, W. D. (1936). *Proc. Inst. Radio Engrs N.Y.* **24**, 1095. [27, 176]

Epstein, W. D. *See also under* Maloff, I. G.

Evans, R. D. *See under* Deutsch, M.

Everhart, T. E. (1967). *J. Appl. Phys.* **38**, 4944. [60, 373]

Ewan, G. T. *See under* Graham, R. L.

Fagot, B. *See under* Durandeau, P.

Fagot, M., Ferré, J. and Fert, C. (1961). *C.R. Acad. Sci.* **252**, 3766. [259]

Fane, R. W. (1958). *Br. J. Appl. Phys.* **9**, 149. [323]

Farley, F. J. *See under* Citron, A.

Farnsworth, P. T. (1934). *J. Franklin. Inst.* **218**, 411. [82]

Feld, R. *See under* Lampert, I. E.

Felici, N. J. (1959). *J. Phys. Rad.* **20**, 97A. [59]

Fernandez-Moran, H. (1965). *Proc. Nat. Acad. Sci. USA* **53**, 445. [136]

Fernandez-Moran, H. (1966). *Argonne Report* ANL-7275. [136]

Ferré, J. *See under* Fagot, M.

Fert, C. *See under* Durandeau, P.; Fagot, M.

Field, L. M., Spangenberg, K. R. and Helm, R. (1947). *Elect. Commun.* **24**, 108. [273, 296]

Field, L. M. *See also under* Spangenberg, K. R.

Firestein, F. and Vine, J. (1963). *Br. J. Appl. Phys.* **14**, 449. [66, 183]

Fishkova, T. Ya. *See under* Dymikov, A. D.

Fowler, R. D. and Gibson, G. E. (1934). *Phys. Rev.* **46**, 1075. [270]

Francken, J. C. and Dorrenstein, R. (1951). *Philips Res. Rep.* **6**, 323. [46]

Francken, J. C., Gier, J. and Nienhuis, W. F. (1956). *Philips Tech. Rev.* **18**, 73. [328]

Françon, M. (1963). *Modern Applications of Physical Optics.* New York. [263]

Frankel, S. (1948). *Phys. Rev.* **73**, 804. [239, 245]

Frost, G. (1958). *Z. angew. Phys.* **10**, 546. [246]

Frost, R. D., Purl, O. T. and Johnson, H. R. (1962). *Proc. Inst. Radio Engrs* **50**, 1800. [365]

Fry T. C. (1932*a*). *Amer. Math. Monthly* **39**, 199. [39]

Fry, T. C. (1932*b*). *Bell. Syst. Monogr.* B**671**. [80]

Fultz, C. S. *See under* Austin, N. A.

Gabor, D. (1937). *Nature, Lond.* **139**, 373. [62]

Gabor, D. (1946). *Nature, Lond.* **158**, 198. [192]

Gambling, W. A. *See under* Czarczynski, W.

Gans, R. (1937). *Z. Tech. Phys.* **19**, 204. [60]

Gardener, M. E. *See under* Jungermann, J. A.

Gaug, H. *See under* Bas, E. B.

Gautier, P. (1954). *J. Phys. et Rad.* **15**, 684. [138]

Gaydou, F. *See under* Bas, E. B.

Geiger, J. *See under* Boersch, H.; Graham, R. L.

Geoffrion, C. (1949). *Rev. Sci. Instr.* **20**, 638. [449, 459, 460]

Gerholm, T. R. (1956). *Springer Encyclopaedia of Physics* **33**, 609. [233]

Gianola, U. F. (1950). *Proc. Phys. Soc.* B**63**, 1037. [192]

Gibson, G. E. *See under* Fowler, R. D.

Gier, J. *See under* Francken, J. C.

Gill, E. W. B. (1925). *Phil. Mag.* **49**, 993. [311]

Ginzton, E. L. and Wadia, B. H. (1954). *Proc. Inst. Radio Engrs* **42**, 1548. [299]

Ginzton, E. L. *See also under* Chodorow, M.

Glaser, A. and Henneberg, W. (1935). *Z. Tech. Phys.* **16**, 229. [76, 80]

Glaser, W. (1933 a). *Z. Phys.* **80**, 450. [8]

Glaser, W. (1933 b). *Z. Phys.* **81**, 649. [212]

Glaser, W. (1933 c). *Z. Phys.* **83**, 104. [160]

Glaser, W. (1935). *Z. Phys.* **97**, 177. [205]

Glaser, W. (1938). *Z. Phys.* **111**, 357. [444]

Glaser, W. (1940). *Z. Phys.* **116**, 19, 56, 737. [226]

Glaser, W. (1941 a). *Z. Phys.* **117**, 285. [118, 193]

Glaser, W. (1941 b). *Z. Phys.* **118**, 264. [118]

Glaser, W. (1949). *Ann. Phys. Lpz.* **4**, 389. [441]

Glaser, W. (1951). *Proc. Phys. Soc.* B**64**, 114. [8]

Glaser, W. (1956). *Handbuch der Physik* vol. 33, pp. 123–395. Berlin. [85]

Glaser, W. and Lammel, E. (1943). *Arch. Elektrotech.* **37**, 347. [212]

Glaser, W. and Schiske, P. (1954). *Optik* **11**, 422. [39]

Gobrecht, G. (1961). *Exper. Tech. Phys.* **9**, 184. [183]

Gobrecht, R. (1941). *Arch. Elektrotech.* **35**, 672. [91, 184]

Gobrecht, R. (1942). *Arch. Elektrotech.* **36**, 484. [203]

Gobrecht, R. (1958 a). *Exper. Tech. Phys.* **5**, 241. [202]

Gobrecht, R. (1958 b). *Exper. Tech. Phys.* **6**, 1, 97. [207, 210, 213]

Goddard, L. S. (1944). *Proc. Phys. Soc.* **56**, 372. [60]

Goddard, L. S. (1946 a). *Proc. Camb. Phil. Soc.* **42**, 106. [66]

Goddard, L. S. (1946 b). *Proc. Camb. Phil. Soc.* **42**, 127. [205]

Goddard, L. S. and Klemperer, O. (1944). *Proc. Phys. Soc.* **56**, 378. [60, 61, 138, 143, 145, 149]

Goudet, G. and Gratzmuller, A. M. (1944). *J. Phys. Rad.* **5**, 142. [291]

Goudet, G. and Gratzmuller, A. M. (1945). *J. Phys. Rad.* **6**, 153. [291]

Grad, E. M. *See under* Liebmann, G.

Graham, R. L. (1949). Thesis, London. [120]

Graham, R. L., Ewan, G. T. and Geiger, J. (1960). *Nucl. Instr. Meth.* **9**, 245. [428, 459]

Graham, R. L. and Klemperer, O. (1952). *Proc. Phys. Soc.* B**65**, 921. [118, 120, 188, 241, 245]

Gratzmuller, A. M. *See under* Goudet, G.

Graut, T. (1966). *Proc. I.E.E.E.* **54**, 801. [268]

Gray, F. (1939). *Bell. Syst. Tech. J.* **18**, 1. [39]

Gregory, B. C. and Beck, A. H. (1966). *Int. J. Electronics* **21**, 561. [366]

Gregory, B. C. and Sander, K. F. (1962). *J. El'onics and Control* **13**, 123. [103]

Grivet, P. (1950). *J. Phys. Rad.* **11**, 582. [239, 241]

Grivet, P. (1951). *J. Phys. Rad.* **12**, 1. [241]

Grivet, P. (1952). *J. Phys. Rad.* **11**, 582; **12**, 1. [26, 119]

Grivet, P. (1965). *Electron Optics* (revised A. Septier, trans. P. W. Hawkes). Oxford. [26, 119]

Grivet, P. and Bernard, M. (1954). In *Electron Physics* (ed. L. Marton), p. 205; *N.B.S. Circular* no. 527. [59]

Grivet, P. and Septier, A. (1960). *Nucl. Instr. Meth.* **6**, 126, 243. [219]

Groendijk, H. *See under* de Beer, A. J.; Hasker, J.

Gruner, H. *See under* Möllenstedt, G.

Gundert, E. (1939). *Z. Phys.* **112**, 689. [183]

Gurney, R. W. (1925). *Proc. Roy. Soc.* A**109**, 540. [449, 456, 459]

Haantjes, J. and Lubben, G. J. (1957). *Philips Res. Rep.* **12**, 46. [446]

Hachenberg, O. (1948). *Ann. Phys. Lpz.* **2**, 225. [416]

Hadfield, D. (1950). *Electronic Engng* **22**, 132. [134]

Haeff, A. V. (1939). *Proc. Inst. Radio. Engrs* **27**, 586. [301, 317]

Haeff, A. V. *See also under* Salzberg, B.

Hafner, H., Simpson, J. A. and Kuyatt, C. E. (1968). *Rev. Sci. Instr.* **39**, 33. [415, 447]

Hahn, E. (1958). *Optik* **15**, 500. [353]

Haine, M. E. and Cosslett, V. E. (1961). *The Electron Microscope.* London. [187, 189, 261]

Haine, M. E. and Einstein, P. A. (1952). *Br. J. Appl. Phys.* **3**, 40. [368, 370]

Haine, M. E., Einstein, P. A. and Borcherds, P. H. (1958). *Br. J. Appl. Phys.* **9**, 482. [371]

Haine, M. E. and Jervis, M. W. (1957). *Proc. Inst. El. Engrs* **104**B, 379, 385. [400]

Haine, M. E. and Mulvey, T. (1954). *J. Sci. Instr.* **31**, 326. [261]

Hall, C. E. (1949). *J. Appl. Phys.* **20**, 631. [177]

Hall, T. A. *See under* Allison, S. K.

Halliday, D. *See under* Quade, E. A.

Hamisch, H. and Oldenburg, K. (1964). *Proc. European Regional Conference on Electron Microscopy, Prague* vol. *A*, p. 41. [68, 76]

Hamza, V. (1966). *I.E.E.E. Trans.* ED-**13**, 551. [360]

Hanley, T. E. (1948). *J. Appl. Phys.* **19**, 583. [322]

Hansen, W. W. and Webster, D. L. (1936). *Rev. Sci. Instr.* **7**, 17.

Hansen, W. W. *See also under* Chodorow, M.

Hanszen, K. J. (1956). *Optik* **13**, 385. [92]

Hanszen, K. J. (1958). *Optik* **15**, 304. [92]

Hanszen, K. J. (1964). *Z. Naturforsch.* **19**a, 896. [360]

Hanszen, K. J. (1966). *Z. angew. Phys.* **20**, 427. [267]

Hanszen, K. J. and Lauer, R. (1967). *Z. Naturforsch.* **22**a, 238. [372]

Hanszen, K. J. and Lauer, R. (1969). *Z. Naturforsch.* **24**a, 97. [266]

Hanszen, K. J. and Morgenstern, B. (1965). *Z. angew. Phys.* **19**, 215. [264]

Hardy, A. E. *See under* Morrell, A. M.

Harris, F. K. (1934). *Bur. Stand. J. Res.* **13**, 391. [167]

Harris, L. A. (1959). *J. Appl. Phys.* **30**, 826. [387]

Harrison, S. W. *See under* Hutter, R. G. E.

Harrower, G. A. (1955). *Rev. Sci. Instr.* **26**, 850. [409]

Hart, P. A. H. and Weber, C. (1961). *Philips Res. Rep.* **16**, 376. [378]

Hartl, W. A. M. (1966). *Z. Phys.* **191**, 487. [250]

Hartman, P. L. *See under* Smith, L. P.

Hartwig, D. and Ulmer, K. (1963*a*). *Z. Phys.* **173**, 294. [250, 467]

Hartwig, D. and Ulmer, K. (1963*b*). *Z. angew. Phys.* **15**, 309. [250, 467]

Hasker, J. and Groendijk, H. (1962). *Philips Res. Rep.* **17**, 401. [330]

Hawkes, P. W. (1966). *Springer Tracts in Modern Physics* vol. 42. *See also*: *Adv. El'onics, Suppl.* 7 (1970). [214, 217]

Haynes, S. K. and Wedding, J. W. (1951). *Rev. Sci. Instr.* 22, 97. [167]

Headrick, L. B. *See under* Thompson, B. J.

Hechtel, J. R. (1962). *Trans. Inst. Radio Engrs* ED-9, 62. [473, 476]

Hechtel, J. R. and Seeger, J. A. (1961). *Proc. Inst. Radio Engrs* 49, 933. [309]

Hedgran, A., Siegbahn, K. and Swartholm, N. (1950). *Proc. Phys. Soc.* A63, 960 [459]

Heffner, H. *See under* Clogston, A. M.

Hehlgans, F. (1935). *Z. Tech. Phys.* 16, 197. [157]

Heinz, W. (1966). *Nucl. Instr. Meth.* 39, 61. [379]

Heinz, W. (1967). *Nucl. Instr. Meth.* 46, 16. [379]

Heinz, W. *See also under* Hoffmann, G.

Heise, F. (1949). *Optik* 5, 479. [179, 187, 199, 209]

Heise, F. and Rang, O. (1949). *Optik* 5, 201. [91]

Hellwig-Zwiessler, O., Stierstadt, O. and Hellwig-Zwiessler, W. (1962). *Optik* 19, 571. [385]

Hellwig-Zwiessler, W. *See under* Hellwig-Zwiessler, O.

Helm, R. *See under* Field, L. M.

Helmer, J. C. (1966). *Am. J. Phys.* 34, 222. [17]

Henneberg, W. (1934a). *Ann. Phys. Lpz.* 19, 335. [435, 438, 452]

Henneberg, W. (1934b). *Ann. Phys. Lpz.* 20, 1. [426]

Henneberg, W. and Recknagel, A. (1935). *Z. Tech. Phys.* 16, 621. [93, 232]

Henneberg, W. *See also under* Glaser, A.

Henry, L. *See under* Castaing, R.

Hernquist, K. G. *See under* Linder, E. G.

Herrmann, G. (1958). *J. Appl. Phys.* 29, 127. [289, 303]

Herrmann, G. and Wagner, S. (1951). *The Oxide Coated Cathode* (2 vols). London, [322]

Herzog, R. (1934). *Z. Phys.* 89, 447. [414, 435, 442]

Herzog, R. (1935). *Z. Phys.* 97, 596 [414, 442]

Herzog, R. (1939). *Z. Phys.* 113, 166. [442]

Herzog, R. (1940). *Phys. Z.* 41, 13. [442]

Herzog, R. F. and Tischler, O. (1953). *Rev. Sci. Instr.* 24, 1000. [138]

Hess, E. (1934). *Z. Phys.* 92, 274. [25]

Hesse, M. B. (1950). *Proc. Phys. Soc.* B63, 386. [137]

Hillier, J. (1940). *Phys. Rev.* 58, 842. [260]

Hillier, J. and Ramberg, E. (1947). *J. Appl. Phys.* 18, 48. [167]

Himmelbauer, E. E. (1969). *Philips Res. Rep. Suppl.* 1, 1–114. [105]

Hinde, R. M. *See under* Taylor, C. A.

Hines, M. E. *See under* Cutler, C. C.

Hintenberger, H. (1948). *Z. Naturforsch.* 3a, 125, 669. [450]

Hintenberger, H. (1949). *Rev. Sci. Instr.* 20, 748. [450]

Hipman, J. *See under* Picht, J.

Hoeft, J. (1959). *Z. angew. Phys.* 11, 380. [79]

Hoffmann, G. and Heinz, W. (1967). *Optik* 26, 211. [379]

Hoffmann, G. and Heinz, W. (1968). *Optik* 26, 221. [379]

Hogan, T. K. (1943). *J. Inst. Engrs Aust.* **15**, 89. [52]

Hollway, D. L. (1955). *Aust. J. Phys.* **8**, 74. [309]

Hollway, D. L. (1962). *J. Br. Radio. Engrs* **24**, 209. [277]

Hopkins, H. H. and Barham, P. M. (1950). *Proc. Phys. Soc.* B **63**, 737. [259]

Hoppe, W. (1963). *Optik* **20**, 599. [195, 263]

Hoppe, W. *See also under* Langer, R.; Möllenstedt, G.

Hornsby, J. S. *See under* Kirstein, P. T.

Hoselitz, K. (1952). *Ferromagnetic Properties of Metals and Alloys.* Oxford. [165]

Hosemann, R. (1955). *Z. angew. Phys.* **7**, 532. [386]

Hottenroth, G. (1936). *Z. Phys.* **103**, 460. [93, 96]

Houtermans, F. G. and Riewe, K. H. (1941). *Arch. Elektrotech.* **35**, 686. [281]

Hubert, P. (1951). *C.R. Acad. Paris* **233**, 943. [120, 195]

Hubert, P. (1952). *Physica* **18**, 1129. [242]

Hudson, R. P. (1949). *J. Appl. Phys.* **26**, 401. [403]

Hughes, A. L. and MacMillan, J. H. (1929). *Phys. Rev.* **34**, 291. [411]

Hughes, A. L. and Rojanski, V. (1929). *Phys. Rev.* **34**, 284. [410]

Hull, A. W. (1921). *Phys. Rev.* **18**, 31. [438]

Hull, C. R. *See under* Dungey, J. W.

Hutter, R. G. E. (1947). *J. Appl. Phys.* **18**, 740. [441]

Hutter, R. G. E. and Harrison, S. W. (1950). *J. Appl. Phys.* **21**, 84. [442]

Iams, H. (1939). *Proc. Inst. Radio Engrs* **27**, 103. [91]

Iams, H. *See also under* Rose, A.

Jacob, L. (1938). *Phil. Mag.* **26**, 570. [55]

Jacob, L. (1939). *Phil. Mag.* **28**, 81. [347, 354, 355]

Jacob, L. (1944). *J. Inst. El. Engrs* pt. III **91**, 512. [344]

Jacob, L. (1948). *Phil. Mag.* **39**, 20, 400. [344]

Jacob, L. (1952). *Proc. Phys. Soc.* B **65**, 421. [342]

Jacob, L. and Mulvey, T. (1949). *Nature, Lond.* **163**, 525. [209]

Jacob, L. *See also under* Einstein, P. A.

Jansen, M. J. *See under* Lemmens, H. J.

Jennings, J. C. E. (1952). *Proc. Phys. Soc.* B **65**, 256. [424, 426, 451]

Jennings, J. C. E. *See also under* Ehrenberg, W.

Jentsch, F. *See under* Wehnelt, A.

Jervis, M. W. *See under* Haine, M. E.

Johannson, H. (1933). *Ann. Phys. Lpz.* **18**, 385. [203, 325, 337, 352]

Johannson, H. (1934). *Ann. Phys. Lpz.* **21**, 275. [203, 325]

Johannson, H. and Scherzer, O. (1933). *Ann. Phys. Lpz.* **18**, 385. [92]

Johannson, H. *See also under* Brüche, E.

Johnson, H. R. *See under* Frost, R. D.

Johnson, K. E. (1962). *Mullard Research Technical Note* no. 533. [105]

Jonker, J. L. H. (1949). *Philips Res. Rep.* **4**, 357. [385]

Jonker, J. L. H. (1953). *Philips Tech. Rev.* **14**, 361. [134]

Judd, F. L. (1950). *Rev. Sci. Instr.* **21**, 213. [430]

Jungermann, J. A., Gardener, M. E., Patten, C. G. and Peek, N. F. (1962). *Nucl. Instr. Meth.* **15**, 1. [242, 245]

Kaashoek, J. (1968). *Philips Res. Rep. Suppl.* **11**, 1–114. [444]

Kaden, G. (1950). *Die electromagnetische Schirmung in fernmelde und hoch-frequenz Technik.* Berlin. [164]

Kamagaito, O. *See under* Ando, K.

Kamiya, Y. *See under* Ando, K.

Kanaya, K., Kawakatsu, H. and Yamazaki, H. (1965). *Br. J. Appl. Phys.* **16**, 991. [295]

Kanaya, K., Kawakatsu, H., Yamazaki, H. and Sibata, S. (1966). *J. Sci. Instr.* **43**, 416. [85, 185]

Kanaya, K., Yamazaki, H., Takano, Y. and Kobayashi, M. (1968). *J. Sci. Instr.* ser. 2 **1**, 289. [376]

Karplus, W. J. (1958). *Analog Simulation.* New York. [47]

Katerbau, K. H. *See under* Möllenstedt, G.

Kawakatsu, H., Vosburgh, K. G. and Siegel, B. M. (1968). *J. Appl. Phys.* **39**, 245, 255. [221]

Kawakatsu, H. *See also under* Kanaya, K.

Keller, J. M., Koenigsberg, E. and Paskin, A. (1950). *Rev. Sci. Instr.* **21**, 713. [241]

Keller, M. (1961). *Z. Phys.* **164**, 292. [374]

Kelman, V. M. *See under* Dolmatova, K. A.; Korsunsky, M.

Kershaw, H. *See under* Cockcroft, J. D.

Kerst, D. W. and Serber, R. (1941). *Phys. Rev.* **60**, 53. [426, 427, 429]

Kerwin, L. (1949). *Rev. Sci. Instr.* **20**, 36. [450]

Kerwin, L. and Marmet, P. (1960). *J. Appl. Phys.* **31**, 2071. [412]

Kerwin, L. *See also under* Marmet, P.

Kessler, J. (1961). *Z. Naturforsch.* **16a**, 1038. [71]

King, P. G. R. *See under* Mathias, L. E. S.

Kino, G. S. (1961). In *Crossed Field Microwave Devices* (ed. Okress) vol. 1, p. 164. [388]

Kino, G. S. and Taylor, N. (1962). *Trans. Inst. El. Engrs* ED-9, 1. [310]

Kino, G. S. *See also under* Kirstein, P. T.

Kirstein, P. T. and Hornsby, J. S. (1964). *Trans. Inst. El. Engrs* ED-11, 196. [309]

Kirstein, P. T., Kino, G. S. and Waters, W. E. (1967). *Space-Charge Flow.* New York. [301]

Kitamura, N., Schulhof, M. P. and Siegel, B. M. (1966). *Appl. Phys. Lett. USA* **9**, 377. [137]

Kittel, C. (1966). *Introduction to Solid State Physics* 3rd ed. New York. [136]

Klanfer, L. *See under* Motz, H.

Klemperer, O. (1935). *Phil Mag.* **20**, 45. [116, 245]

Klemperer, O. (1937). *Brit. Pat.* 498 511. [397]

Klemperer, O. (1941). *Brit. Pat.* 534 215. [293]

Klemperer, O. (1942). *Brit. Pat.* 568 572. [184, 294]

Klemperer, O. (1944). *J. Sci. Instr.* **21**, 88. [50]

Klemperer, O. (1947a). *Proc. Phys. Soc.* **59**, 302. [281, 286, 383]

Klemperer, O. (1947b). *Proc. Roy. Soc.* A**190**, 376. [335]

Klemperer, O. (1951). *Proc. Phys. Soc.* B**64**, 790. [180]

Klemperer, O. (1953). Originally published in *Electron Optics* 2nd ed. Cambridge. [66, 90, 94, 99, 104, 182, 220, 233, 287, 347, 357]

Klemperer, O. (1965). *Rep. Progr. Phys.* **28**, 77. [233]
Klemperer, O. and Klinger, Y. (1951). *Proc. Phys. Soc.* B**64**, 231. [333]
Klemperer, O. and Mayo, B. J. (1948). *J. Inst. El. Engrs* pt. III **95**, 135. [296, 333, 350, 356, 357]
Klemperer, O. and Miller, H. (1939). *J. Sci. Instr.* **16**, 121. [138]
Klemperer, O. and Shepherd, J. P. G. (1963). *Br. J. Appl. Phys.* **14**, 85. [424, 459]
Klemperer, O. and Wright, W. D. (1939). *Proc. Phys. Soc.* **51**, 296. [27, 28, 176, 204, 361]
Klemperer, O. *See also under* Dolder, K. T.; Goddard, L. S.; Graham, R. L.
Klinger, Y. *See under* Klemperer, O.
Knoll, M. (1934). *Z. Tech. Phys.* **15**, 584. [385]
Knoll, M. (1939. *Telefunken Hausmit.* **20**, no. 81, 72. [393, 394, 404]
Knoll, M. and Ruska, E. (1931). *Z. Tech. Phys.* **12**, 394. [1]
Knoll, M. and Ruska, E. (1932). *Ann. Phys. Lpz.* **12**, 607, 641. [1, 98]
Knoll, M. and Weichardt, H. (1938). *Z. Phys.* **110**, 233. [99]
Kobayashi, M. *See under* Kanaya, K.
Koenigsberg, E. *See under* Keller, J. M.
Kollath, R. (1936). *Ann. Phys. Lpz.* **27**, 721. [468]
Konopinski, E. *See under* Beiduck, F. M.
Korsunsky, M. (1945). *J. Phys. USSR* **9**, 14. [450]
Korsunsky, M., Kelman, V. and Petrov, B. (1945). *J. Phys. USSR* **9**, 7. [450]
Korsunsky, M. *See also under* Zashkavara, V. V.
Kosmachev, O. S. *See under* Zashkavara, V. V.
Kramer, J. *See under* Bok, A.
Krasnow, M. E. *See under* Lassettre, E. N.
Krumm, J. *See under* Borries, B. v.
Kulsrud, H. E. (1967). *R.C.A. Review* **28**, 351. [45, 61, 309]
Kunz, C. (1962). *Z. Phys.* **167**, 53. [412]
Kuyatt, C. E. and Simpson, J. A. (1967). *Rev. Sci. Instr.* **38**, 103. [415]
Kuyatt, C. E. *See also under* Hafner, H.; Simpson, J. A.
Kyhl, R. L. *See under* Chodorow, M.
Lafferty, J. M. (1951). *J. Appl. Phys.* **22**, 299. [322]
Lammel, E. *See under* Glaser, W.
Lampert, I. E. and Feld, R. (1946). *Proc. Inst. Radio Engrs* **34**, 432. [399]
Langer, L. M. and Cook, C. S. (1948). *Rev. Sci. Instr.* **19**, 257. [459]
Langer, L. M. and Scott, F. R. (1950). *Rev. Sci. Instr.* **21**, 522. [138]
Langer, R. and Hoppe, W. (1967). *Optik* **25**, 413, 507, 599. [195]
Langer, R. *See also under* Möllenstedt, G.
Langmuir, D. B. (1937a). *Nature, Lond.* **139**, 1067. [62]
Langmuir, D. B. (1937b). *Proc. Inst. Radio Engrs* **25**, 977. [15, 231, 353]
Langmuir, D. B. (1950). *R.C.A. Review* **11**, 143. [62]
Langmuir, I. (1913). *Cf.* Langmuir and Compton (1931). [300]
Langmuir, I. and Blodgett, K. B. (1924). *Phys. Rev.* **24**, 49. [363]
Langmuir, I. and Compton, K. T. (1931). *Rev. Mod. Phys.* **3**, 191. [300, 344]

Langner, G. (1955). *Optik* **12**, 554. [135]

Laplume, J. (1947). *Cahiers Physique* nos. 29, 30, 55. [92]

Lassettre, E. N., Berman, A. S., Silverman, S. M. and Krasnow, M. E. (1964). *J. Chem. Phys.* **40**, 1232. [409]

Laudet, M. *See under* Durandeau, P.

Lauer, R. *See under* Hanszen, K. J.

Law, H. B., Davne, L. and Ramberg, E. G. (1961). *R.C.A. Review* **22**, 603. [399]

Law, R. R. (1937). *Proc. Inst. Radio Engrs N.Y.* **25**, 954. [332, 353, 354, 355]

Lawson, J. L. and Tyler, A. W. (1940). *Rev. Sci. Instr.* **11**, 14. [449, 459]

Lee, R. N. (1968). *Rev. Sci. Instr.* **39**, 1306. [378]

Legler, W. (1963). *Z. Phys.* **171**, 424. [435, 437]

Leisegang, S. (1953). *Optik* **10**, 5. [266]

Leisegang, S. (1954). *Optik* **11**, 397. [227, 266]

Lemmens, H. J., Jansen, M. J. and Loosjes, R. (1950). *Philips Tech. Rev.* **11**, 341. [323]

Lenz, F. (1951). *Ann. Phys. Lpz.* **9**, 245, [119]

Lenz, F. (1952). *Optik* **9**, 3. [119]

Lenz, F. (1964). *Optik* **21**, 489. [195]

Lenz, F. (1965). *Optik* **22**, 270. [262]

Lenz, F. *See also under* Möllenstedt, G.

Le Poole, J. B. (1948). *J. Sci. Instr.* **25**, 24. [163]

Le Poole, J. B. *See also under* Bok, A.; Ments, M.

Lerch, H. *See under* Bas, E. B.

Le Rutte, W. A. (1948). *Nature, Lond.* **161**, 392. [253]

Levi, R. (1953). *J. Appl. Phys.* **24**, 233. [323]

Levi, R. (1955). *J. Appl. Phys.* **26**, 639. [323]

Li, K. T. (1937). *Proc. Camb. Phil. Soc.* **33**, 164. [448]

Liebmann, G. (1946). *Phil. Mag.* **37**, 677. [189]

Liebmann, G. (1948). *Phil. Mag.* **39**, 281. [76]

Liebmann, G. (1949*a*). *Proc. Phys. Soc.* B**62**, 213. [90, 92, 99, 175, 179, 184, 199]

Liebmann, G. (1949*b*). *Proc. Phys. Soc.* B**62**, 753. [52, 60, 62, 143]

Liebmann, G. (1950*a*). *Br. J. Appl. Phys.* **1**, 92. [53]

Liebmann, G. (1950*b*). *Phil. Mag.* **41**, 1143. [140]

Liebmann, G. (1952*a*). *Proc. Phys. Soc.* B**65**, 94. [30, 193, 212, 213]

Liebmann, G. (1952*b*). *Proc. Phys. Soc.* B**65**, 188. [193]

Liebmann, G. (1955). *Proc. Phys. Soc.* B**68**, 737. [126, 189]

Liebmann, G. and Grad, E. M. (1951). *Proc. Phys. Soc.* B**64**, 956. [126, 128, 199]

Linder, E. G. and Hernquist, K. G. (1950). *J. Appl. Phys.* **21**, 1088. [297, 298]

Lippert, W. (1955). *Optik* **10**, 467. [243]

Lippert, W. and Pohlit, W. (1952). *Optik* **9**, 456. [89, 228]

Lippert, W. and Pohlit, W. (1953). *Optik* **10**, 447. [89]

Lippert, W. and Pohlit, W. (1954). *Optik* **11**, 253. [89]

Lipson, H. *See under* Taylor, C. A.

Lischke, B. *See under* Boersch, H.

Livingstone, M. S. *See under* Courant, E. D.

Loeb, J. (1947). *Onde Electrique*, **27**, 27. [148]

Lohff, J. (1963). *Z. Phys.* **171**, 442. [413]

Löhner, H. (1930). *Ann. Phys. Lpz.* **6**, 66. [409]

Loosjes, R. *See under* Lemmens, H. J.

Lubben, G. J. *See under* Haantjes, J.

Lubszynski, G. *See under* McGee, J. D.

Lubszynski, H. G., Mayo, B. J., Wardley, J. and Barford, N. C. (1969). *Proc. Inst. El. Engrs* **116**, 339. [31, 205, 401]

Lyddane, R. H. and Ruark, E. A. (1939). *Rev. Sci. Instr.* **10**, 253. [139, 167]

McGee, J. D. *et al.* (eds) (1958, 1966, 1969). *Adv. El'onics* vols. 12, 22, 28A. [367]

McGee, J. D. and Lubszynski, G. (1939). *J. Inst. El. Engrs* **84**, 468. [150]

McGregor-Morris, J. T. and Mines, R. (1925). *J. Inst. El. Engrs* **63**, 1065. [192, 270]

MacMillan, J. H. *See under* Hughes, A. L.

Magnan, C. (1961). *Traité de Microscopie Electronique*. Paris. [266]

Mahl, H. and Pendzich, A. (1943). *Z. Tech. Phys.* **24**, 38. [95, 97].

Mahl, H. and Recknagel, A. (1944). *Z. Phys.* **122**, 60. [177]

Maloff, I. G. and Epstein, W. D. (1934). *Proc. Inst. Radio Engrs N.Y.* **23**, 1386. [59, 357]

Maloff, I. G. and Epstein, W. D. (1938). *Electron Optics in Television*. New York. [59, 357, 358, 393]

Maloney, C. E. *See under* Beck, A. H.

Manifold, M. B. and Nicoll, F. H. (1938). *Nature, Lond.* **142**, 39. [49]

Markovitch, M. G. and Zuckerman, I. I. (1960). *Zh. Tekh. Fiz.* **30**, 1362. [218]

Markovitch, M. G. and Zuckerman, I. I. (1961). *Sov. Phys. Tech. Phys.* **5**, 1292. [218]

Marmet, P. and Kerwin, L. (1960). *Canad. J. Phys.* **38**, 787. [378]

Marmet, P. *See also under* Kerwin, L.

Marschall, H. (1944). *Phys. Z.* **45**, 1. [416]

Marschall, H. and Schröder, W. (1942). *Z. Tech. Phys.* **23**, 297. [405, 444]

Marton, L. (1950). *J. Opt. Soc. Amer.* **40**, 269. [163]

Marton, L. and Bol, K. (1947). *J. Appl. Phys.* **8**, 522. [195]

Marton, L. and Reverdin, D. L. (1950). *J. Appl. Phys.* **21**, 842. [254]

Marton, L., Simpson, J. A., Wagner, M. D. and Watanabe, H. (1958). *Phys. Rev.* **110**, 1057. [431]

Marton, L. *See also under* Simpson, J. A.

Mathias, L. E. S. and King, P. G. R. (1957). *Inst. Radio Engrs Trans.* (*Electron Devices*) ED-**4**, 280. [304]

Mayer, L. (1955). *J. Appl. Phys.* **26**, 1228. [96, 267]

Mayer, L. (1957). *J. Appl. Phys.* **28**, 975. [267]

Mayo, B. J. *See under* Klemperer, O.; Lubszynski, H. G.

Mecklenburg, W. (1942). *Z. Phys.* **120**, 21. [325]

Mendel, J. T., Quate, C. F. and Yocom, W. H. (1954). *Proc. Inst. Radio Engrs* **42**, 800. [305]

Ments, M. and Le Poole, J. B. (1947). *Appl. Sci. Res.* B**1**, 3. [118, 119, 139]

Metherell, A. J. F. and Whelan, M. J. (1965). *Br. J. Appl. Phys.* **16**, 1038. [243]

Meyer, S. and Schweidler, E. (1899). *Phys. Z.* **1**, 90. [419]

Meyer, W. E. (1961). *Optik* **18**, 69, 101. [223]

Michaelis, E. *See under* Citron, A.

Miessner, H. *See under* Boersch, H.

Mileikowsky, C. M. (1952). *Ark. Fys.* **4**, 337. [431]

Miller, H. *See under* Klemperer, O.

Millet, W. E. (1948). *Phys. Rev.* **74**, 1058. [439]

Mines, R. *See under* McGregor-Morris, J. T.

Möllenstedt, G. (1949). *Optik* **5**, 499. [243, 245]

Möllenstedt, G. (1952). *Optik* **9**, 473. [243]

Möllenstedt, G. (1956). *Optik* **13**, 209. [223]

Möllenstedt, G. and Düker, H. (1953). *Z. Naturforsch.* **8a**, 79. [380]

Möllenstedt, G. and Düker, H. (1956). *Z. Phys.* **145**, 377. [374]

Möllenstedt, G. and Gruner, H. (1968). *Optik* **27**, 602. [252]

Möllenstedt, G. and Lenz, F. (1963). *Adv. El'onics* **18**, 251. [230, 326]

Möllenstedt, G., Thon, F., Speidel, R., Hoppe, W., Katerbau, K. H. and Langer, R. (1969). *Siemens Rev.* **36**. [263]

Möllenstedt, G. *See also under* Sakaki, Y.

Montegue, J. H. *See under* Allison, S. K.

Moon, P. B. and Spencer, D. G. (1961). *Field Theory Handbook*. Berlin. [38]

Morgenstern, B. *See under* Hanszen, K. J.

Morito, N. *See under* Watanabe, A.

Morrell, A. M. and Hardy, A. E. (1964). *I.E.E.E. Trans.* Publication ST-2814. [329]

Morrish, A. H. *See under* Allison, S. K.

Morton, G. A. (1946). *Rev. Mod. Phys.* **18**, 362. [327]

Morton, G. A. and Ramberg, E. G. (1936). *Physics* **7**, 451. [202, 209]

Morton, G. A. *See also under* Zworykin, V. K.

Moss, H. (1945). *Wireless Engr.* **22**, 316. [292]

Moss, H. (1946). *J. Br. Inst. Radio Engrs* **6**, 99. [341, 342, 344]

Moss, H. (1961). *J. El'onics and Control* **11**, 289. [359]

Moss, H. (1968). *Adv. El'onics, Suppl.* **3**, 1–244. [327, 342]

Motz, H. and Klanfer, L. (1946). *Proc. Phys. Soc.* **58**, 30. [45, 60]

Müller, H. *See under* Borries, B. v.

Müller, M. (1955). *Arch. El. Übertrag.* **9**, 20. [365]

Müller, M. (1956). *J. Br. Inst. Radio Engrs* **16**, 83. [365]

Mulvey, T. and Wallington, M. J. (1969). *J. Sci. Instr.* **2**, 466. [126, 129, 190, 226, 265]

Mulvey, T. *See also under* Haine, M. E.; Jacob, L.

Neal, R. B. *See under* Chodorow, M.

Neumann, G. (1914). *Ann. Phys. Lpz.* **45**, 529. [392]

Nicoll, F. H. (1938). *Proc. Phys. Soc.* **50**, 888. [95, 192]

Nicoll, F. H. (1955). *Rev. Sci. Instr.* **12**, 1206. [321]

Nicoll, F. H. *See also under* Manifold, M. B.

Niemeck, F. W. and Ruppin, D. (1954). *Z. angew. Phys.* **6**, 1. [322]

Nienhuis, W. F. *See under* Francken, J. C.

Nixon, W. C. *See under* Barnett, M. E.; Cosslett, V. E.; Swift, D. C.
Nottingham, W. B. (1956). In *Springer Encyclopaedia of Physics* vol. 21. [321]
Oatley, C. W. *See under* Smith, K. C. A.
Oldenburg, K. *See under* Hamisch, H.
Opatowski, I. (1944). *Phys. Rev.* **65**, 45. [8]
Openshaw, I. K. *See under* Barnes, R. L.
Orr, D. (1963). *Nucl. Instr. Meth.* **24**, 377. [220]
Øverås, H. *See under* Citron, A.
Ovsyanikova, L. P. *See under* Dymikov, A. D.; Yavor, S.
Owen, G. E. (1949). *Rev. Sci. Instr.* **20**, 916. [449]
Packh, D. G. de (1947). *Rev. Sci. Instr.* **18**, 798. [52]
Palmer, J. L. *See under* Bevc, V.
Panofsky, W. K. H. and Baker, W. R. (1950). *Rev. Sci. Instr.* **21**, 445. [157]
Panofsky, W. K. H. *See also under* Chodorow, M.
Parker, R. S. and Studders, R. J. (1962). *Permanent Magnets and their Applications.* New York. [131]
Paskin, A. *See under* Keller, J. M.
Patten, C. G. *See under* Jungermann, J. A.
Pearson, A. (1962). *J. Sci. Instr.* **39**, 8. [138]
Peek, N. F. *See under* Jungermann, J. A.
Pendzich, A. *See under* Mahl, H.
Petrov, B. *See under* Korsunsky, M.
Picht, J. (1939). *Theorie der Elektronenoptik.* Berlin. [58]
Picht, J. and Hipman, J. (1941). *Ann. Phys. Lpz.* **39**, 401, 436, 448. [441]
Pierce, J. R. (1940). *J. Appl. Phys.* **11**, 548. [50, 300, 362, 385]
Pierce, J. R. (1941). *Proc. Inst. Radio Engrs* **29**, 28. [398]
Pierce, J. R. (1949). *Theory and Design of Electron Beams.* New York. [50, 301, 304]
Pierce, J. R. (1953). *J. Appl. Phys.* **24**, 1247. [306]
Pizer, H. I., Yates, J. G. and Sander, K. F. (1956). *J. Electronics* **2**, 65 [62]
Placius, R. C. *See under* Schreck, R. A.
Plocke, M. (1952). *Z. angew. Phys.* **4**, 1. [358]
Pohlit, W. *See under* Lippert, W.
Post, R. F. (1956). *Rev. Mod. Phys.* **28**, 338. [151]
Preuss, L. and Bas, E. B. (1966). *Z. angew. Math. Phys.* **17**, 168. [53]
Preuss, L. *See also under* Bas, E. B.
"Processes" (1960, 1961, 1962, 1963). *Proceedings of the 2nd, 3rd, 4th and 5th Symposia on Electron Beam Processes.* Boston, Mass. [379, 380]
"Processes" (1965). *International Conference on Electron and Ion Beam Science.* New York. [379, 380]
Purcell, E. M. (1938). *Phys. Rev.* **54**, 818. [413]
Purl, O. T. *See under* Frost, R. D.
Quade, E. A. and Halliday, D. (1948). *Rev. Sci. Instr.* **19**, 234. [194]
Quate, C. F. *See under* Mendel, J. T.
Rajchman, J. (1938). *Arch. Sci. Phys. Nat. (Genève).* **20**, 9. [57]

Ramberg, E. G. (1942). *J. Appl. Phys.* **13**, 582. [92, 191, 227, 253]
Ramberg, E. G. (1949). *J. Appl. Phys.* **20**, 183. [193, 228]
Ramberg, E. G. *See also under* Hillier, J.; Law, H. B.; Morton, G. A.
Ramo, S. *See under* Bachman, C. H.
Rang, O. (1948). *Optik* **4**, 251. [91, 209]
Rang, O. (1949). *Optik* **5**, 518. [168]
Rang, O. *See also under* Heise, R.
Rawlinson, W. E. *See under* Rutherford, E.
Read, F. H. (1969a). *J. Sci. Instr.* (*E.*) **2**, 165. [75, 184]
Read, F. H. (1969b). *J. Sci. Instr.* (*E.*) **2**, 679. [86, 184, 185]
Recknagel, A. (1937). *Z. Phys.* **104**, 381. [60]
Recknagel, A. (1938). *Z. Phys.* **111**, 61. [392, 407, 408, 442]
Recknagel, A. (1941). *Z. Phys.* **117**, 689. [352]
Recknagel, A. *See also under* Henneberg, W.; Mahl, H.
Regenstreif, E. (1951). *Ann. Radioelect.* **6**, 51, 114, 164, 244, 299. [60, 84,
    91, 246]
Reisman, J. *See under* Siegel, B. M.
Reisner, J. H. (1951). *J. Appl. Phys.* **22**, 561. [134]
Reisner, J. H. and Dornfield, E .G. (1950). *J. Appl. Phys.* **21**, 11, 31. [134]
Reverdin, D. L. *See under* Marton, L.
Riewe, J. H. *See under* Houtermans, F. G.
Roberts, A. S., Cox, J. L. and Bennett, W. H. (1966). *J. Appl. Phys.* **37**,
    3231. [381]
Rogers, F. T. (1946). *Phys. Rev.* **69**, 537. [414]
Rogers, F. T. (1951). *Rev. Sci. Instr.* **22**, 723. [414]
Rogowski, W. and Thielen, H. (1939). *Arch. Elektrotech.* **33**, 411.
Rohr, M. von. (1920). *Formation of images in Optical Instruments.* London.
    [12]
Rojanski, V. *See under* Hughes, A. L.
Romani, L. *See under* Bruck, H.
Robinson, H. *See under* Rutherford, E.
Rose, A. and Iams, H. (1939). *Proc. Inst. Radio Engrs N.Y.* **27**, 547.
    [437]
Rose, H. (1967). *Optik* **26**, 289. [254]
Rosenbruch, K. J. *See under* Rosenhauer, K.
Rosenfeld, J. L. *See under* Danielson, W. E.
Rosenhauer, K. and Rosenbruch, K. J. (1967). *Rep. Progr. Phys.* **30** pt. 1,
    1. [266]
Ruark, E. A. *See under* Lyddane, R. H.
Ruppin, D. *See under* Niemeck, F. W.
Ruska, E. (1933). *Z. Phys.* **83**, 684. [331]
Ruska, E. (1934). *Z. Phys.* **89**, 90. [116, 189]
Ruska, E. (1944). *Arch. Elektrotech.* **38**, 102. [147]
Ruska, E. (1965). *Optik* **22**, 319. [190]
Ruska, E. *See also under* Borries, B. v.; Knoll, M.
Ruthemann, G. (1948). *Ann. Phys. Lpz.* **2**, 113. [459]
Ruthemann, G. *See also under* Voges, H.
Rutherford, E., Robinson, H. and Rawlinson, W. E. (1914). *Phil. Mag.*
    28, 281. [419]

Ryley, J. E. *See under* Czarczynski, W.

Sakaki, Y. and Möllenstedt, G. (1956). *Optik* **13**, 193. [322]

Saloom, J. A. *See under* Danielson, W. E.

Salzberg, B. and Haeff, A. V. (1937). *R.C.A. Review* **2**, 336. [311]

Samuel, A. L. (1945). *Proc. Inst. Radio Engrs N.Y.* **33**, 233. [50, 363]

Samuel, A. L. (1949). *Proc. Inst. Radio Engrs N.Y.* **37**, 1252. [301]

Sander, K. F. and Yates, J. G. (1956). *I.E.E. Monograph* 195M. [48, 62]

Sander, K. F. *See also under* Barber, M. R.; Gregory, B. C.; Pizer, H. I.

Sandor, A. (1941). *Arch. Elektrotech.* **35**, 217, 259. [139, 149]

Schade, O. H. (1938). *Proc. Inst. Radio Engrs* **26**, 137. [385]

Schagen, P. and Woodhead, A. D. (1967). *Philips Tech. Rev.* **28**, 161. [367]

Scherzer, O. (1939). *Z. Phys.* **114**, 427. [259]

Scherzer, O. (1947). *Optik* **2**, 114. [195, 217, 222, 223]

Scherzer, O. (1949). *J. Appl. Phys.* **20**, 20. [195, 222, 223]

Scherzer, O. and Typke, D. (1967). *Optik* **26**, 564. [223, 224]

Scherzer, O. *See also under* Brüche, E.; Johannson, H.

Schiske, P. *See under* Glaser, W.

Schlesinger, K. (1949). *Electronics* **22**, 102. [406]

Schlesinger, K. (1956). *Proc. Inst. Radio Engrs* **44**, 659. [399]

Schlesinger, K. (1961). *Proc. Inst. Radio Engrs* **49**, 1538. [60]

Schreck, R. A. and Placius, R. C. (1956). *Rev. Sci. Instr.* **27**, 412. [322]

Schröder, W. *See under* Marschall, H.

Schulhof, M. P. *See under* Kitamura, N.

Schweidler, E. *See under* Meyer, S.

Schwartz, E. (1938). *Fernseh Mitt.* **1** no. 2, 19. [398]

Schwartz, J. W. (1958). *Proc. Inst. Radio Engrs* **46**, 1846. [387]

Schwarz, H. (1962). *Rev. Sci. Instr.* **33**, 688. [380]

Schwarz, H. (1964). *J. Appl. Phys.* **35**, 2020. [380]

Schwarzschild, K. (1903). *Nachr. Ges. Wiss. Göttingen. Phys. Kl.* **3**, 126. [7]

Scott, F. R. *See under* Langer, L. M.

Seeger, J. A. *See under* Hechtel, J. R.

Seeliger, R. (1949). *Optik* **5**, 490. [223]

Seeliger, R. (1951). *Optik* **8**, 311. [217, 223]

Seidel, L. (1856). *Astr. Nachr.* **43** nos. 1027-9. [160]

Septier, A. (1955). *C.R. Acad. Sci.* **240**, 1200. [352]

Septier, A. (1961). *Adv. El'onics El. Phys.* **14**, 85. [30, 36, 37, 101, 154, 214, 221]

Septier, A. and Van Acker, J. (1961). *Nucl. Instr. Meth.* **13**, 335. [220]

Septier, A. *See also under* Grivet, P.

Serber, R. *See under* Kerst, D. W.

Shepherd, J. P. G. *See under* Klemperer, O.

Shull, F. G. and Dennison, D. M. (1947). *Phys. Rev.* **71**, 681. [428, 451]

Shutt, R. P. and Whittemore, W. L. (1951). *Rev. Sci. Instr.* **22**, 73. [404]

Sibata, S. *See under* Kanaya, K.

Siday, R. E. (1942). *Proc. Phys. Soc.* **54**, 266. [142]

Siday, R. E. (1947). *Proc. Phys. Soc.* **59**, 905. [424, 426, 450]

Siday, R. E. and Silverston, D. A. (1952). *Proc. Phys. Soc.* A**65**, 328.

Siday, R. E. *See also under* Ehrenberg, W.

Siegbahn, K. (1942). *Ark. Mat. Astr. Fys.* A28 no. 17. [245]

Siegbahn, K. (1944). *Ark. Mat. Astr. Fys.* A30 no. 20. [245, 459]

Siegbahn, K. (1946). *Phil. Mag.* **37**, 162. [194, 242, 245]

Siegbahn, K. (ed.) (1966). *Alpha, Beta and Gamma Ray Spectrography* vol. 1. Amsterdam. [233]

Siegbahn, K. and Edwardson, K. (1956). *Nucl. Phys.* **1**, 137. [428]

Siegbahn, K. and Swartholm, N. (1946a). *Nature, Lond.* **157**, 872. [426, 459]

Siegbahn, K. and Swartholm, N. (1946b). *Ark. Mat. Astr. Fys.* A33 no. 21 [426, 428, 459]

Siegbahn, K. *See also under* Hedgran, A.; Slätis, H.

Siegel, B. M. and Reisman, J. (1957). *J. Appl. Phys.* **28**, 1379. [221]

Siegel, B. M. *See also under* Kawakatsu, H.; Kitamura, N.

Silverman, S. M. *See under* Lassettre, E. N.

Silverston, D. A. *See under* Siday, R. E.

Simpson, J. A. and Kuyatt, C. E. (1963). *Rev. Sci. Instr.* **32**, 265. [378]

Simpson, J. A. and Marton, L. (1961). *Rev. Sci. Instr.* **32**, 802. [251]

Simpson, J. A. *See also under* Hafner, H.; Kuyatt, C. E.; Marton, L.

Slätis, H. and Siegbahn, K. (1949). *Ark. Mat. Astr. Fys.* **1**, 339. [242, 245]

Smith, K. C. A. and Oatley, C. W. (1955). *Br. J. Appl. Phys.* **6**, 391. [374]

Smith, L. P. and Hartman, P. L. (1940). *J. Appl. Phys.* **11**, 220. [286, 301, 317]

Snyder, H. S. *See under* Courant, E. D.

Soa, E. A. (1959). *Jena Jb.* **1**, 115. [321, 326, 337, 339]

Southwell, R. V. (1946). *Relaxation Methods in Theoretical Physics.* Oxford. [44]

Spangenberg, K. R. (1948). *Vacuum Tubes.* New York. [66, 276]

Spangenberg, K. R. and Field, L. M. (1942a). *Elect. Commun.* **20**, 305. [28, 179]

Spangenberg, K. R. and Field, L. M. (1942b). *Proc. Inst. Radio Engrs N.Y.* **30**, 138. [28, 179]

Spangenberg, K. R. and Field, L. M. (1943). *Elect. Commun.* **21**, 194. [68, 70, 75]

Spangenberg, K. R. *See also under* Field, L. M.

Spanner, D. C. (1967). *Br. J. Appl. Phys.* **18**, 773. [18]

Speidel, R. *See under* Möllenstedt, G.

Spencer, D. G. *See under* Moon, P. B.

Spitzer, L. (1956). *Physics of Fully Ionized Gases.* New York.

Stabenow, G. (1935). *Z. Phys.* **96**, 634. [119]

Standley, K. J. (1962). *Oxide Magnetic Materials.* Oxford. [136]

Steffen, K. G. (1965). *High Energy Ion-Beam Optics.* New York. [446]

Steigerwald, K. H. (1949). *Optik* **5**, 469. [375]

Steigerwald, K. H. (1953). *Ver. Deu. Phys. Ges.* **4**, 123. [379]

Stephens, W. E. (1934). *Phys. Res.* **45**, 513. [419]

Stickel, W. *See under* Boersch, H.

Störmer, C. (1933). *Ann. Phys. Lpz.* **16**, 685. [109, 426]

Strashkevitch, A. (1940). *J. Tech. Phys. USSR.* **10**, 91. [80]

Studders, R. J. *See under* Parker, R. S.

Sturrock, P. A. (1954). *Proc. London. Conf. on Electron Microscopy* (Royal Microscopical Society). [59]

Susskind, C. *See under* Bevc, V.

Swartholm, N. (1946). *Ark. Mat. Astr. Fys.* A**33**, no. 24. [451]

Swartholm, N. (1948). *Phys. Rev.* **74**, 108. [440]

Swartholm, N. (1950). *Ark. Mat. Astr. Fys.* A**2**, no. 14. [431]

Swartholm, N. *See also under* Hedgran, A.; Siegbahn, K.

Sweeny, L. *See under* Terrill, H. M.

Sweer, J. H. *See under* Brown, W. F.

Swift, D. C. and Nixon, W. C. (1962). *Br. J. Appl. Phys.* **13**, 292. [322, 371]

Synge, J. L. (1937). *Geometrical Optics.* Cambridge. [160]

Takano, Y. *See under* Kanaya, K.

Takashi, S. *See under* Ando, K.

Taylor, C. A., Hinde, R. M. and Lipson, H. (1951). *Acta Cryst.* **4**, 261. [263]

Taylor, N. *See under* Kino, G. S.

Teasdale, R. (1953). *Tele-Tech and Electronic Industries* December, p. 74. [165]

Teng, L. C. (1954). *Rev. Sci. Instr.* **25**, 265. [106]

Terrill, H. M. and Sweeny, L. (1944*a*). *J. Franklin Inst.* **237**, 495. [272]

Terrill, H. M. and Sweeny, L. (1944*b*). *J. Franklin Inst.* **238**, 220. [272]

Thielen, H. *See under* Rogowski, W.

Thompson, B. J. and Headrick, L. B. (1940). *Proc. Inst. Radio Engrs N.Y.* **28**, 319. [279, 280]

Thomson, J. J. (1897). *Phil. Mag.* **4**, 293. [3, 434]

Thon, F. (1965). *Z. Naturforsch.* **20**a, 154. [262]

Thon, F. (1966). *Z. Naturforsch.* **21**a, 476. [262]

Thon, F. *See also under* Möllenstedt, G.

Tischler, O. *See under* Herzog, R. F.

Tricker, R. A. (1924). *Proc. Camb. Phil. Soc.* **22**, 454. [233]

Tyler, A. W. *See under* Lawson, J. L.

Typke, D. (1967). *Optik* **24**, 1. [68]

Typke, D. *See also under* Scherzer, O.

Ulmer, K. *See under* Hartwig, D.

Uyeda, R. *See under* Ando, K.

Van Acker, J. *See under* Septier, A.

Van Duzer, T. and Brewer, G. R. (1959). *J. Appl. Phys.* **30**, 291. [309]

Verhof, J. A. (1954). *Philips Res. Rep.* **15**, 214. [136]

Verster, J. L. (1963). *Philips Res. Rep.* **18**, 456. [99]

Verster, J. L. *See also under* de Beer, A. J.

Verster, N. F. (1950). *Appl. Sci. Res. Netherlands* B**1**, 363. [241]

Vine, J. (1959). *Computer J.* **2**, 134. [62]

Vine, J. (1960). *Br. J. Appl. Phys.* **11**, 408. [83, 89]

Vine, J. (1966). *I.E.E.E. Trans.* ED-**13**, 544. [359]

Vine, J. *See also under* Firestein, F.

Voges, J. and Ruthemann, G. (1939). *Z. Phys.* **114**, 709. [449]

Vosburgh, K. G. *See under* Kawakatsu, H.

Wadia, B. H. (1958). *J. El'onics and Control* **6**, 305. [296]
Wadia, B. H. *See also under* Ginzton, E. L.
Wagner, M. D. *See under* Marton, L.
Wagner, S. *See under* Herrmann, G.
Walcher, W. (1949). *Nucleonics* **5**, 42. [421]
Walcher, W. (1951). *Z. angew. Phys.* **3**, 189. [311]
Wall, I. *See under* Crewe, A. V.
Wallington, M. J. *See under* Mulvey, T.
Wallraff, A. (1935). *Arch. Elektrotech.* **29**, 351. [441]
Wallraff, A. *See also under* Becker, H.
Walter, L. M. *See under* Crewe, A. V.
Wang, C. C. T. (1950). *Proc. Inst. Radio Engrs* **38**, 135. [304]
Wang, C. C. T. (1967). *J. Appl. Phys.* **38**, 4938. [445]
Wardley, J. *See under* Lubszynski, H. G.
Warshaw, S. D. *See under* Allison, S. K.
Watanabe, A. and Morito, N. (1955). *Optik* **12**, 166. [227]
Watanabe, H. *See under* Marton, L.
Waters, W. E. (1960). *J. Appl. Phys.* **31**, 1814. [307]
Waters, W. E. *See also under* Kirkstein, P. T.
Watson, E. E. (1927). *Phil. Mag.* **3**, 849. [270, 275]
Weber, C. (1963). *Proc. I.E.E.E.* **51**, 252. [53]
Weber, C. (1967). *Philips Res. Rep. Suppl.* **6**, 3, 23, 42, 64, 69. [45, 46, 53, 62, 277, 289, 309]
Weber, C. *See also under* Hart, P. A. H.
Weber, E. (1950). *Electromagnetic Fields* vol. 1. New York. [38]
Webster, D. L. *See under* Hansen, W. W.
Wedding, J. W. *See under* Haynes, S. K.
Wehnelt, A. (1903). *Ber. Dtsch. Phys. Ges.* **5**, 29. (A proper description of the Wehnelt cylinder was only given very much later. See for instance Wehnelt, A. and Jentsch, F. (1909). *Ann. Phys. Lpz.* **28**, 541.) [321]
Weichardt, H. *See under* Knoll, M.
Weissenberg, G. *See under* Bartz, G.
Wendt, G. (1942). *Z. Phys.* **119**, 423. [444]
Wendt, G. (1943). *Z. Phys.* **120**, 720. [416]
Wendt, G. (1953). *Onde Elect.* **33**, 92. [445]
Wendt, G. *See also under* Diels, K.
Whelan, M. J. *See under* Metherell, A. J. F.
Whittemore, W. L. *See under* Shutt, R. P.
Wiedmer, H. B. *See under* Bas, E. B.
Wien, W. (1898). *Ann. Phys. Lpz.* **65**, 440. [434]
Wierl, R. (1931). *Ann. Phys. Lpz.* **8**, 533. [380]
Williamson, K. I. (1947). *J. Sci. Instr.* **24**, 242. [138]
Williamson, R. M. *See under* Browne, C. P.
Wiskott, D. *See under* Bartz, G.
Witcher, C. M. (1941). *Phys. Rev.* **60**, 32. [233, 239, 245]
Wohlfarth, E. P. (1959). *Adv. Phys.* **8**, 87. [131]
Wolf, E. *See under* Born, M.
Woodhead, A. D. *See under* Schagen, P.
Wooster, W. A. (1927). *Proc. Roy. Soc.* **114**, 729. [448]

Wreathall, W. M. (1966). *Adv. Electronics* **22**A, 583. [46, 61, 367]

Wreathall, W. M. *See also under* Carré, B. A.

Wright, W. D. *See under* Klemperer, O.

Yamazaki, H. *See under* Kanaya, K.

Yarnold, G. D. and Bolton, H. C. (1949). *J. Sci. Instr.* **26**, 38. [408]

Yates, J. G. *See under* Pizer, H. I.; Sander, K. F.

Yavor, S., Dymikov, A. D. and Ovsyanikova, L. P. (1964a). *Nucl. Instr. Meth.* **26**, 13. [229]

Yavor, S., Dymikov, A. D. and Ovsyanikova, L. P. (1964b). *Zh. Tekh. Fiz.* **34**, 99; *Sov. Phys. Tech. Phys.* **9**, 76. [229]

Yavor, S. *See also under* Dymikov, A. D.

Yocom, W. H. *See under* Mendel, J. T.

Young, D. M. (1954). *Trans. Am. Math. Soc.* **76**, 92. [46]

Zashkavara, V. V., Korsunsky, M. I. and Kosmachev, O. S. (1966). *Zh. Tekh. Fiz.* **36**, 132; *Sov. Phys. Tech. Phys.* **11**, 96. [414]

Zuckerman, I. I. *See under* Markovitch, M. G.

Zworykin, V. K. (1933). *J. Inst. El. Engrs* **73**, 437. [327]

Zworykin, V. K. and Morton, G. A. (1940). *Television.* New York. [443]

SUBJECT INDEX

Abbe's sine law, 13, 16
aberrations
  anisotropic, 212
  chromatic, 225
  classification of, 159, 161
  combination of, 264
  electronic, 225
  in emission systems, 325
  field, 196
  geometrical, 161
  relativistic, 253
  by space charge, 285
  spherical, *see* spherical aberration
  see also deflexion error; distortion
acceleration of emitted electrons, 320
action integral, 3
Airy disc, 258
alignment errors, 163
alternating gradient focusing, 106
alternating magnetic field, 306
analogue methods, 47, 52, 62
angular momentum, 110
anisotropic aberrations, 212 ff.
anisotropic coma, 214
annular beams, 385, 387
aperture, focal length of, 79
  potential distribution, 39
aperture aberration of line focus
  lenses, 215
aperture error, *see* spherical aberration
aperture lens, *see* diaphragm lens;
  single aperture
aperture, numerical, 256
astigmatism, anisotropic, 214
astigmatism
  axial, 162
  third order, 200
astigmatism, test of Fresnel fringes,
  261

baffle coil, 195
beam spread
  in accelerating fields, 292
  in emission systems, 348
beam-spread coefficient, 279 ff., 356
beam-spread, curve, 274
beam-spread equation, 272
beam transmission through tunnel,
  277, 281

beam transport, 157
bell-shaped magnetic field distribution, 118
bell-shaped potential distribution, 39,
  59, 85, 126
betatron, 426, 429–30
black-out voltage, 340
Boersch effect, 250, 467
bolt cathode, 322, 377
bore-to-gap ratio of microscope lens,
  129
brightness, 16, 17
Brillouin flow, 301, 318
Busch short lens formula, 115
Busch theorem, 110

cardinal points, 10, 20 ff., 26
  of line focus lenses, 30 ff.
cathode-ray oscillograph, 398, 400
cathode-ray tube, 136, 175, 344, 398,
  408, 443; *see also* television
cathodes, 321 ff.
caustic, 266, 360
caustic baffle, 242
Child's equation, 312, 343
chromatic aberration, 225 ff.
  of emission systems, 352
classification of aberrations, 159, 161
coherence
  of electron beam, 258
  of illumination and resolving power,
  259
coil form factor, 116
collecting efficiency, 234
collective of paths, 19
collinear projection, 9 ff., 14
colour television picture tube, 329
coma, 196
combination, of aberrations, 264
  of lenses, 26
combined electrostatic – magnetic
  lenses, 148 ff.
computation of trajectories, 473 ff.
computational ray tracing, 145
computer, *see* analogue; digital
concave cathodes, 361 ff.
condenser lens, 129
congradient, 12
contrast, 266